# Into the Sunlit Splendor

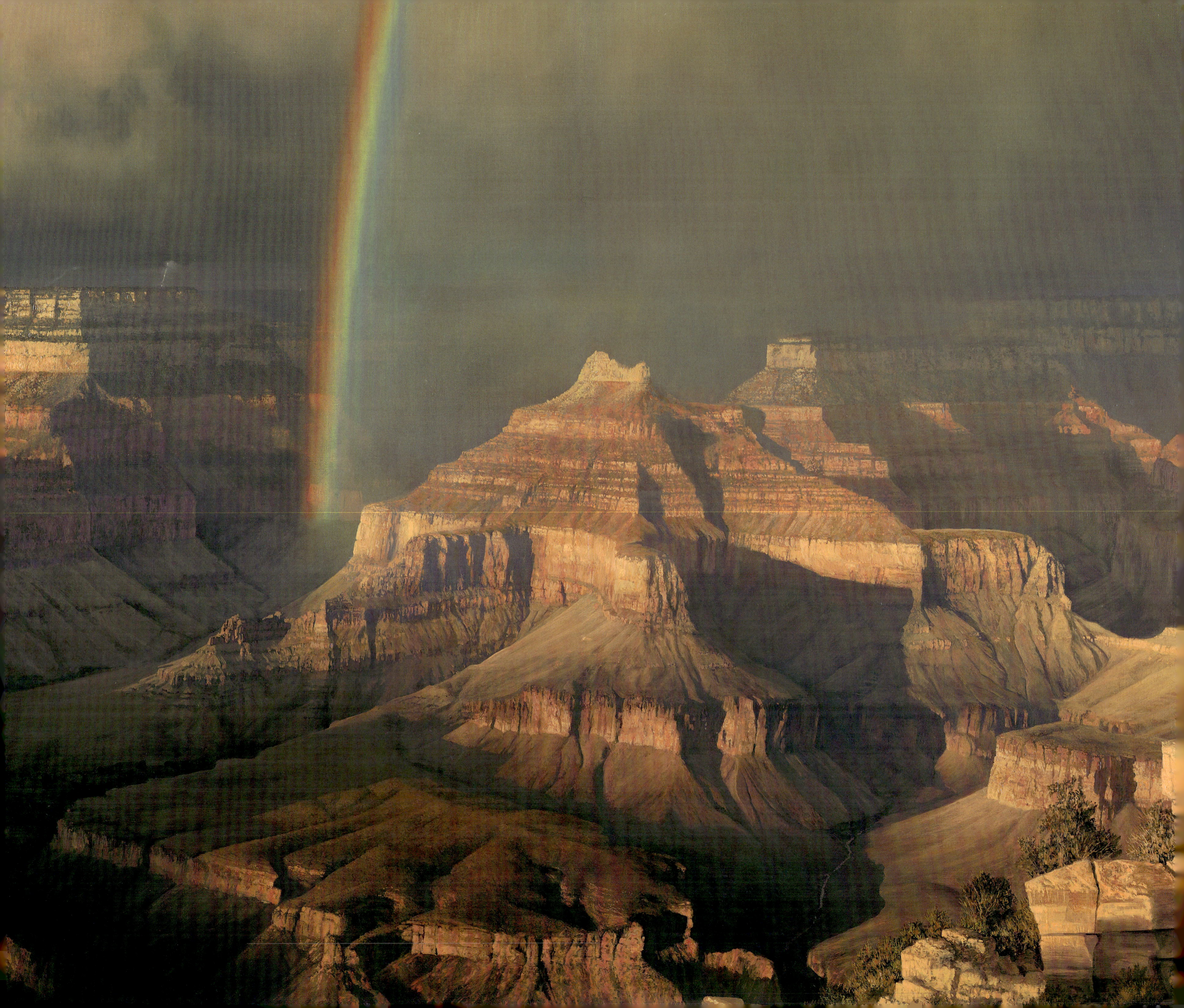

# Into the Sunlit Splendor

## The Aviation Art of William S. Phillips

By Ann and Charlie Cooper

Foreword by Apollo Astronaut Captain Alan Bean

Introduction by Artist Wilson Hurley

THE GREENWICH WORKSHOP PRESS

A GREENWICH WORKSHOP PRESS BOOK



Published by The Greenwich Workshop Press
151 Main St. P.O. Box 231, Seymour, CT 06483
(203) 881-3336
www.greenwichworkshop.com

Library of Congress Cataloging-in-Publication Data is available from the publisher upon request.
ISBN-10: 0-86713-093-8
ISBN-13: 978- 0-86713-093-5
Frontis art: *Revelation*. Isis Temple from the South Rim.
p. 6: *And the Light Shall Prevail*
pp. 12–13: *Sleigh Ride at Apple Creek* (detail)
pp. 50–51: *The Dream Fulfilled, Where the Love Light Shines* (detail)
pp. 82–83: *The Storm Watchers* (detail)
pp. 118–119: *Moonwatchers* (detail)
pp. 160–161: *Winter Symphony* (detail)

Greenwich Workshop Press books are distributed to the trade by Artisan, a division of Workman Publishing.

Book design: Steven Schoenfelder
Manufactured in China by Imago
First Printing 2005
1 2 3 4 08 07 06 05

**TO KRISTI FOR ALL OF HER LOVE AND SUPPORT, MY MOTHER AND FATHER FOR LAYING THE FOUNDATION AND TO DAVE USHER FOR SEEING THE POTENTIAL.**

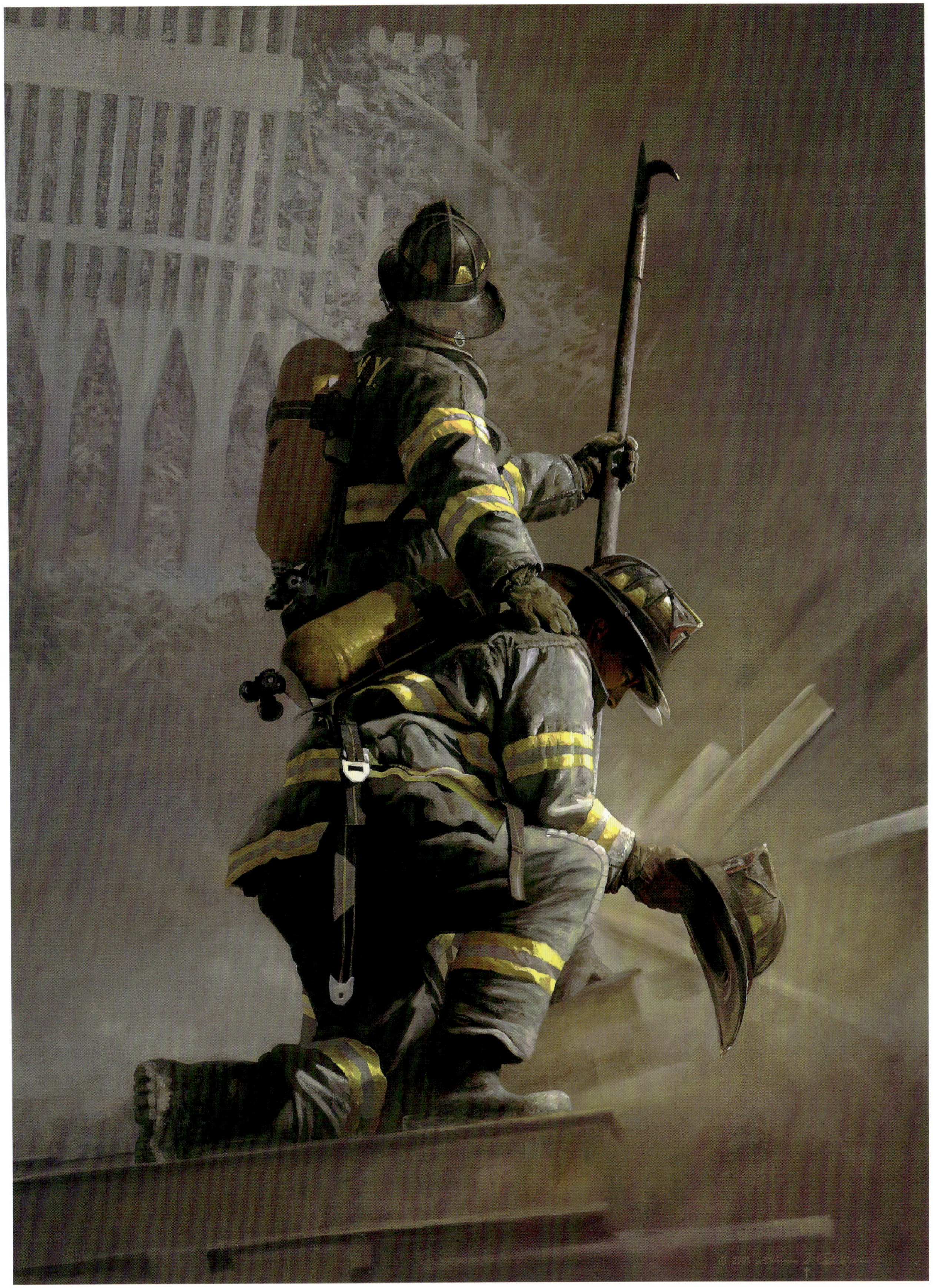

A PRAYER FOR MY BROTHER

## ARTIST ACKNOWLEDGEMENT

For the Fire Fighters, Police Officers, Military Personnel and their families whose sacrifices and service must never be taken for granted by those who derive the benefits of their efforts.

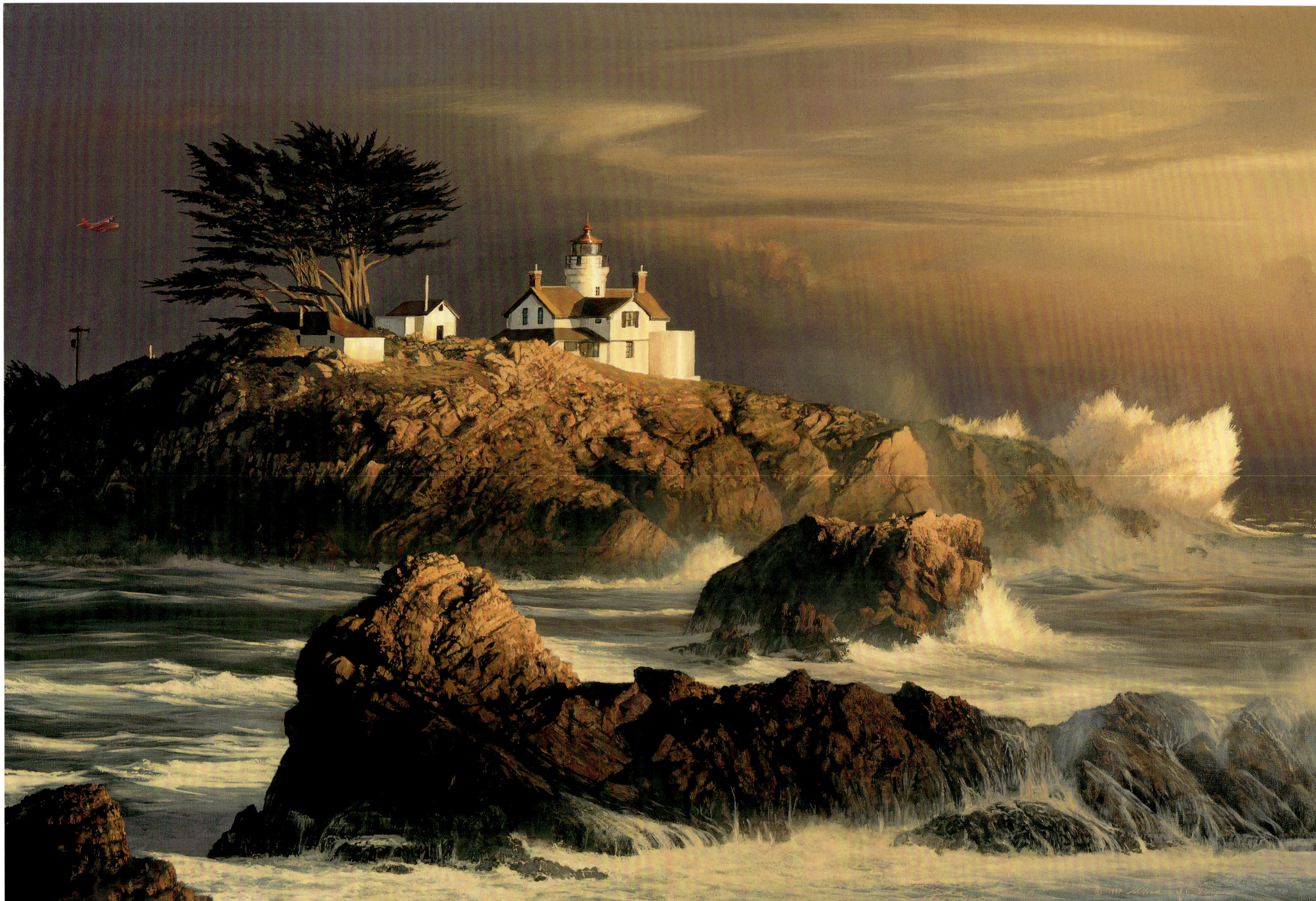

**And the Light Shall Prevail**

# CONTENTS

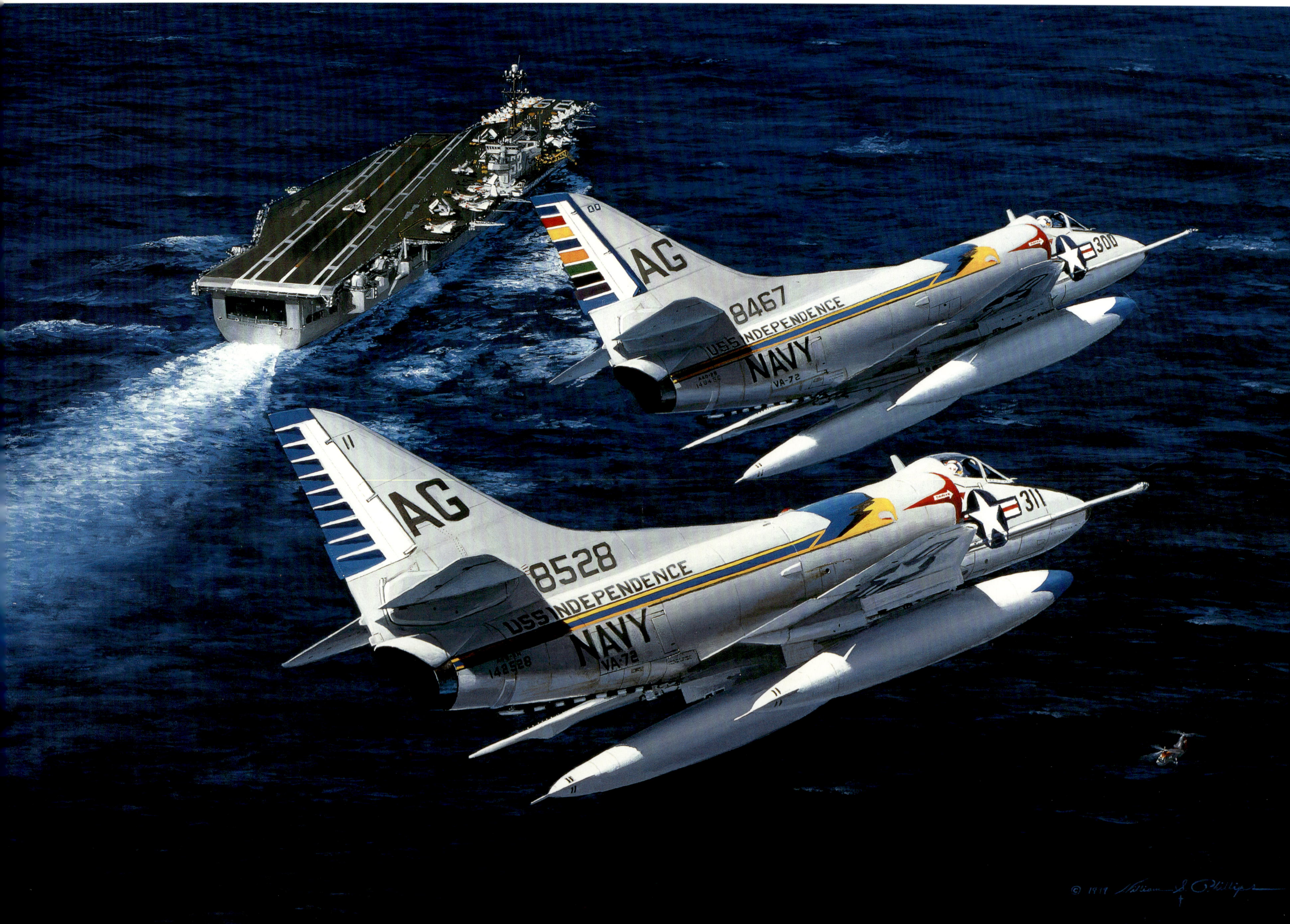


**Return to the Independence:** VA-72 Skyhawks off USS *Independence* CV-62 in the Mediterranean Sea.

# Foreword

## CAPTAIN ALAN BEAN

On November 19, 1969, I stepped out of the Lunar Module *Intrepid* onto the Ocean of Storms. My friend and Apollo 12 Mission Commander, Captain Pete Conrad, had already descended the ladder. "You look great," he said. "Welcome aboard." I'll never forget that moment. I became the fourth man to set foot on the moon's rocky and cratered surface.

We had been chosen by NASA to train to become astronauts because of our ability to fly high performance airplanes. Attitude, mindset, and the technical skills of flying are the necessary foundation for learning to operate spaceships.

Like Bill Phillips, I had spent much of my young life looking at planes, mostly at a distance from the ground. I could identify them all at a glance. Later, as a Navy pilot, I was able to observe planes in flight close up. I loved their unique aerodynamic shapes, their special military markings, and the way they moved through the air as if propelled by some supernatural forces.

Look at the facing page! Now that's what it really looks and feels like when you're flying number three in probably a four-plane flight. You're circling to get into position for formation break up prior to approaching and landing aboard your ship. Boy, does this beautiful painting bring back a bunch of fun (and exciting) memories.

Why do I enjoy looking at Bill Phillips' paintings so much? I think it is because he captures memorable moments in time that we can study at length. Pilots don't have the luxury of appreciating visual detail that doesn't directly contribute to the success of their mission. For example, if we were flying #3 in Bill's wonderfully accurate painting, we would need to concentrate on flying precise formation. We would be thinking about our remaining fuel and how much flying time we have available at this low altitude to get aboard. Are we getting ready to switch our radios to the landing frequency? What is our interval in the break? When do we put our hook down? And these thoughts are just a few of those that must go through our minds to complete a safe and successful flight.

The right artist is needed to preserve these special moments. Not just any artist, but Bill Phillips, who knows about airplanes and flying, and cares enough to do the difficult and time consuming research to make his paintings real and to show what it was really like to be there.

And Bill creates paintings that help us in another important way. Many of us wonder what it would be like to fly and fight in World War II airplanes like the P-51, or the P-38, or the B-17, or how it would seem to us to break the sound barrier in the X-1. Bill helps us better imagine experiences we can never have and he does it masterfully. Almost every page in this book answers that need. Maybe we just want to enjoy a beautiful memory of ours, or of America's past, or of the earth and sky.

We can be thankful that a young boy from Southern California grew up with both a passion for airplanes and an exceptional talent for painting. Bill Phillips paints planes that *look* like they are climbing, or turning, or landing. He paints paintings that look just right. With his brush he captures moments in time when sunlight, sky, and clouds combine to form a magic no earthbound vision can touch.

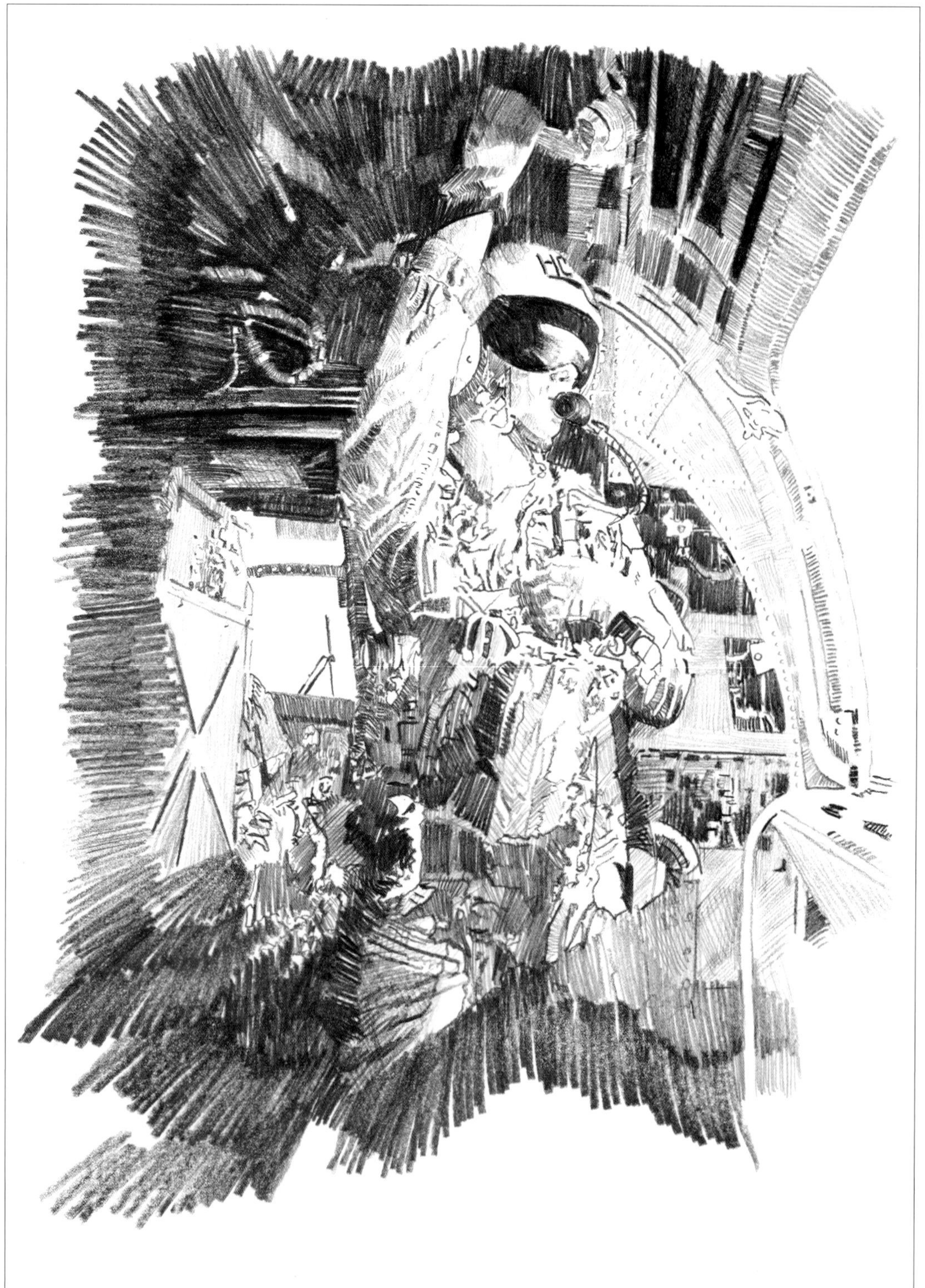

**On Alert**
A crewman aboard a CH-46 helicopter scans the horizon for any sign of enemy small boat activity over the northern waters of the Persian Gulf. Artist Bill Phillips sketched this and many other moments of Naval aviation activity in the Persian Gulf as a Navy Combat Artist.

# Introduction

## WILSON HURLEY

Over a century ago, to become an artist, a person was required to master drawing, understand the geometry of accurate perspective, cope with the effects of atmosphere, and be able to control light and shadow on a complex shape to make it look real. Modern Art disregarded all that. Today, the advocates of the abstract movement write books on supposed contraptions of Renaissance painters that helped them create the vivid images of their world, because modern artists have not been trained in these skills. It takes time, intelligence and effort to be able to paint an object that looks real. Bill Phillips has taken that trouble and mastered it well. Therefore, the next step is to see what he, having achieved the status of an artist, does with his ability. For he was born to difficult times.

Not only has abstract art been the dominant fashion, but there have also arisen a number of groups who prefer nature entirely without human presence. Some look at creation from the viewpoint of dogs, cats, whales, dolphins, trees, birds or bugs. These excursions in fashion are not unique to our times. Consider Reginald Heber's line, "Though every aspect pleases and only man is vile."

No one who ventures to paint can conceal who he is. Look at these paintings and see that Bill Phillips is a member and admirer of our human race. He has a deep reverence for the beauty of the world, the sky, the sea and the glory of the sun that lights it all. He sees the aesthetic beauty of the natural universe. Yet he is also aware of the aesthetic beauty of the works of man. Look how his eye and hand caress the grill, hood, and flowing full fenders of a 1950s automobile! See how functionally pleasing are the masses of a mighty ship driving over a wide sea at sunset. Notice how lovingly he shows the clean, functional lines of a jet fighter blasting over the chaotic beauty of a cliff.

And when he paints about a place of repose, notice the familiar garden wall and low steps down from a manicured garden into a wild valley with massive walls of cumulus clouds piling up in the east at twilight, with glittering lightning showing in their depths. This blending of the works of human beings into the majesty of nature is of much greater weight and origin than those who now seek to avoid it. You need only look at the murals of Ostia, Herculano or Pompeii to see the same theme. Or look at the Renaissance, the best paintings up through history and see that all of us were considered a part of creation. Maxfield Parrish, the extraordinary illustrator, most often opened his landscapes from a man-made environment.

When Bill Phillips uses his great artistic ability as he does, he is obviously at peace with his and our natural place in creation and finds it all beautiful. That is what makes viewing his work such a pleasure.

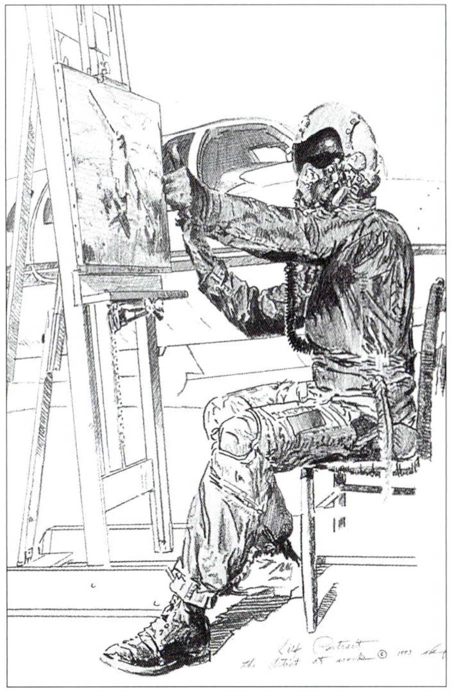

**Self Portrait,** always ready to fly.

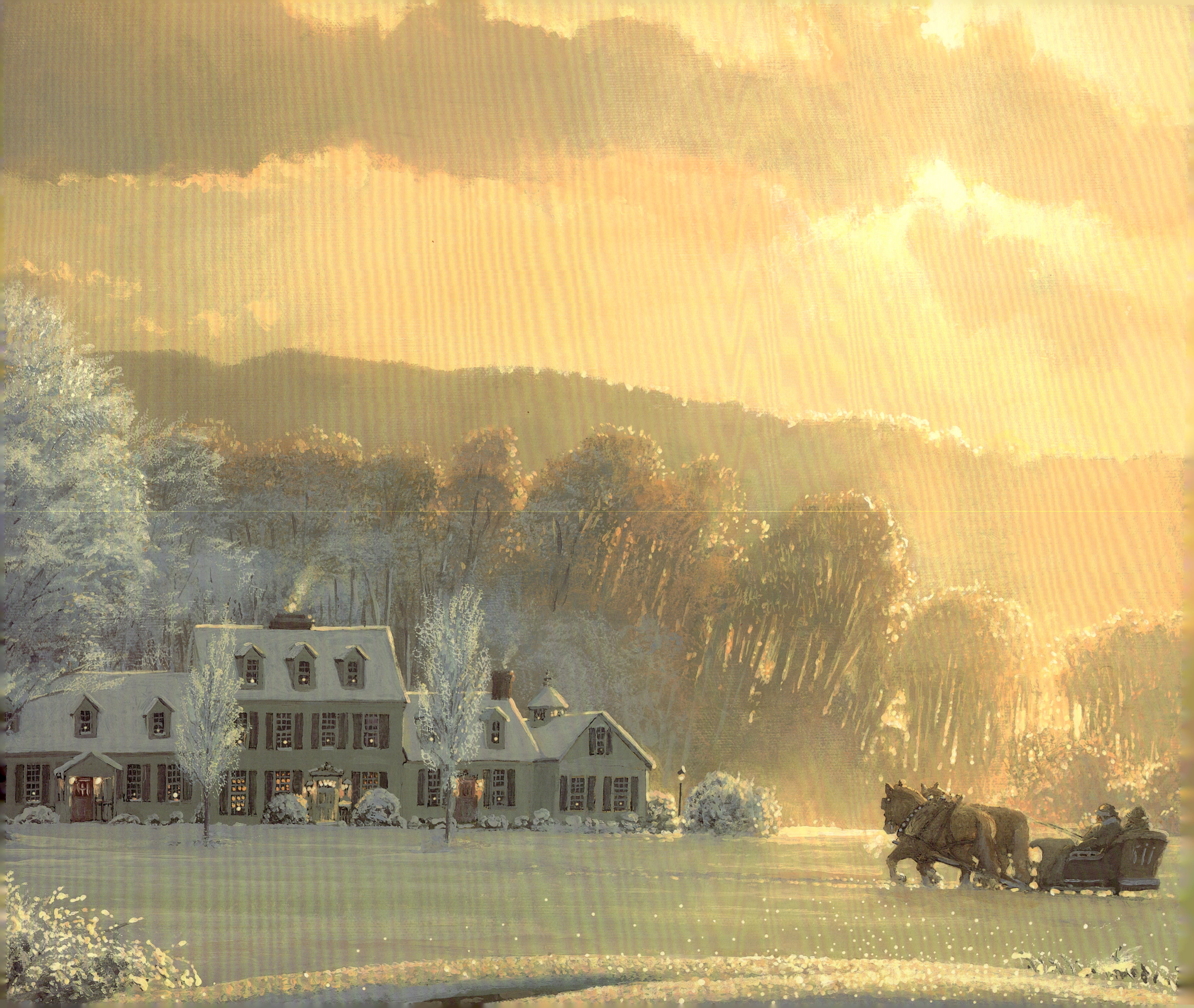

# INSPIRATION

## WHEN PRAYERS ARE ANSWERED

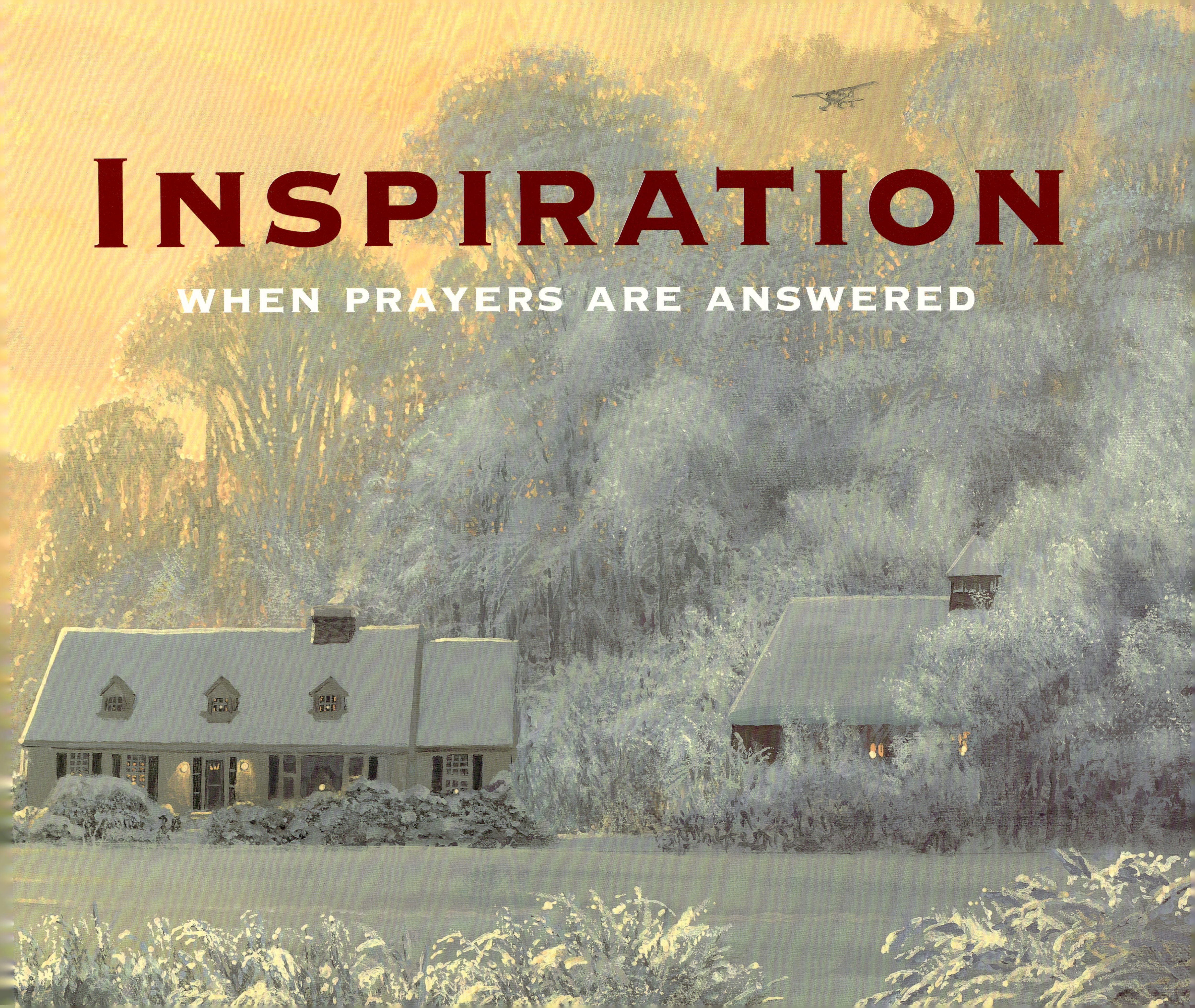

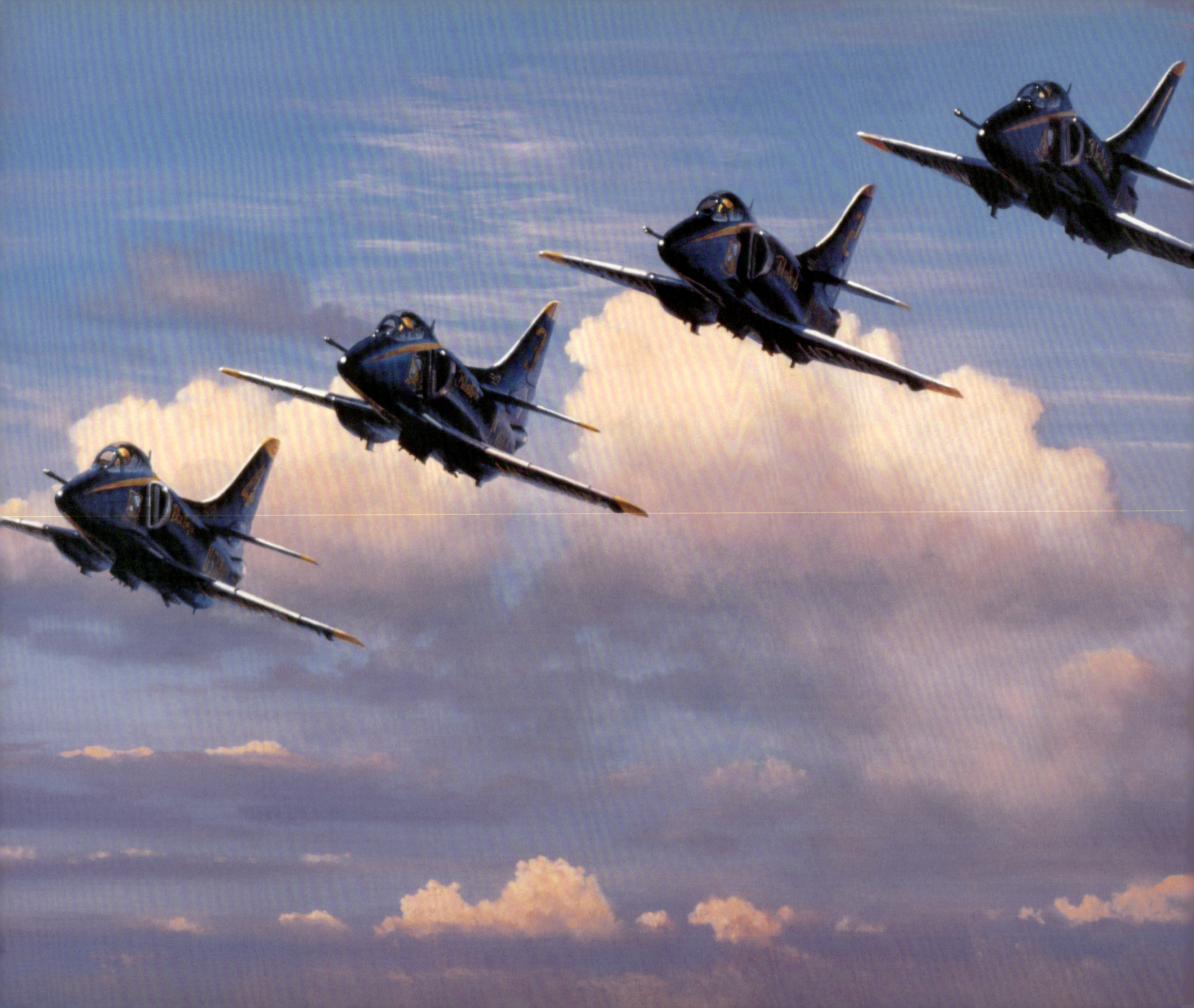

*An artist observes, selects, guesses and synthesizes.*

ANTON CHEKHOV

What shapes a child's heart and mind, unleashing the talent that lies within and inspiring a future that cannot be known until it is lived? What influenced a young William S. "Bill" Phillips, born in Culver City, California in 1945, and kindled his passion for artistic expression?

Surrounded with breathtaking natural beauty and vital, creative people, Bill's early years were filled with wonder. The warm, sunny climate of Culver City and Santa Monica lured stars like Lucille Ball, Paul Newman, Spencer Tracy, Katharine Hepburn, and Jimmy Stewart and pulsated with creative energy in rhythm with the restless Pacific Ocean pounding its shore. Culver City was home to Metro-Goldwyn-Mayer and Bill's father, a character actor, was under contract to the movie studio. He performed in ninety-six movies between 1941 and 1957, playing in such films as *30 Seconds Over Tokyo, Dive Bomber, The Red Badge of Courage* and *High Noon.* In *On The Town,* he performed with Frank Sinatra and Gene Kelly and, in *Holiday In Mexico,* with Jane Powell and Roddy McDowell. In 1951, when Bill was six years old, his father had second billing—a major role—as Sergeant Kirby in *A Yank In Korea.*

Bill received more than his name from his father, William S. "Bill" Phillips. The impressionable boy listened to his father practice his lines and watched his father perform. He used his father's discarded scripts as an endless supply of paper for his insatiable urge to draw. He drew strength from his father, a strict disciplinarian, and his mother, Mary Davis Phillips, a loving, but firm parent.

Unlike his parents and Mary Jane, his younger sister, Bill was entranced with airplanes. At four, he lined toy airplanes along the hall of their home until his mother told him to clean them up. Bill told her, "No. You step over them," and received a swift swat on the behind as punishment. His parents were similarly critical of his first wall mural—a freestyle airplane drawn on his parents' newly dry-walled master bedroom wall. Disrespect wasn't tolerated.

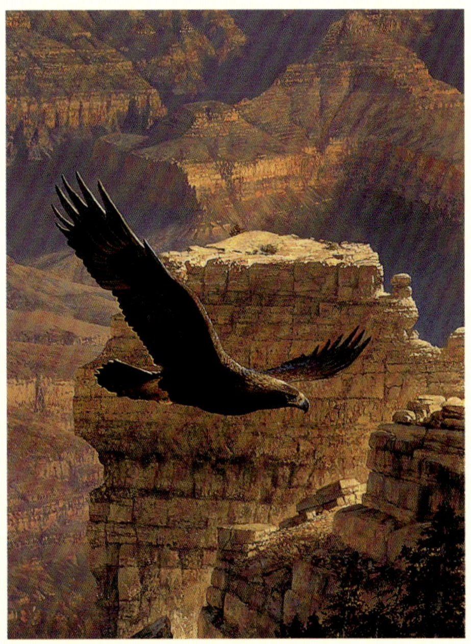

**SOARING THE WALL**
A golden eagle catches thermal updrafts and glides across the majestic gorges of Grand Canyon National Park in this painting that was among the Top 100 Winners in the 2003 Arts for the Parks international painting competition.

Santa Monica Airport was located not far from the Phillips' home and was the aircraft manufacturing site of Donald Douglas, the aircraft engineer and designer. "Aviation had come into its own after its Golden Age of barnstormers and record-setters," recalled Bill. "It flourished with the huge build-up of our armed forces for World War II. Airplanes always circled overhead. One of my earliest recollections is of lying on the grass and watching a yellow biplane—it had to be a Stearman—fly past in a bright blue sky with one puffy white cloud. I was captivated."

Memories of several pivotal events with his father remain with Bill. The senior William Phillips had been bitterly disappointed to be sidelined from World War II because of knees badly injured in football. He turned his own regrets into support for his son's love of aviation and actively encouraged in Bill an appreciation for service and for the armed forces.

Bill reminisced, "When I was about four, I won a drawing held at a local drive-in theater. The prize was a ride in a four-place taildragger. I rode in the front with the pilot and my dad took a seat in the back. I remember seeing my mother and sister as we swooped in over the taxiway to a landing. I think Dad stuffed the contest box to make sure I was the one to win the plane ride."

The Phillipses relocated from Culver City to the San Fernando Valley (then a rural area filled with orange and walnut groves and protected by a wreath of mountains to the west, northwest and northeast) and built a home on one acre in Sepulveda. Again, an airport was nearby—the Van Nuys Airport. "Aviation was as important in our valley as were the farms and the numerous movie studios that grew up around us. P-51 Mustangs, F-94 Starfires, and F-86 Sabres crisscrossed the sky. The years I spent there were some of the most carefree, wonderful years in my life. I made model airplanes; I flew model airplanes. I spent endless hours holding model airplanes in my hands and watching the sunlight on their wings. To this day, my paintings reflect what I learned as a child. Aviation was in the real world for all to see, but it was also in my head. Studying that interplay of light and shadow developed something inside of me."

When Bill was eleven, his father took him to lunch to tell him the facts of life. He was direct and positive—and wise. Bill recalled, "Feeling that I might be traumatized (after all, I was Catholic and these were the Fifties) Dad had the good sense to take me to the airport directly afterward so we could watch airplanes together.

"Another time, my father and I dressed up in our best clothes and he took me to lunch in the Sky Room of the old Lockheed Hollywood Burbank Airport terminal. The Sky Room boasted a showcase full of Lockheed airplane models. My dad was far more interested in acting, yet he sat with me watching DC-3s load and unload passengers. It was a special day."

Little did Bill know such outings were soon to end. The elder William S. Phillips died suddenly of a heart attack on June 27, 1957. Bill was twelve years old. Life changed drastically for the family. His mother, whose losses were devastating, faced an uncertain future with limited finances. She could no longer keep her children in Catholic school as she had only Social Security and meager savings.

Bill's deepest regret was that he'd disappointed his father in his struggles to pass math courses.

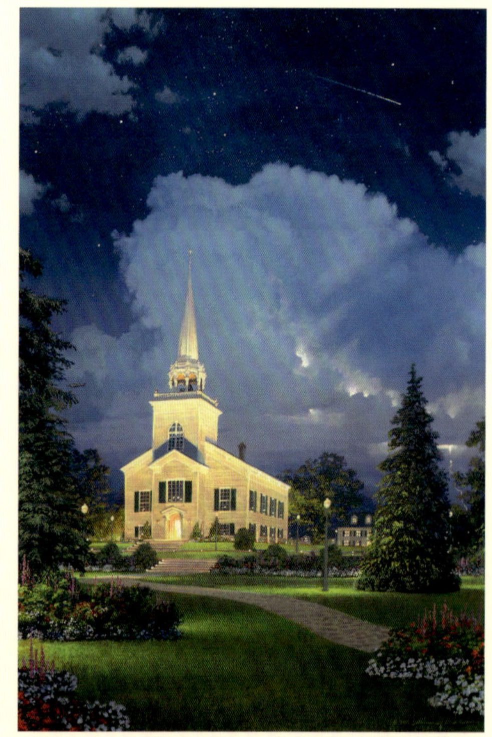

**The Heavens Proclaim His Glory**
One can almost hear the peal of bells from the steeple of this gleaming church, erected on the highest point in the artist's (fictional) Dogwood County. Man's works, however beautiful, pale in comparison to God's creation and, truly, the heavens proclaim His glory.

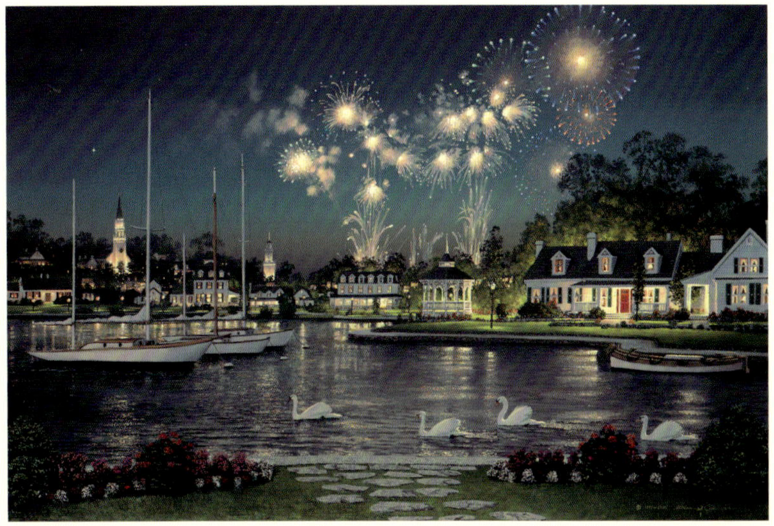

**Rejoice**
Phillips Bay, the charming, fictional locale created by the artist, is lit by a full moon that bathes the town in soft twilight, and sparkling fireworks rival the luminous stars and lights of the town.

"Much later, I learned from Mr. Barnes, one of my teachers, that I had mathematical dyslexia and I regularly transposed numbers. I'm appreciative of Mr. Barnes, but I'll never forget my father's disappointment when I couldn't comprehend fractions and equations. In Catholic school, Sister Mary Regina shared my father's frustration as she tutored me in math. No matter how hard I tried, I couldn't pass tests and Dad dismissed me as lazy. I began to think of myself as stupid. Between my failures in school and the shocking reality of my father's death, I started into a downward spiral. I chose the wrong friends; I got into trouble. I dropped behind in my school work and, most importantly, I didn't care. After having been extremely religious, I began to deny the existence of God. My mother, bless her, recognized the direction I was headed and changed the course."

Mary Phillips was gutsy. In 1960, she sold their home, loaded Bill, his sister and their belongings into the family car and started north to find a more positive future for her children. She drove into the Willamette Valley of Oregon, found a 10-acre farm in Silverton, a rural community in the foothills of the Cascade Mountains, and settled her children far from the influences she rightfully recognized as unhealthy.

"During my second day at Silverton High School," explained Bill, "an authoritative voice over the room's loudspeaker announced, 'William Phillips, come to the Dean's office!' I was astounded. I hadn't been in the school long enough to get into trouble. I couldn't imagine why I was being singled out.

"In the office, Tom Cowan, Vice Principal, Dean of Boys and Coach of the football team, was holding my transcripts. He was a tough guy, short-haired, stocky, with a military bearing. He challenged me, saying he would do everything in his power to see me graduate from high school, but that the responsibility was mine. He noted the loss of my father and immediately began to assume a parental role for me. He asserted that I *was* going out for football and that I would *not* miss any practices. It was a start and, although I wasn't a great football player, I did well in wrestling. Mr. Barnes and Tom Cowan took me under their wings and I owe them both my deepest gratitude."

Pablo Picasso has said, "The artist is a receptacle for emotions that come from all over the place: from the sky, from the earth, from a scrap of paper, from a passing shape, from a spider's web." Bill Phillips was inspired by members of his family, by his experiences, his environs and by particular people who believed in him. He, in turn, inspires others through the magic of his brush.

## Dusting the Ridgeline

Until man flew, only mountain climbers reveled in the glorious vistas of the heavens that wrapped rugged mountains in azure. Here, on negative staggered wings and glowing with the long rays of sunlight, a Beechcraft D17 soars mere feet above the ridge. It was designed in 1932, the same year that Walter Beech created his own company from what had been the triumvirate of aviation pioneering at Travel Air in Wichita, Kansas, with Clyde Cessna and Lloyd Stearman. Initially created in a small portion of Cessna's manufacturing space, the Model 17 outgrew its temporary site. Beech eventually returned to Travel Air's large building, renaming it Beechcraft.

Louise Thaden and Blanche Noyes flew a stock Beech D17 Staggerwing in one of the earliest Bendix Trophy Races in which women were allowed to compete. Judges, expecting the winning prize to be awarded to a male pilot, had a separate and smaller monetary prize for the first woman pilot. The Bendix Trophy of 1936, however, went to Thaden and Noyes. In their historic victory, they captured both prizes and made an east-west speed record as well. The Beechcraft Model 17 Staggerwing set a standard for high performance, single-engine aircraft. It served in World War II and remains one of the most coveted classic aircraft in existence.

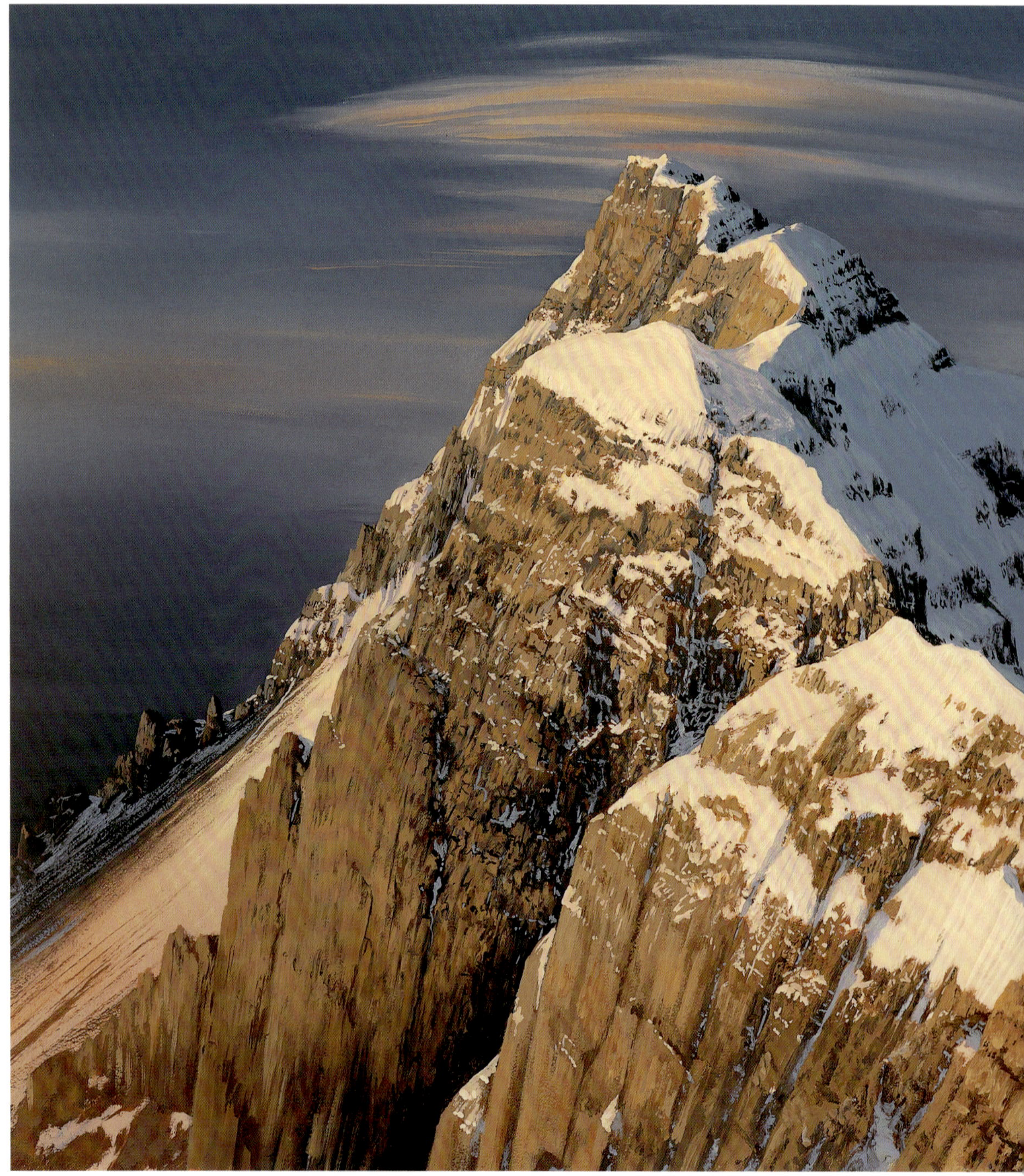

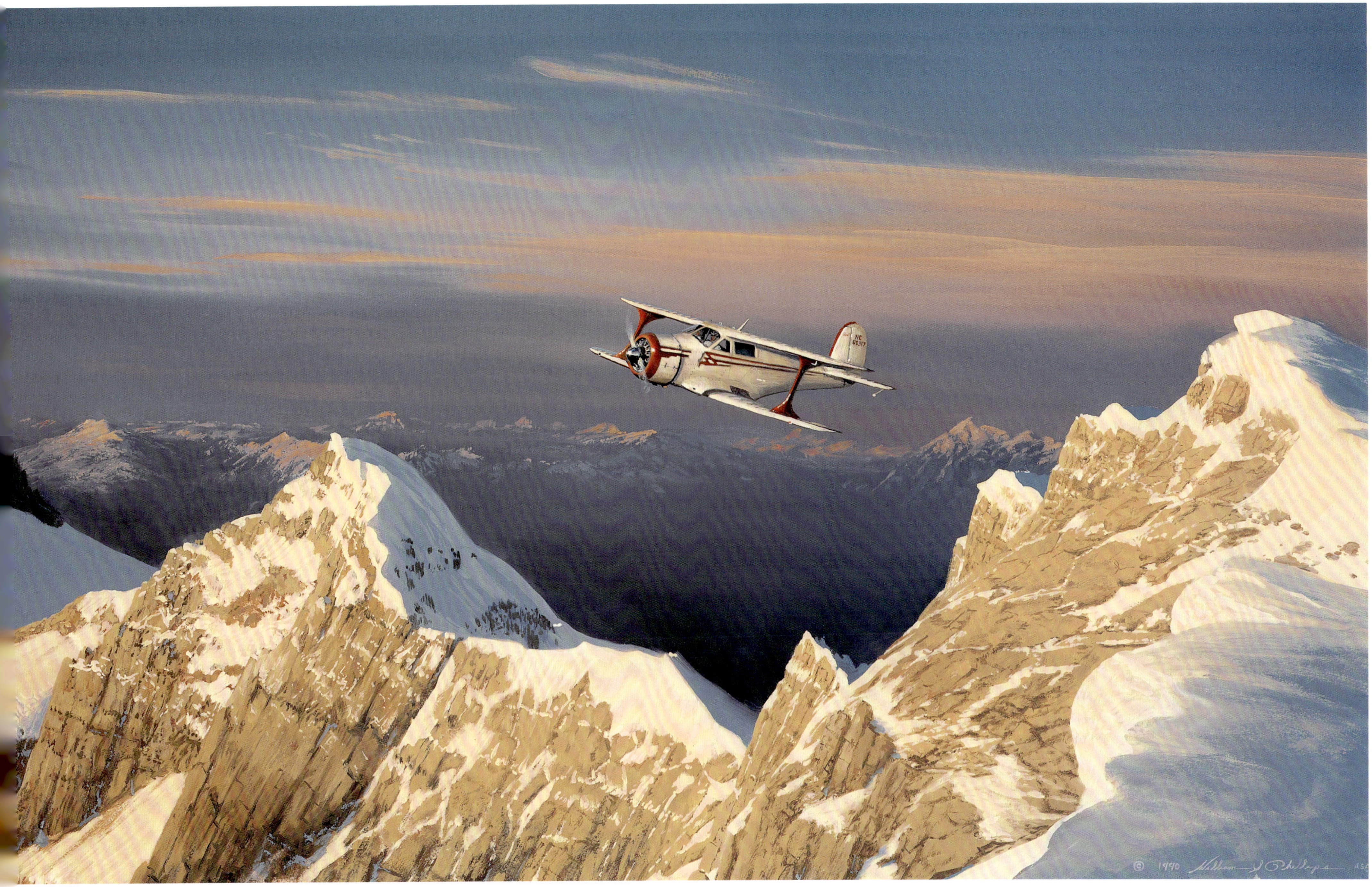
© 1990

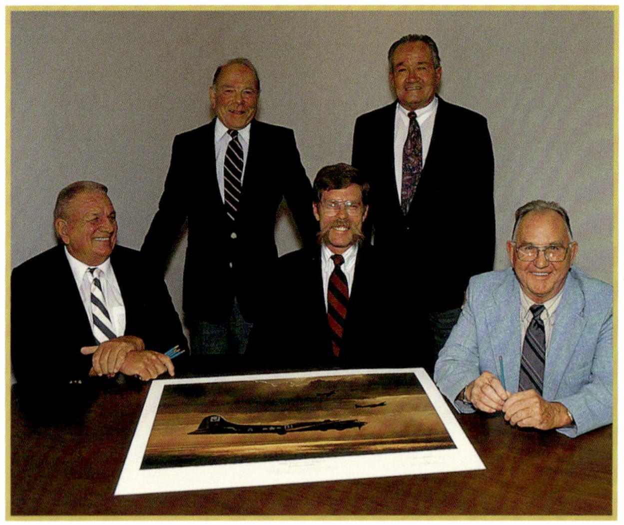

The lead crew of the B-17F as she flew the first daylight raid on Munster was that of James R. Geary with Nathaniel "Gus" Mencow, Navigator; Gene Wilms, Bombardier; and gunners Shirl Hoffman, Clarence Medeck, Clifford Puckett, Paul Morris and Donald May. Shown here during the counter-signing of the Greenwich Workshop Limited Edition Fine Art Print of the painting are Colonel Geary, Lt. Colonel Mencow, artist Bill Phillips, Technical Sergeant Shirl Hoffman, and Master Sergeant Clifford Puckett.

## When Prayers are Answered

Target: Munster. Operating out of Framlingham, England on October 10, 1943, *Betty Boop/Pistol Packin' Mama*, a B-17F of the 390th Bomb Group, 570th Bomb Squadron, was one of 133 B-17s dispatched for a daylight run following the previous night's mission by the RAF—the first wing of Eighth Air Force B-17Fs to target the railroad station and yards of Munster, Germany. Takeoff went smoothly and twelve P-47 fighter escorts arrived to provide top cover, until reaching the limit of the fighters' range. At that crucial juncture, the fighters turned for England and the bombers faced unrestrained German fighters while making their runs—a concentrated and violent attack. The sky was blackened with flak, and peppered with over 200 German fighters including Focke-Wulf 190s, Junkers 88s, and Messerschmitt 109s, 110s and 210s. For the 390th, eight crews and aircraft, out of their departing nineteen, were lost. Of the fourteen in the 100th Bomb Group, twelve were lost and of the 95th Bomb Group, five of the twenty were lost. The mission was grueling; the losses were brutal.

The painting at right depicts *Pistol Packin' Mama* with flak damage to the wing and tail, approaching England's chalk white cliffs that welcomed her crew to safety. As she nears the coast, the danger diminishes and the escort fighters break away to recover at their base. The clouds part and rays of sunlight and hope stream down on the B-17 and her crew as "their prayers are answered."

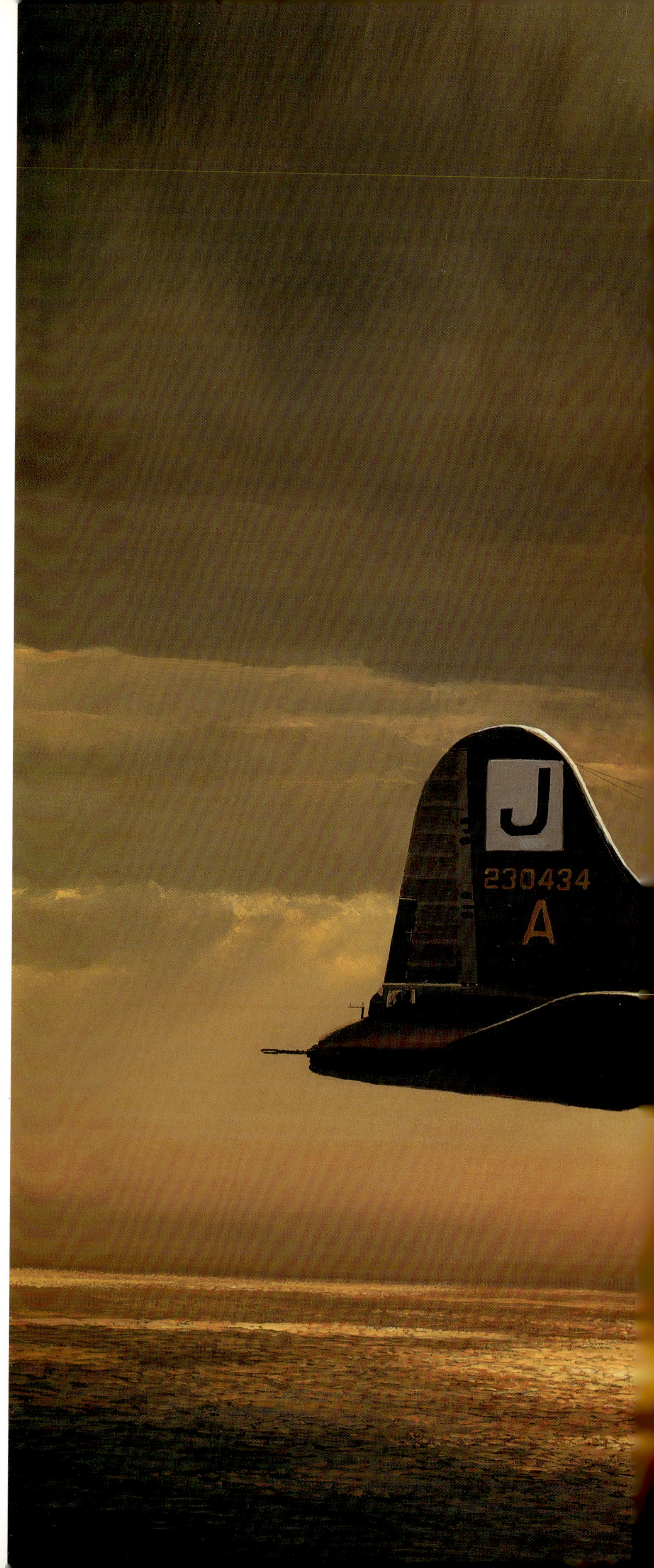

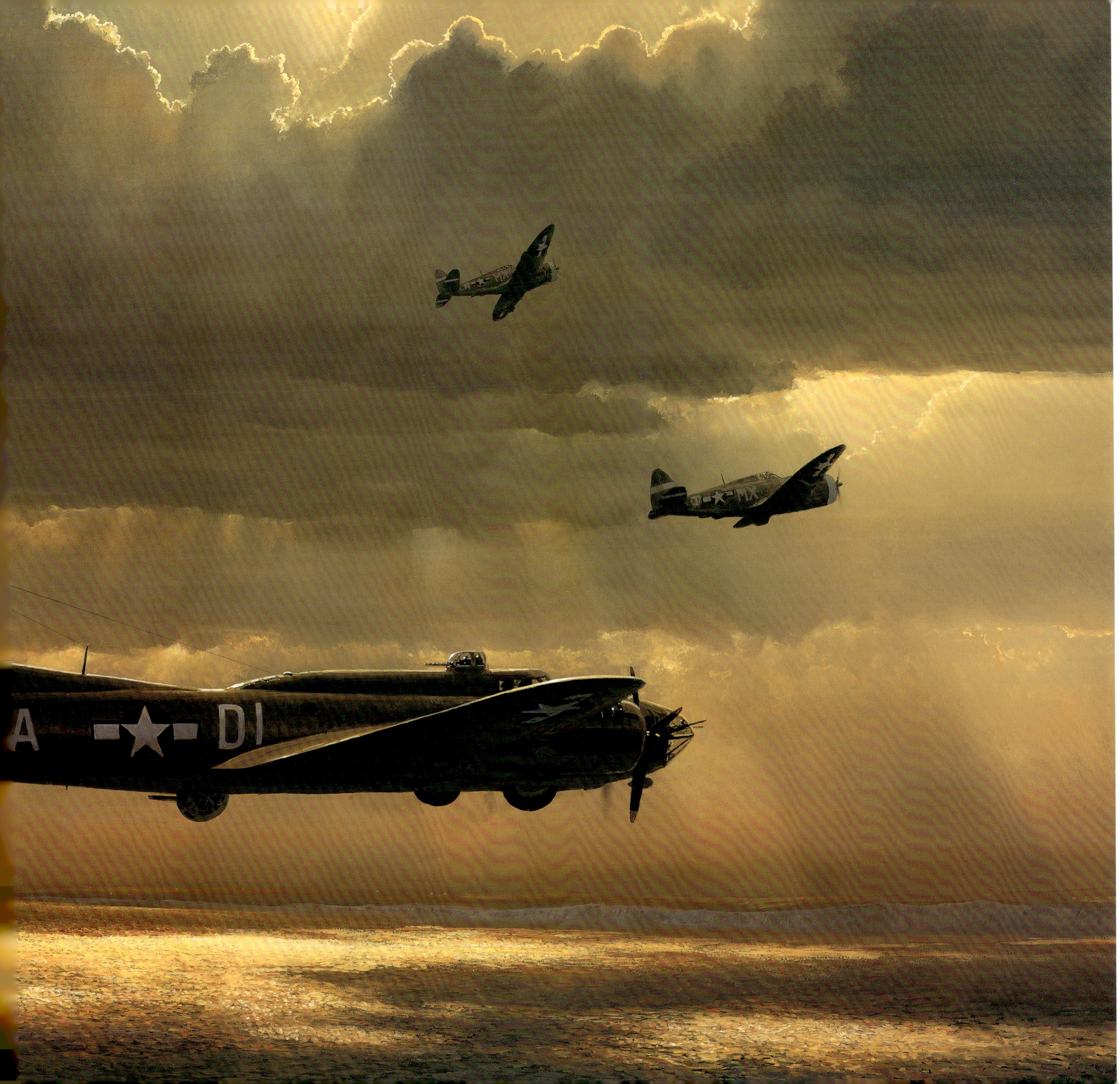

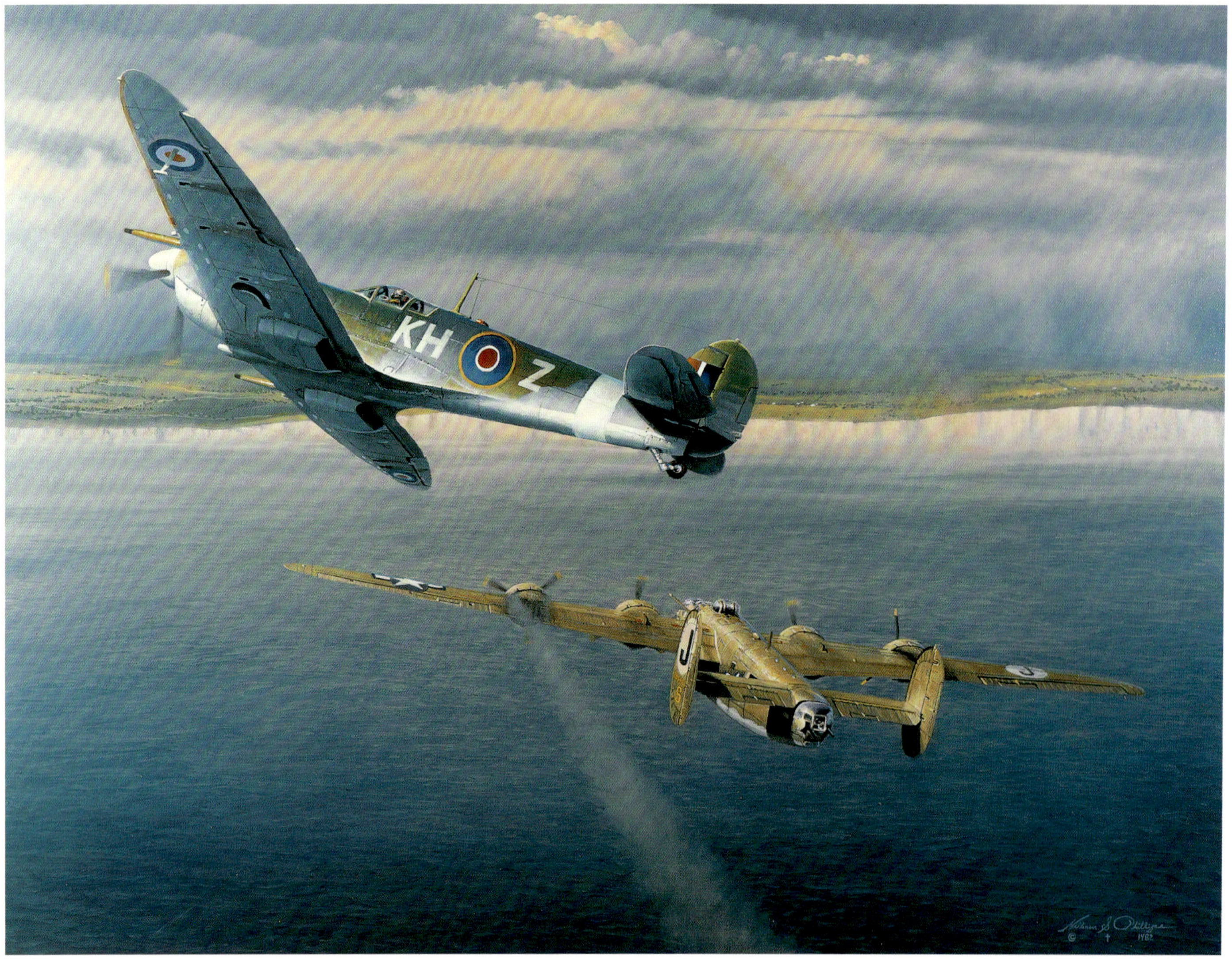

## "Welcome Home, Yank"

Air power from the United States first thundered into European skies during the summer of 1942. Departing Egypt for the first of many daring long-distance raids, thirteen B-24 Liberators carried their bombs to wreak havoc on oil refineries at Ploesti, Rumania. The 1000-mile aerial surprise awakened the Nazi command to America's presence. In this scene, a troubled Liberator limps toward the safety of its English base that is heralded from miles away by a white-striped horizon, the chalk cliffs of England. Did this embattled bomber escape the fire and fury of enemy skies only to fail to make it over those cliffs? Is the crew scrambling to brace for ditching? But hopes rise with the appearance of the beautiful elliptical-winged Supermarine Spitfire circling overhead. The Spit pilot takes a protective position, his radio on and ready to place the call for Air-Sea rescue, if need be. In the meantime, he radios, "Welcome home, Yank."

## Give Us this Day

After the ravages of Nazi assaults on England in the early years of World War II, the bombings in the Pacific by the Japanese on December 7, 1941 (followed four days later by the declaration of war on America by Germany and Italy) the United States ceased its isolationist stand and came to the aid of its allies. By 1942, its military men, women and machines were stationed throughout Britain. Growls of departing aircraft engines broke the stillness of each British morning as men and their machines headed toward dangerous skies over occupied Europe. In the hazy golden light of dawn, this U.S. 8th Air Force B-17G from the 398th Bomb Group, 602nd Bomb Squadron out of Northamsted, carried its crewmembers away from the sanctity of the tranquil countryside into the fires of war. The ardent prayer was shared by all, "Give us this day. . . ."

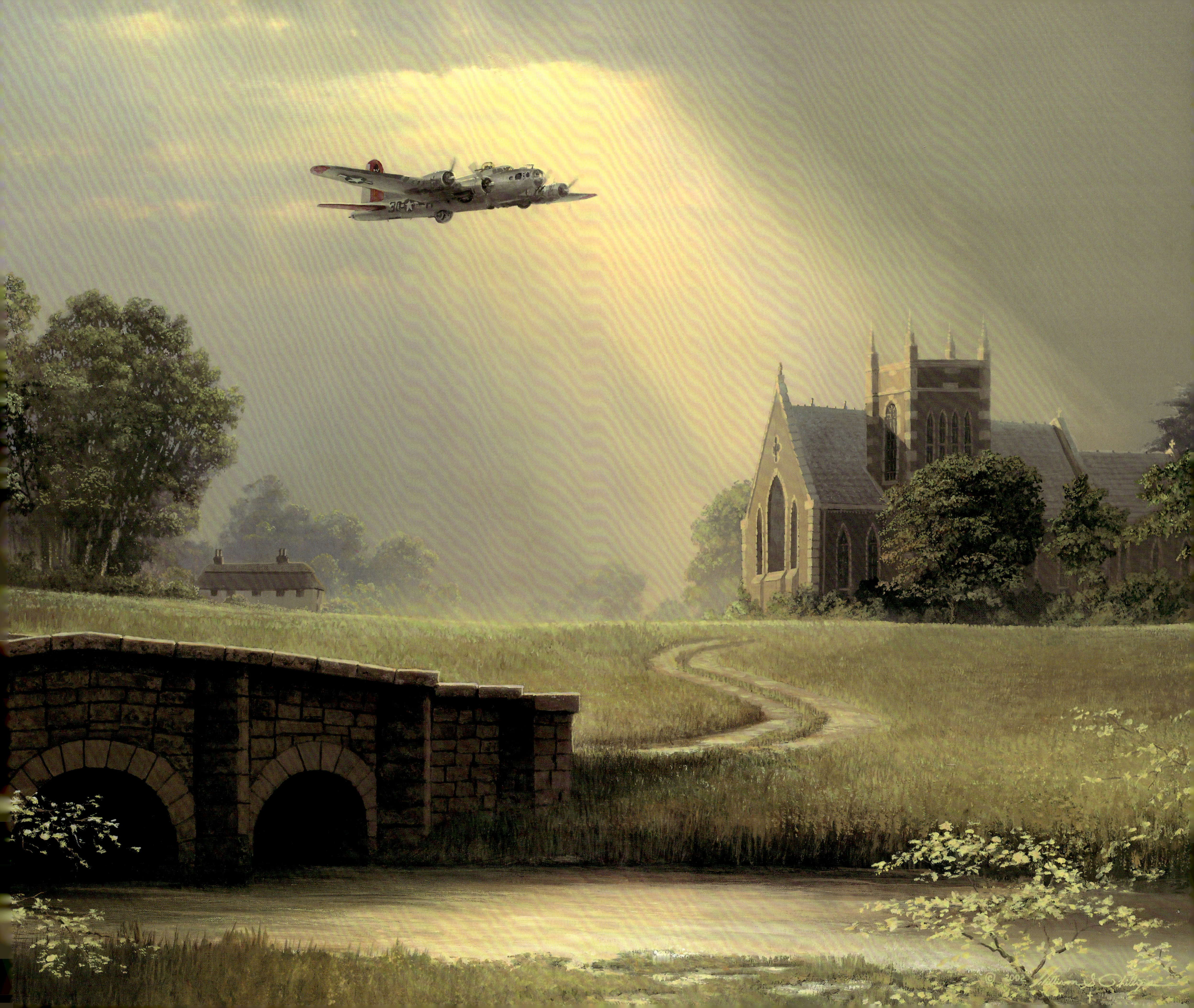

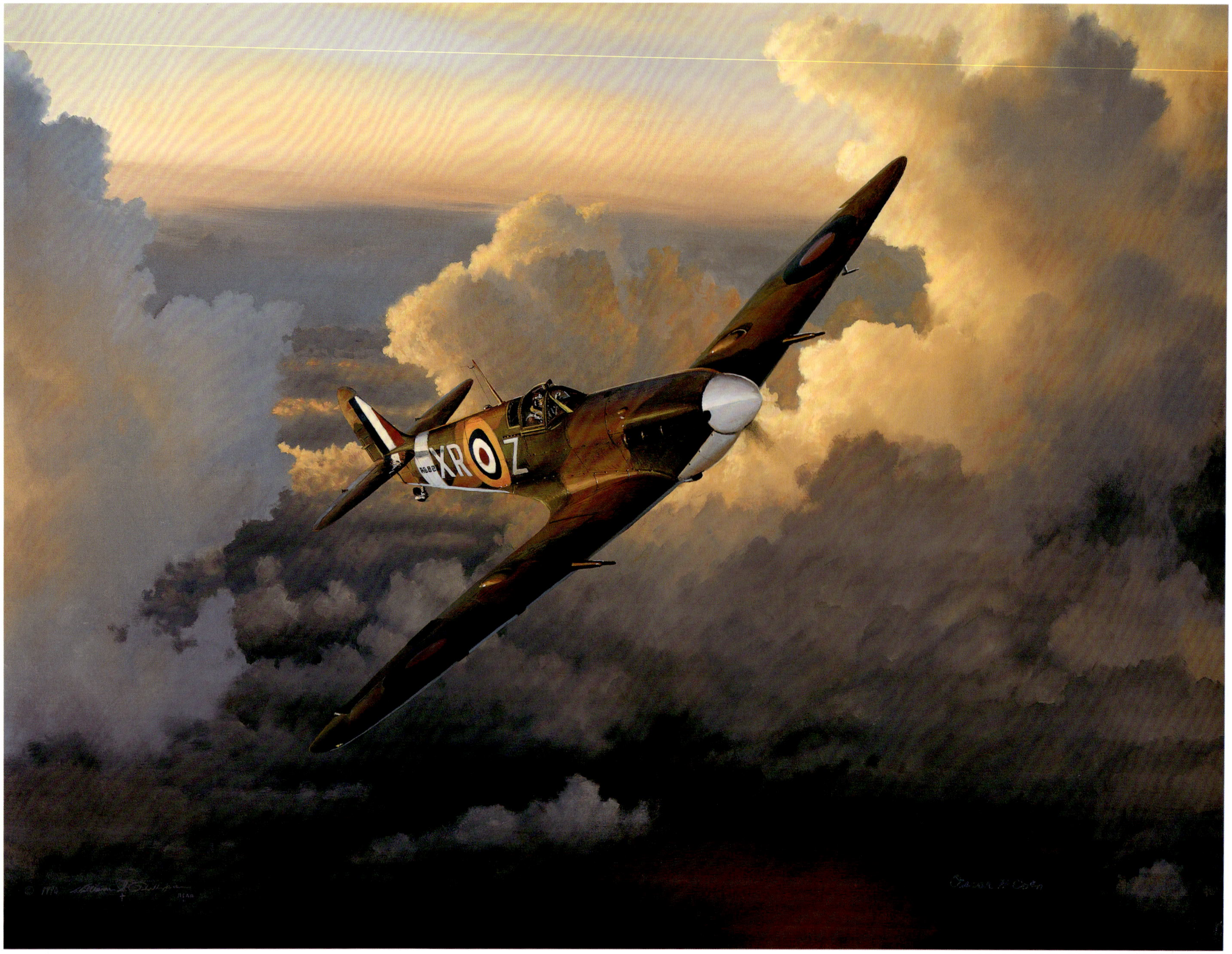

## A Time of Eagles

Following the decisive Battle of Britain in the autumn of 1940, which forced Hitler from invading the United Kingdom, American pilots flocked to volunteer for the Royal Air Force. They formed into three Eagle Squadrons, the first of which was Squadron 71 created in September 1940 at Church Fenton. American volunteer Pilot Officer Oscar Coen was assigned to this squadron. Bill Phillips was honored to meet with Oscar and Jenny Coen at their Oregon ranch and learned first hand what it was like for an American to lead his fellow pilots into World War II. Oscar told him, "A lot of guys went because the British Spitfire was the hottest thing in the air and they wanted to fly it." During his service to the RAF, Oscar downed six enemy planes, took a hit, bailed out, and survived to relish his memories. In capturing Coen's achievements, Bill placed the craft in its element—the vast dome of moisture-laden air filled with billowing cumulus clouds and tinged with lateral streams of early morning light.

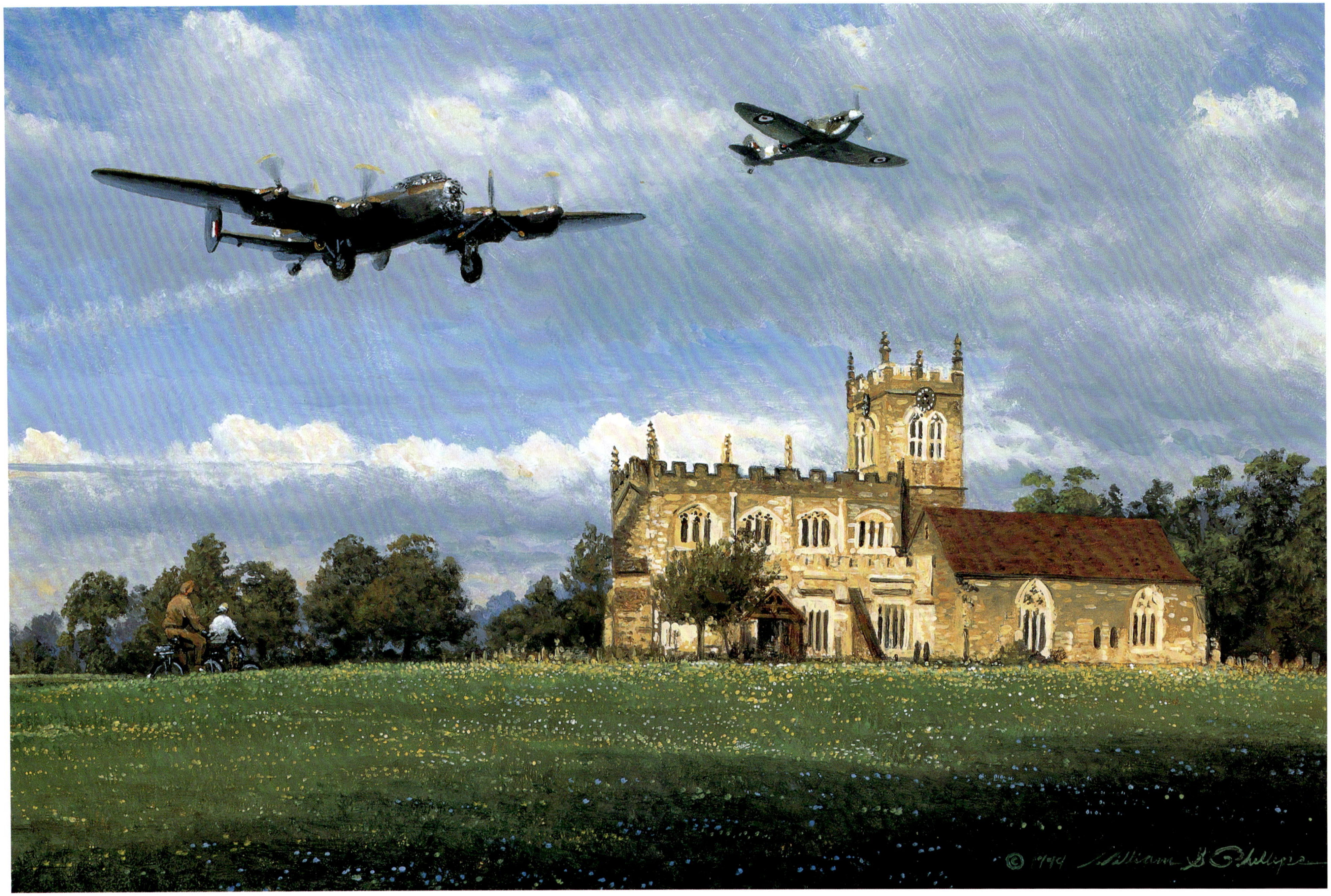

## Almost Home

Having survived a night raid over enemy territory, one of the most successful of the Royal Air Force's night bombers, an Avro Lancaster, limps home on three of its engines. Introduced in early 1941, a total of 7,300 Lancasters were eventually produced. Their most famous mission was the May 1943 skip bombing by 617 Squadron that burst Germany's Moehne, Eder and Sorpe Dams, flooding the industrial Ruhr Valley. Relentless nighttime runs against Germany were instrumental in accelerating war's end. The Avro Lancaster shown here, a source of British pride, is escorted by a RAF Supermarine Spitfire, another aircraft that earned Allied admiration.

# I Could Never Be So Lucky Again

Target: Japan. At 8:20 A.M. on April 18, 1942, the first of sixteen North American B-25B Mitchell bombers took off from U.S. Navy carrier, USS *Hornet*. Piloting the lead aircraft was Lt. Col. James H. Doolittle, General "Hap" Arnold's personal choice to direct the mission. Japanese onslaughts in the Pacific, begun on December 7, 1941, had continued unabated and President Franklin Roosevelt pressed his military commanders to launch a bombing raid against Japan. Plans were discussed and discarded before Navy Chief Admiral King and Air Force Commander Arnold accepted the proposal of Navy Captains Francis S. Low and Donald B. Duncan to launch Army medium bombers from a carrier and have them recover in China. To coordinate Navy activities, King chose Duncan, and Arnold chose Doolittle, a persuasive leader willing to accept risks. Jimmy Doolittle was a race pilot, a test pilot, an aeronautical engineer and later, a living legend who is quoted as having said, "I would never want to relive my life; I could never be so lucky again." In absolute secrecy and working with Admirals Nimitz and Halsey, Duncan assembled a task force of ships, and Doolittle chose twenty-four volunteer crews with B-25 experience. Doolittle enlisted a contractor to modify the Mitchells for the mission, and then selected Eglin Field, Florida for practice in short takeoffs, overwater navigation, and crew coordination.

The crews and aircraft eventually arrived at Alameda Naval Air Station to be taken aboard the *Hornet*, now part of an eight-ship task force. Departing on April 2, 1942 to rendezvous with Halsey and head west, the mission was underway. As they neared seas totally controlled by the Imperial Japanese Navy, Capt. Marc Mitscher, the *Hornet's* skipper, advised Doolittle to be ready to depart on short notice. When two Japanese patrol boats were spotted, the Doolittle Raiders, as they came to be known, prepared to launch despite the fact that they were a greater distance from the target than planned. Doolittle departed, circled back over the carrier to verify his compass accuracy and headed west, followed by Lt. Travis Hoover in plane number two. At 9:20, the last aircraft lifted off the *Hornet* and the Navy task force turned back to Hawaii at full speed.

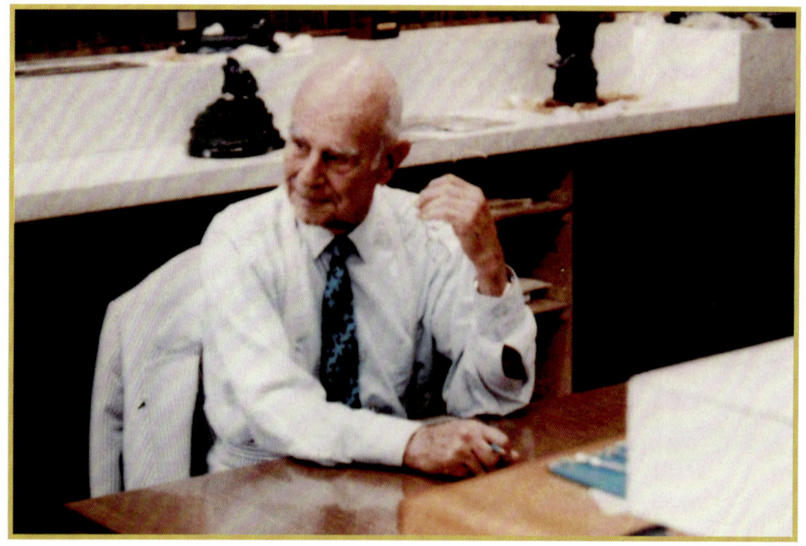

Jimmy Doolittle takes a break from counter-signing the Greenwich Workshop Fine Art Limited Edition Print, *The Giant Begins to Stir*, in Los Angeles, California in 1983.

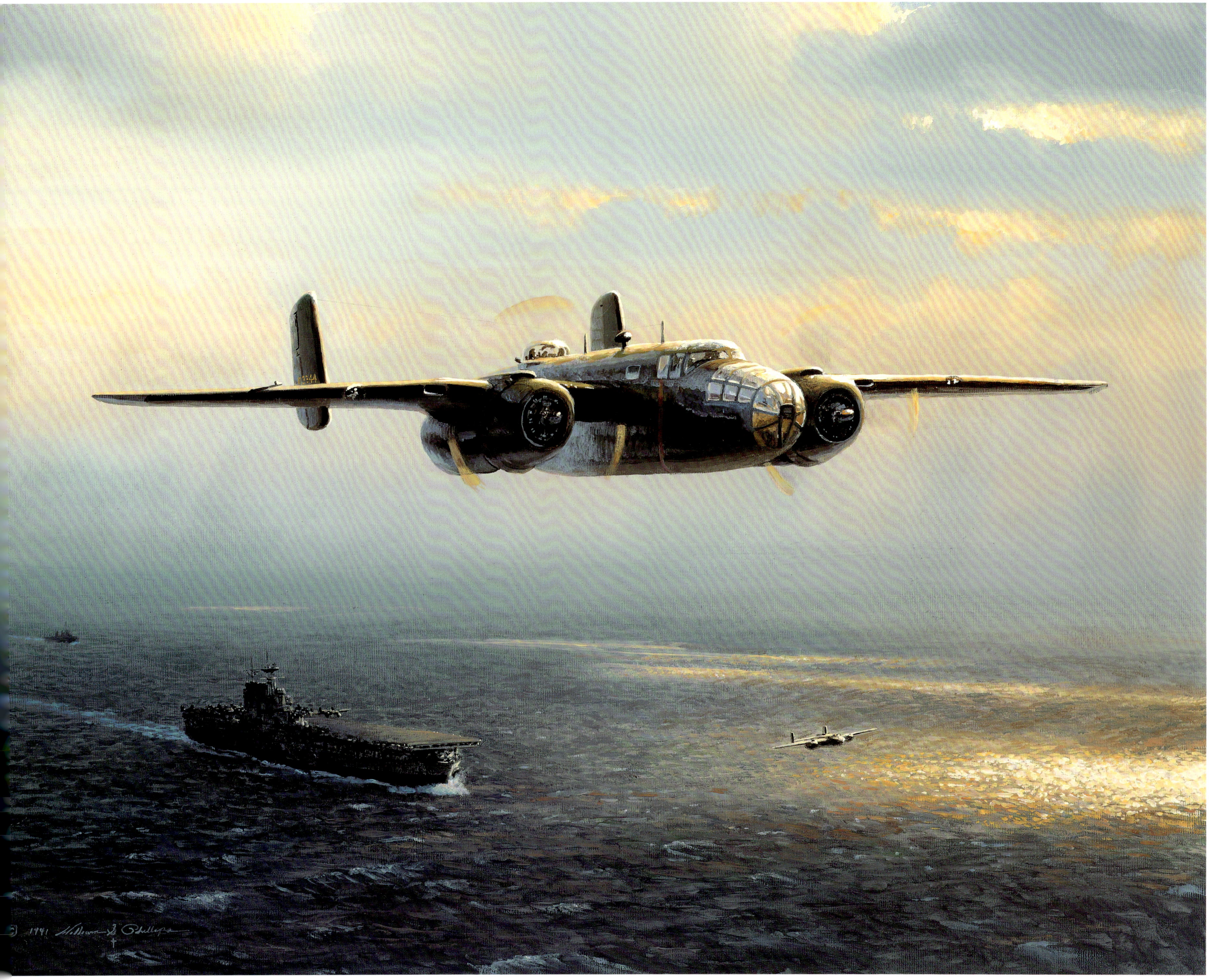
© 1991 William S. Phillips

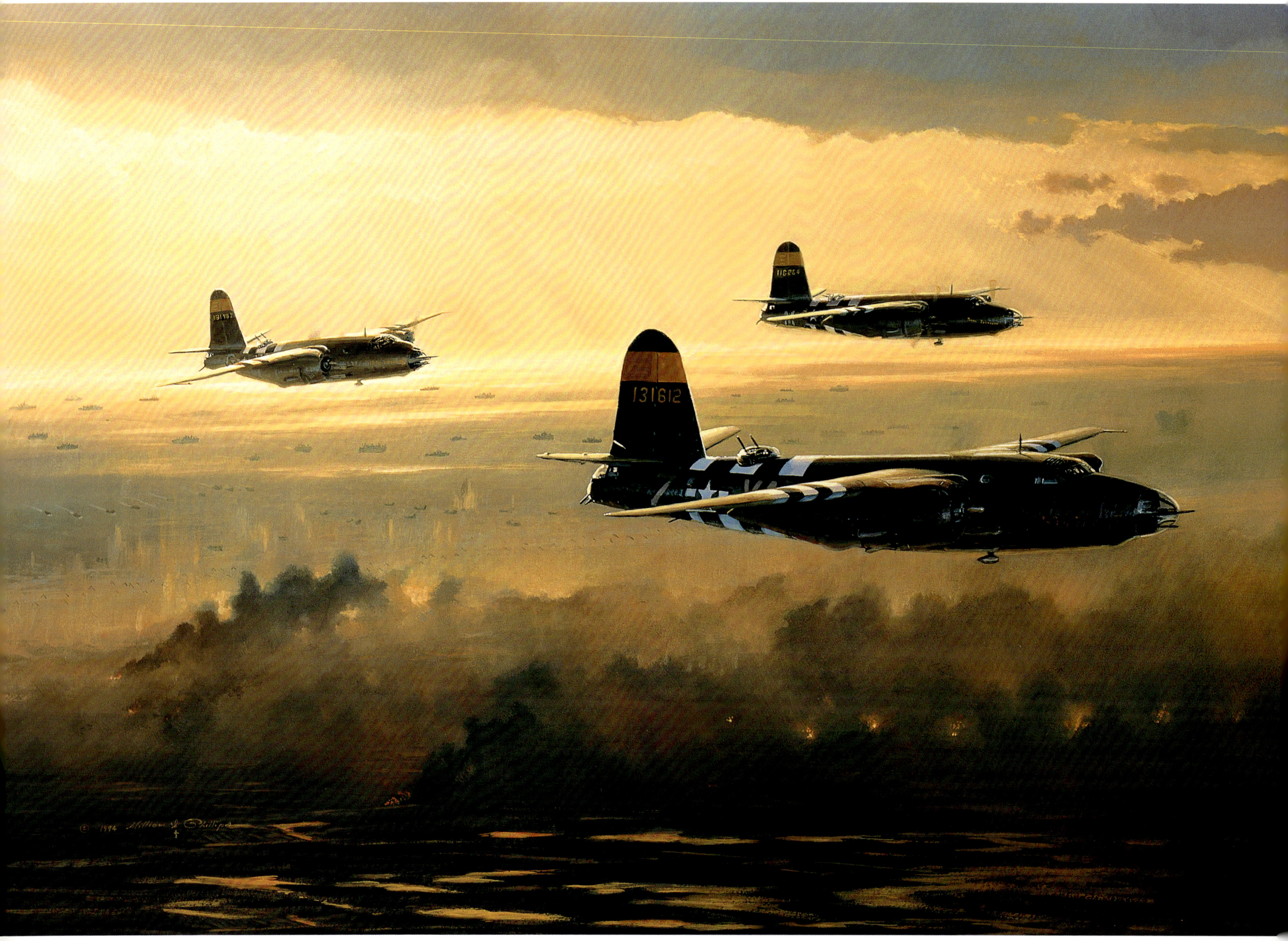
131612

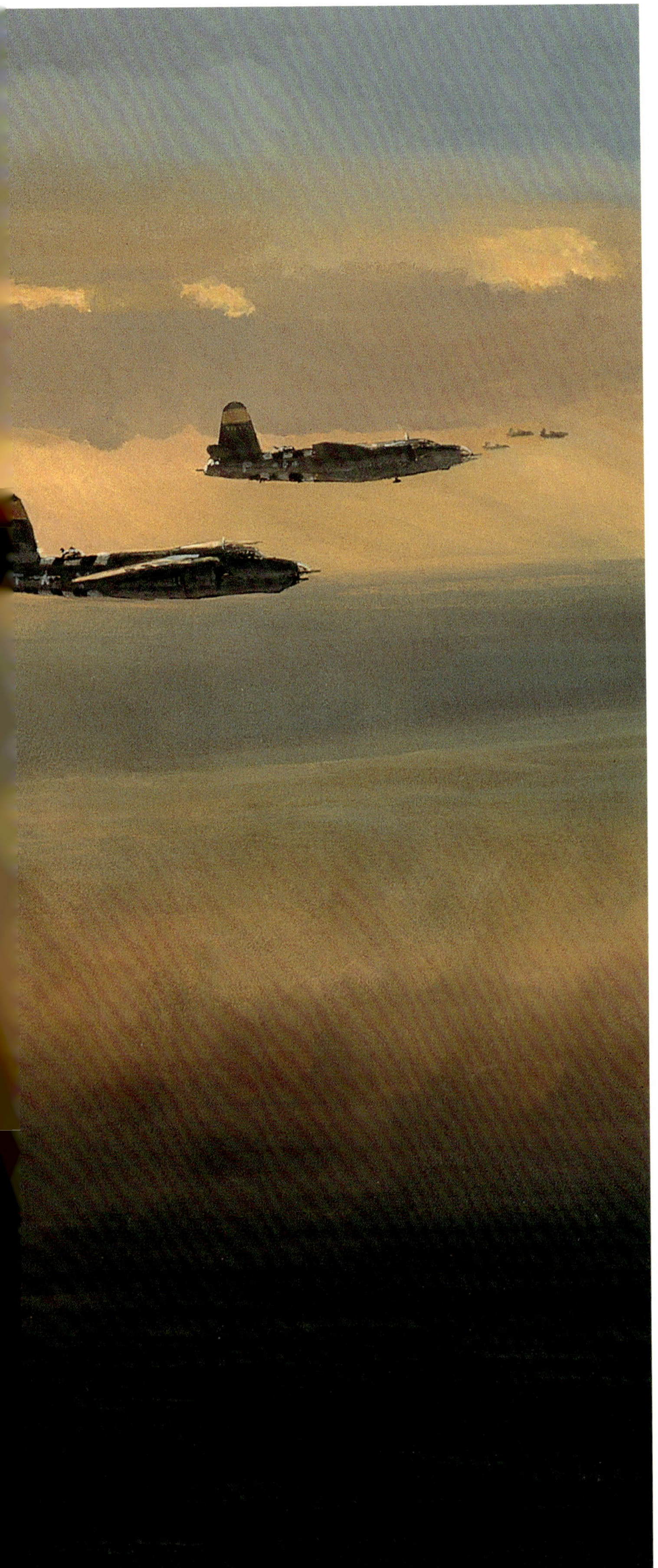

## The Storm Breaks Over Normandy

Before the sun dawned on June 6, 1944—D-Day!—the invasion of Normandy had begun. In an awesome display of airpower, 900 aircraft and 100 gliders dropped paratroopers behind enemy lines. P-38 Lightnings of the Eighth and Ninth Air Forces covered convoys approaching the beaches, screaming low above long rows of German-erected angular steel barricades that rose from the beach like defiant clenched fists and dared any enemy of the Germans to attempt to cross. More than 2,300 B-17 and B-24 heavy bombers blackened the sky and were tasked with strategic bombing of enemy defenses.

Shown here in a painting commissioned by *Smithsonian* magazine in honor of the 50th Anniversary of the invasion, the 800 Ninth Air Force medium bombers, Martin B-26 Marauders, stream in from over the water with their arsenal of bombs to supplement the heavies. The Martin B-26 made its first appearance in November 1940 and its design had shown such promise that the U.S. Army Accessions Command ordered 1,131 the September prior to the bomber's first flight. The B-26s were initially assigned to the Southwest Pacific in 1942 but, two years later, they flew primarily out of England and the Mediterranean theater, providing valuable tactical support on D-Day against their targets: the coastal defense batteries, rail and road crossings, bridges, and marshalling yards.

With six nations involved—Germany, against Australia, New Zealand, Canada, Britain, and the United States—instant aircraft identification was essential. To avoid confusion and to ensure proper recognition, the 13,000 Allied aircraft participating in the invasion bore prominent black and white invasion stripes on the aircraft fuselages and wings. The D-Day operation, called "The Great Crusade," by General Dwight Eisenhower, brought the largest invasion force in human history—by land, sea, and air—to liberate an occupied France, and crush Adolf Hitler's Nazi forces.

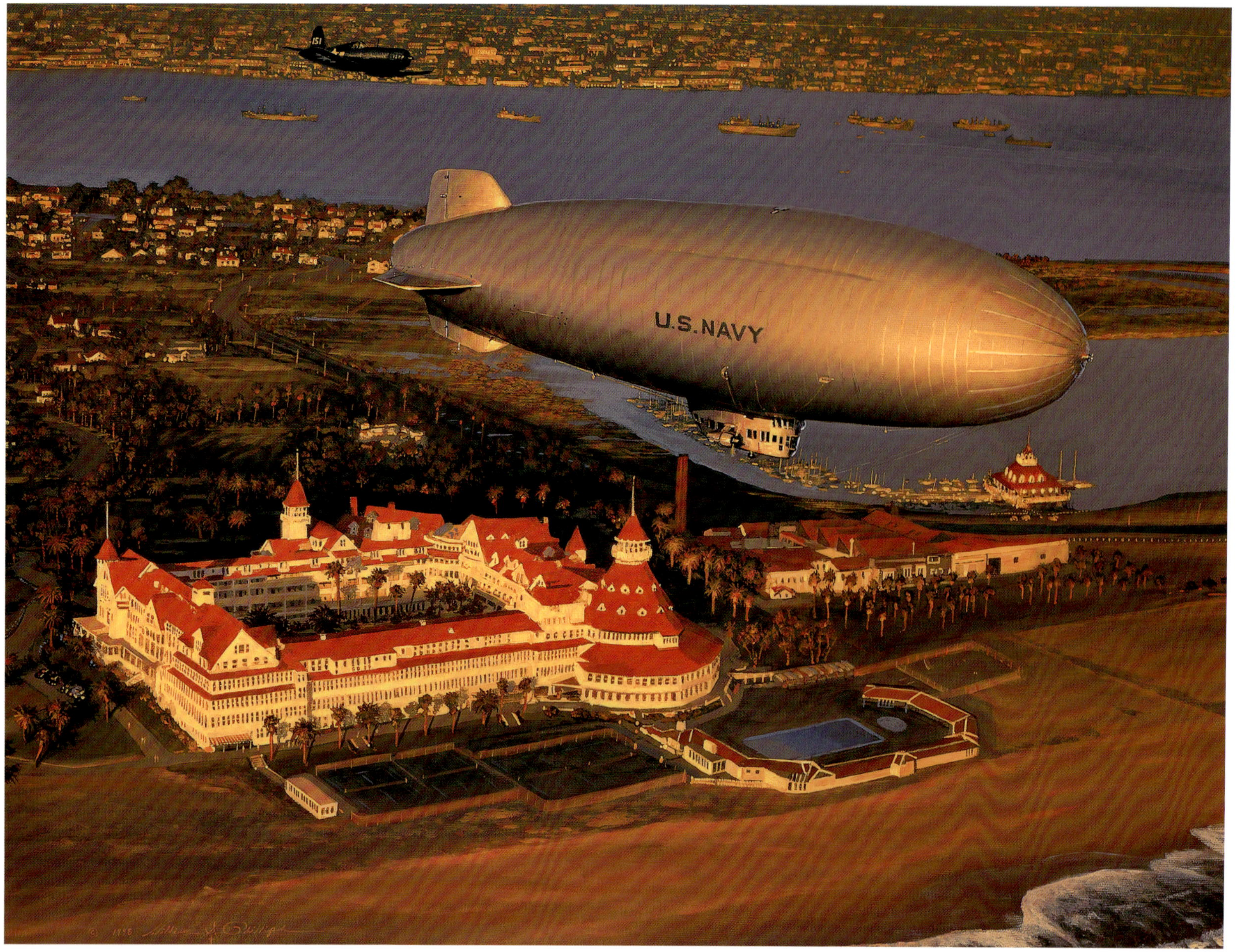


## K-ship Over Hotel del Coronado

First tested in 1931, the prototype K-1, a nonrigid airship, neared 252 feet in length. World War II gas-enveloped K-ships logged 550,000 flight hours and escorted 89,000 naval surface ships; they carried radios, radar, sonobuoy receivers, magnetic detection equipment, depth charges, contact bombs, a machine gun and homing torpedoes. The heavily armored Corsair, with its distinctive inverted gull-wing, was powered by Pratt and Whitney's 2800 radial engine, and could out-climb, out-fly, and achieve a better roll rate than the Japanese Zero. In the background is California's historic Hotel del Coronado. "The Del," built in 1888 and designated as a National Historic Landmark in 1977, is among the world's best resorts. It is shown here as it appeared in the 1940s. The painting was commissioned by the wife of a pilot who flew both the Chance Vought F4U Corsair and the Goodyear-built U.S. Navy K-ship.

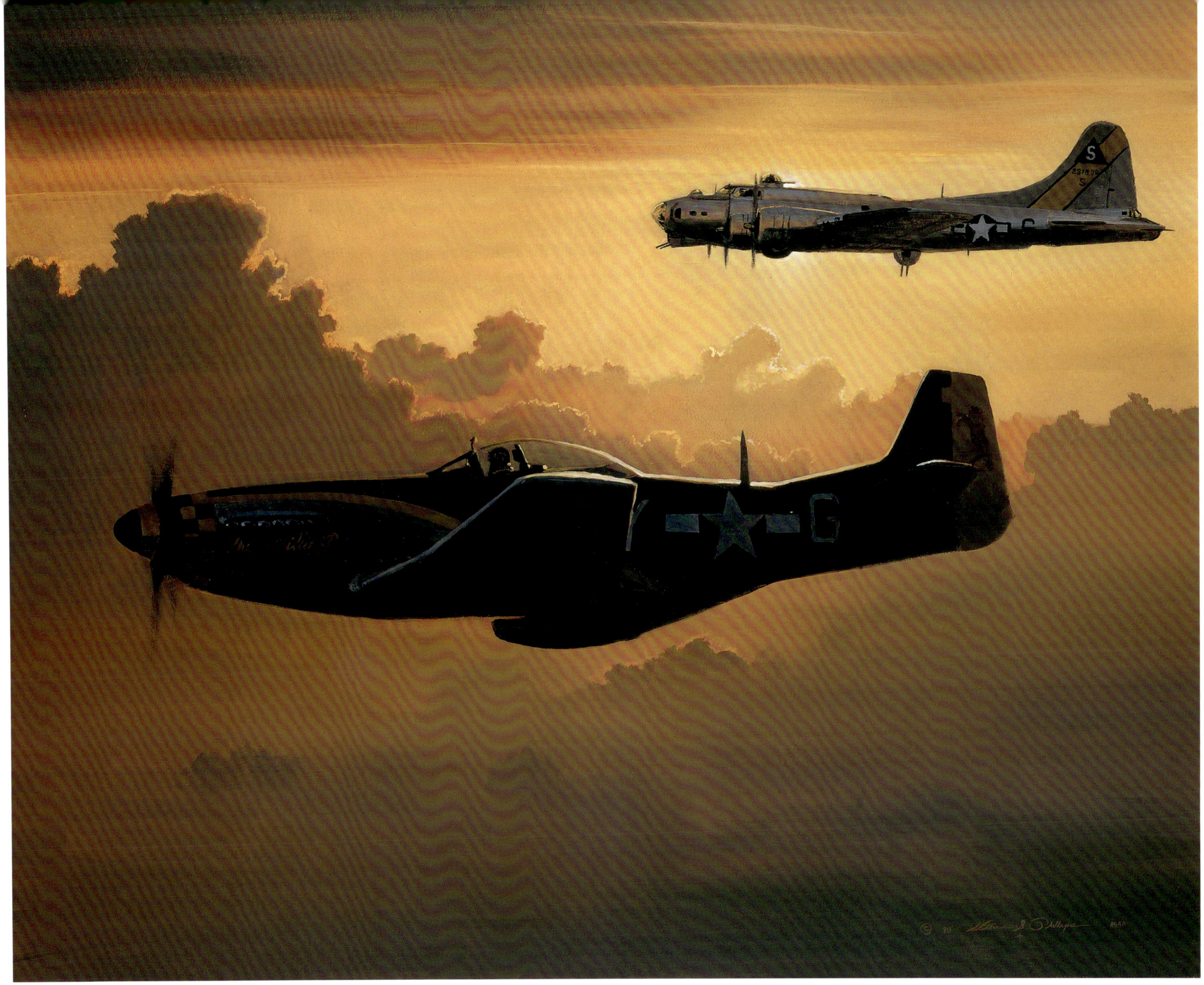

## Sunset Serenade

The air war over Europe brightened considerably for the 8th Air Force with the arrival of P-51 Mustangs in July 1944. Earlier losses of young bomber crewmembers to enemy fighter aircraft had been horrific. Designed with long-range capabilities, P-51s provided protective cover throughout entire bombing runs for beleaguered strategic bombers and finally, losses could be curbed. In a new tactic, fighter pilots were "turned loose" to engage and destroy as many Luftwaffe fighters as possible. This reduced the number of experienced enemy pilots, while bombers attacked the enemy's aircraft production. Fighter escorts with the bombers' range signaled a change in the tides of the air war. Here, a B-17G, having completed the mission safely, cruises in companionable appreciation beside *The Millie G*, named for the wife of the pilot Major Ed Giller, 343rd FS, 55th FG, who was credited with having destroyed a German Me-262, the first combat jet.

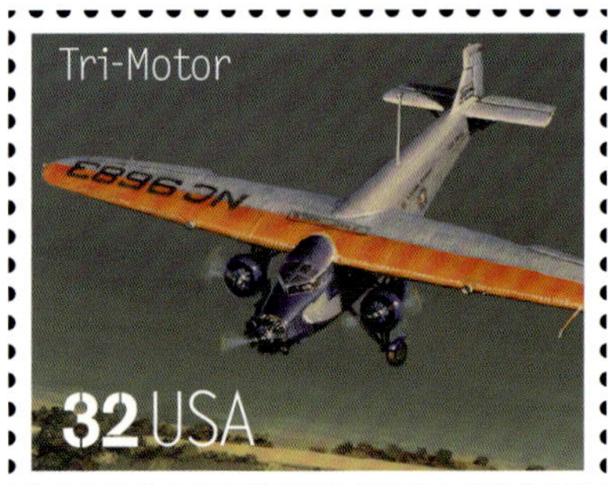


## Ford Tri-Motor

U.S. Postal Service, Classic American Aircraft series, 1997

This legendary transport, Henry Ford's all-metal contribution to the quest to offer the public trustworthy air transportation, rode the aviation boom of the late 1920s. Ford sponsored National Air Tours and the "Tin Goose" was largely responsible for bolstering public confidence in the air carrier industry. It became the first successful airliner in the United States.

## Grand Lady of the Canyon

Promoting a bird's eye view of the awe-inspiring Grand Canyon to fifteen lucky passengers, Scenic Airways was founded in 1927 and the first commercial air tour was born. Taking to the air (originally from the site that transformed cotton fields into what is now the bustling Phoenix, Arizona Sky Harbor Airport) Scenic Airways evolved into Grand Canyon Airlines. In May 1928 the airline purchased the first of several Tri-Motors from Henry Ford, which carried on the Ford legacy of reliability. The Tri-Motor, one of almost 200 built by Ford with engineering that began with William Stout, sought to emphasize safety and encourage the public's trust in aerial transportation. Taking its thrust from three Pratt & Whitney radial engines and gaining lift with a wingspan of seventy-seven feet, ten inches, the "Tin Goose" boasts a unique corrugated aluminum skin that covers a forty-nine foot, ten inch fuselage. Grand Canyon Airlines continues to nurture nostalgia by periodically winging over the majesty of the Grand Canyon in one of two vintage Ford Tri-Motors still owned by the company.

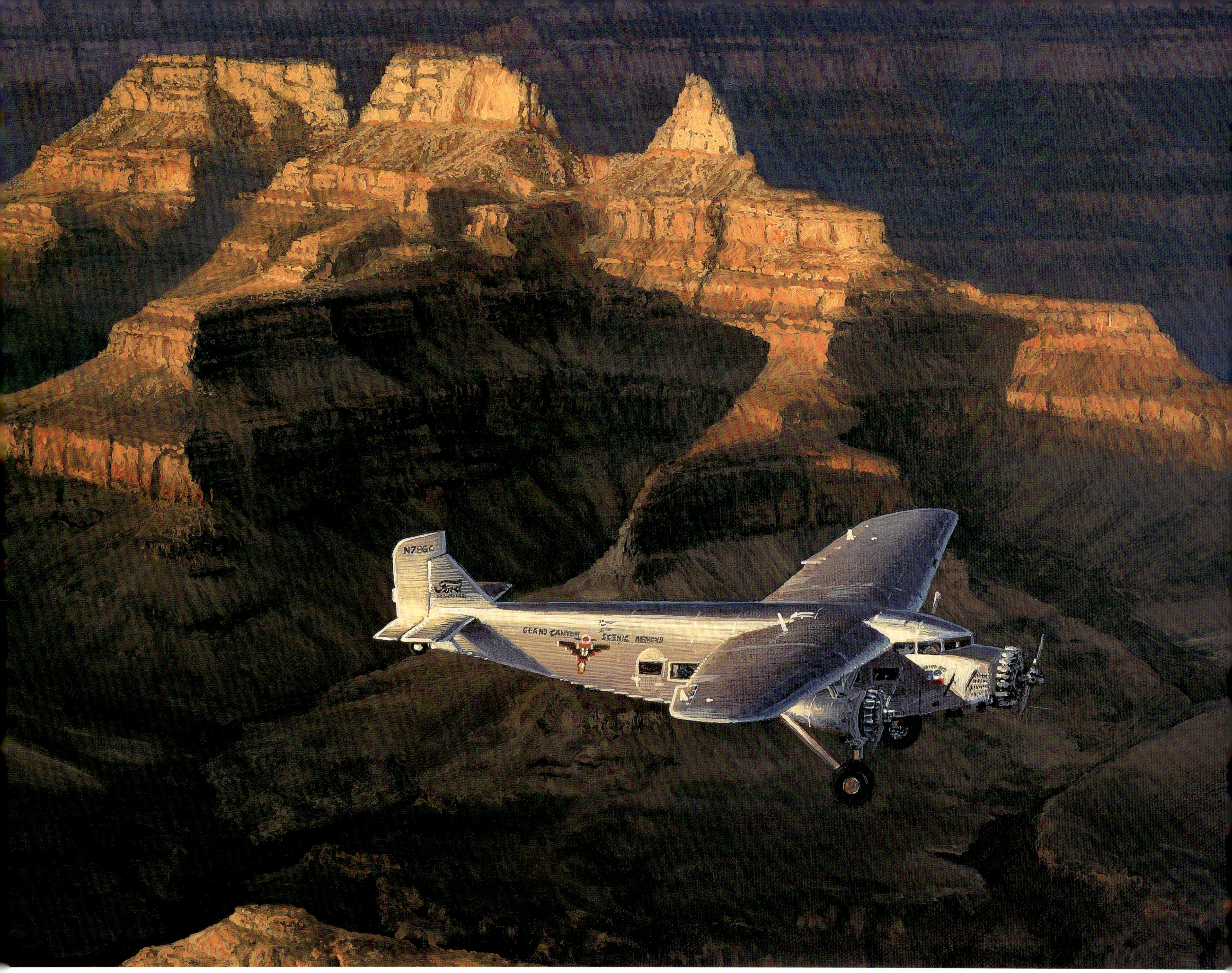
Ford
GRAND CANYON SCENIC AIRWAYS

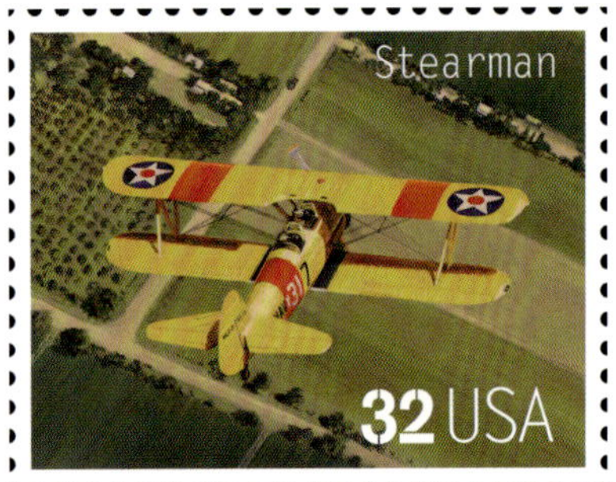


## Boeing Stearman N2S-3

U.S. Postal Service, Classic American Aircraft series, 1997

Described as a superb, kindly and forgiving trainer, the PT-13 Kaydet and its variants were aerial classrooms to thousands of young flight students of the U.S. Army Air Corps and as the N2S-3 for the U.S. Navy. Designed by Lloyd Stearman, more than 8,000 were manufactured and many are still flying as trainers, crop dusters, and as personal aircraft.

## Chasing the Daylight

From the beginnings of powered flight, pilots have dropped low to follow the "iron compass" when away from airfields and seeking directional guidance. Bill Phillips has depicted two U.S. Navy trainees in their N2S-2 Stearman trainers along the southern coast of California. They'd been forced close to the ground by ominous clouds and their joy was complete when the railroad tracks they'd sought happened to be the rails upon which the Southern Pacific's premier Coast Daylight locomotive was charging. The chase was on!

No young man could resist the adrenaline rush of pitting his winged racer against the speeding steam engine. Halfway back along the rushing train, a second Navy trainee banked to join the race. The pilots had the added advantage of the wind whipping past to convince them of their good fortune to be young, to be Navy pilots and to be buzzing a train that had fascinated them from childhood.

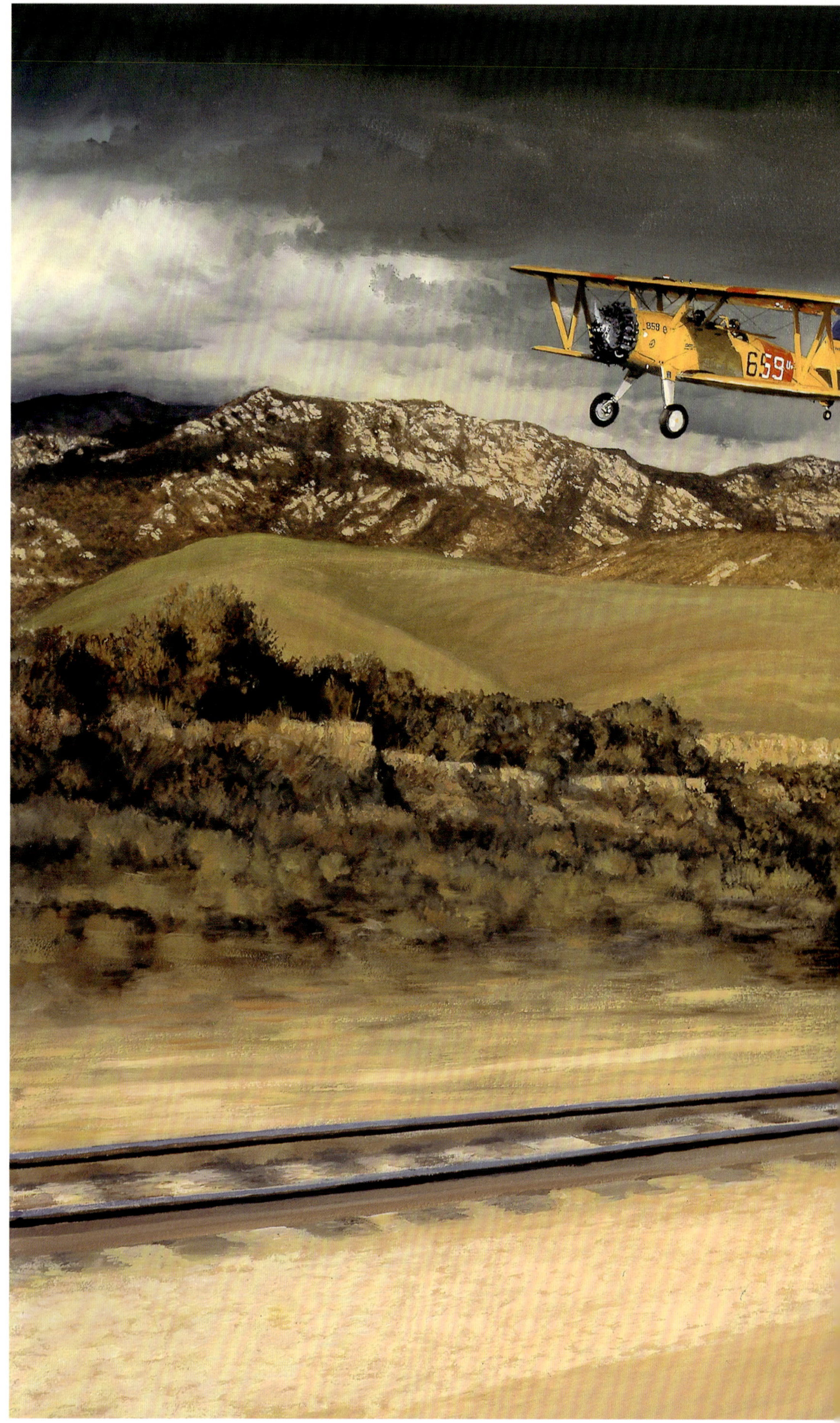

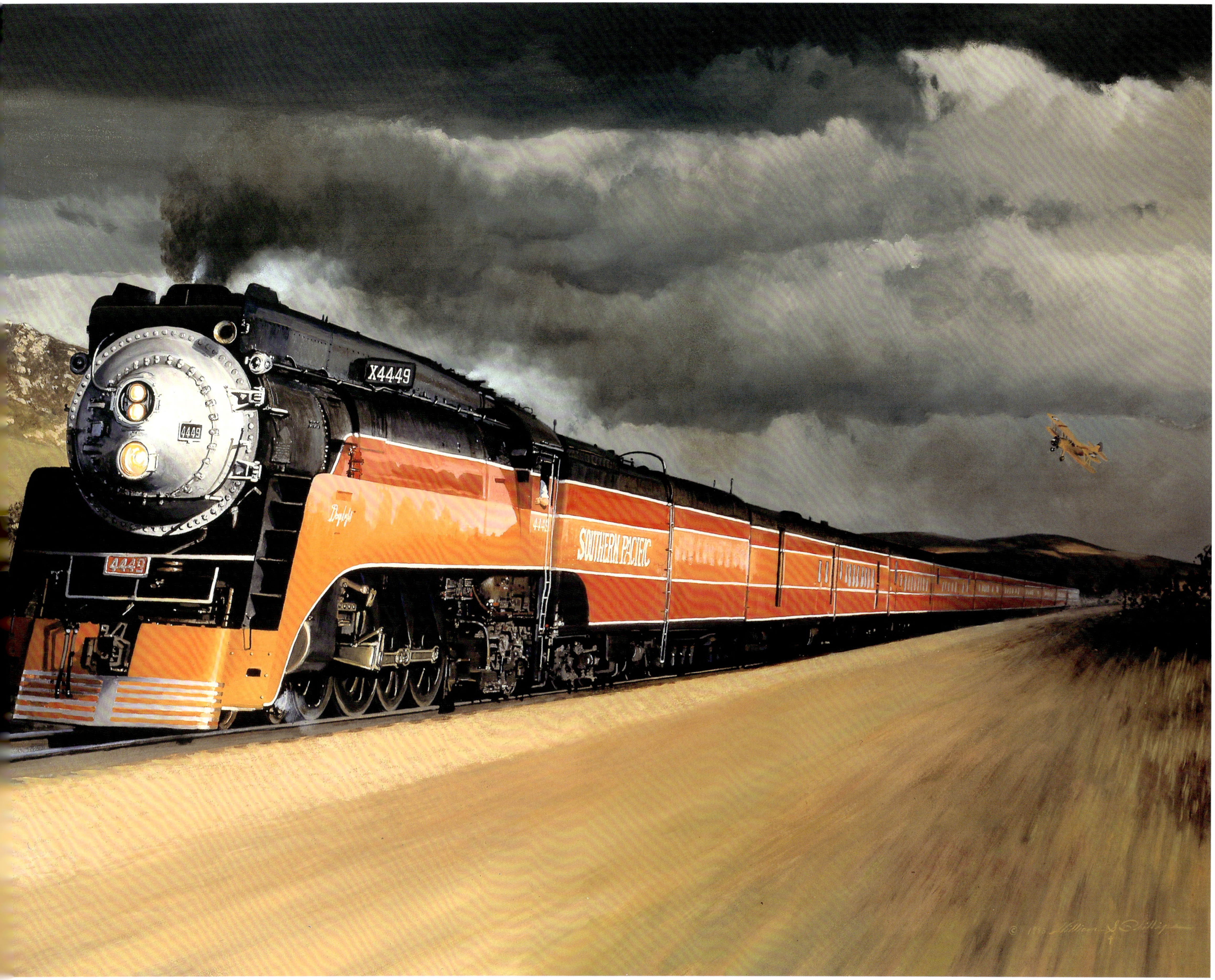
X4449
4449
4449
SOUTHERN PACIFIC

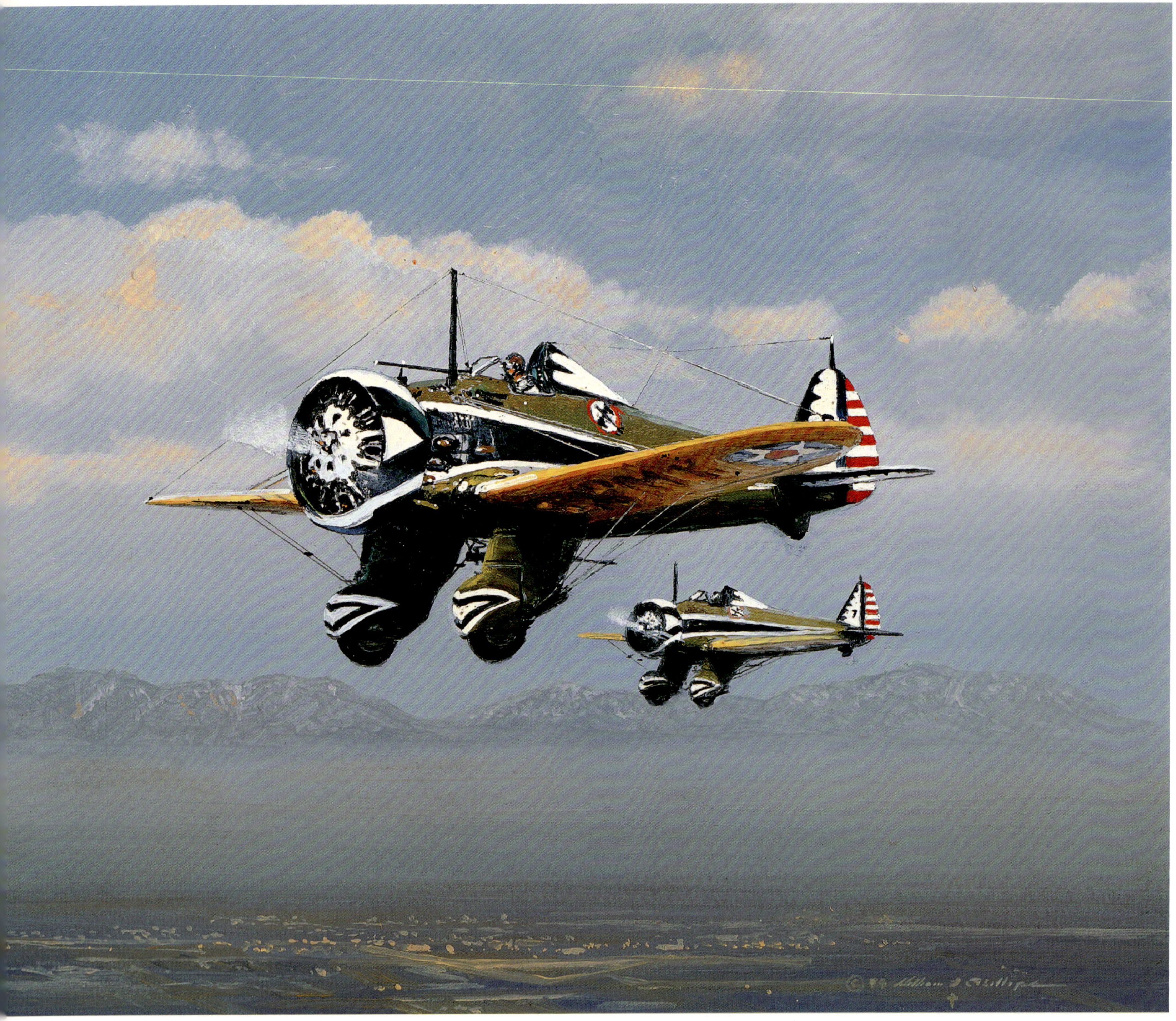

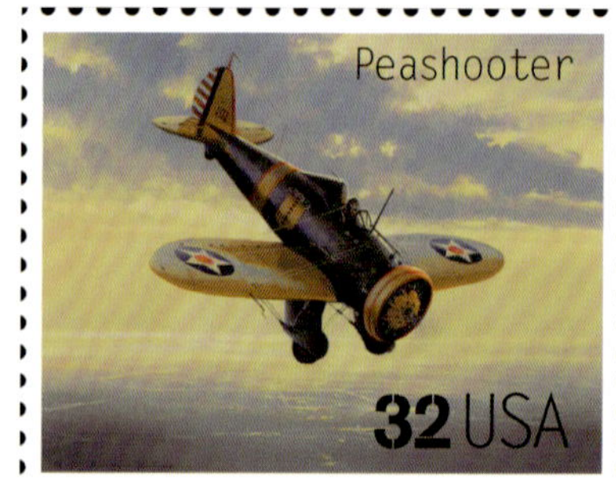


## Peashooter

U.S Postal Service, Classic American Aircraft series, 1997

In marking aircraft so as to identify the unit to which they belonged, distinctive painting of the P-26 Peashooter continued a tradition established by the Army Air Service during WWI. Although the Peashooter resembled 1930s aircraft that competed in Bendix, Thompson, or Greve Trophy races, it was a radical design for the U.S. military.

## P-26 Peashooter

Working on a prototype throughout 1931 to meet specifications determined by the U.S. Army Air Corps, the Boeing Company designed a fighter aircraft that bridged a gap between World War I biplanes and modern metal monoplanes. The resulting P-26, the sports car of the air, was first flown in March 1932. It bore fixed landing gear, its drag reduced by smooth metal fairings, and external bracing wires to further strengthen cantilevered wings. The P-26, called the "Peashooter," was powered by Pratt and Whitney's closely-cowled 600-horsepower Wasp radial engine and was the first among U.S. Army Air Corps' aircraft to be equipped with landing flaps. In addition to the standard Army chrome-yellow wings and blue fuselage, P-26s colorfully depicted their origins—flight, squadron, and group. These Peashooters proudly hail from the 17th Pursuit Group, 34th Pursuit Squadron based at March Field, Riverside, California.

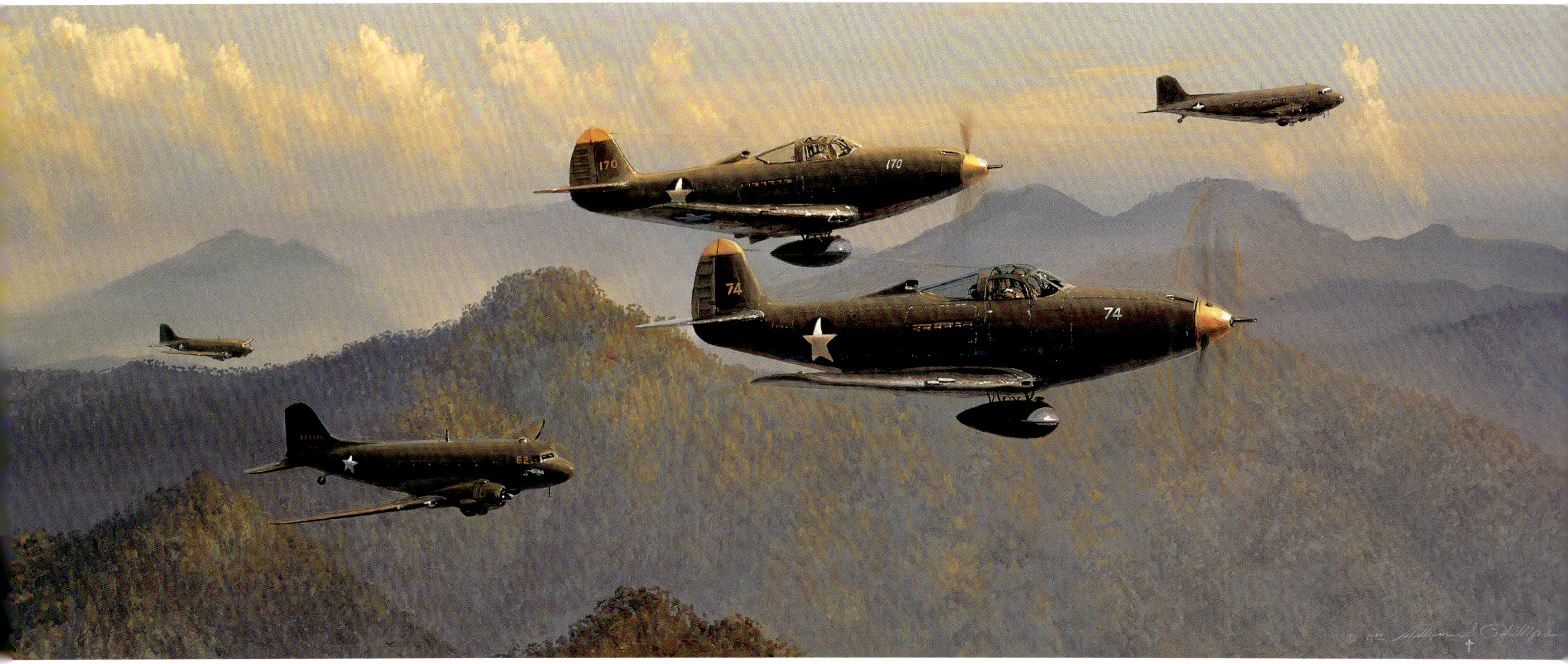

## NANETTE

Edwards Park, the P-39 pilot who flew this "sexy, dicey, ever-lovin' harlot of a plane" out of New Guinea during WWII, titled his biography in her honor. "So here's *Nanette* (number 74)," he wrote, "of the grand old 41st Fighter Squadron, on a milk-run to Wau, a gold-mining settlement in the Owen Stanley Mountains, besieged by enemy infantry, that had inched over the hills from their base at Salamaua. C-47s will land uphill on Wau's tiny dirt strip (it's in a box canyon so no one can go around) disgorge tough Australian infantry ('Diggers'), who'll hit the dirt shooting. And *Nanette* and her sisters will stooge around above, their pilots drenched with sweat, fretting about fuel, endlessly squinting into the sun from whence cometh the Zero. The cagey Zeros generally waited until the short-range '39s were committed for the homeward flight—too low on fuel to turn back. Then they bombed and strafed. But Japan never took Wau. The dainty, nearly useless P-39s finally gave way to P-47s. And *Nanette's* skinny, raunchy pilot lived to tell her story in his book."

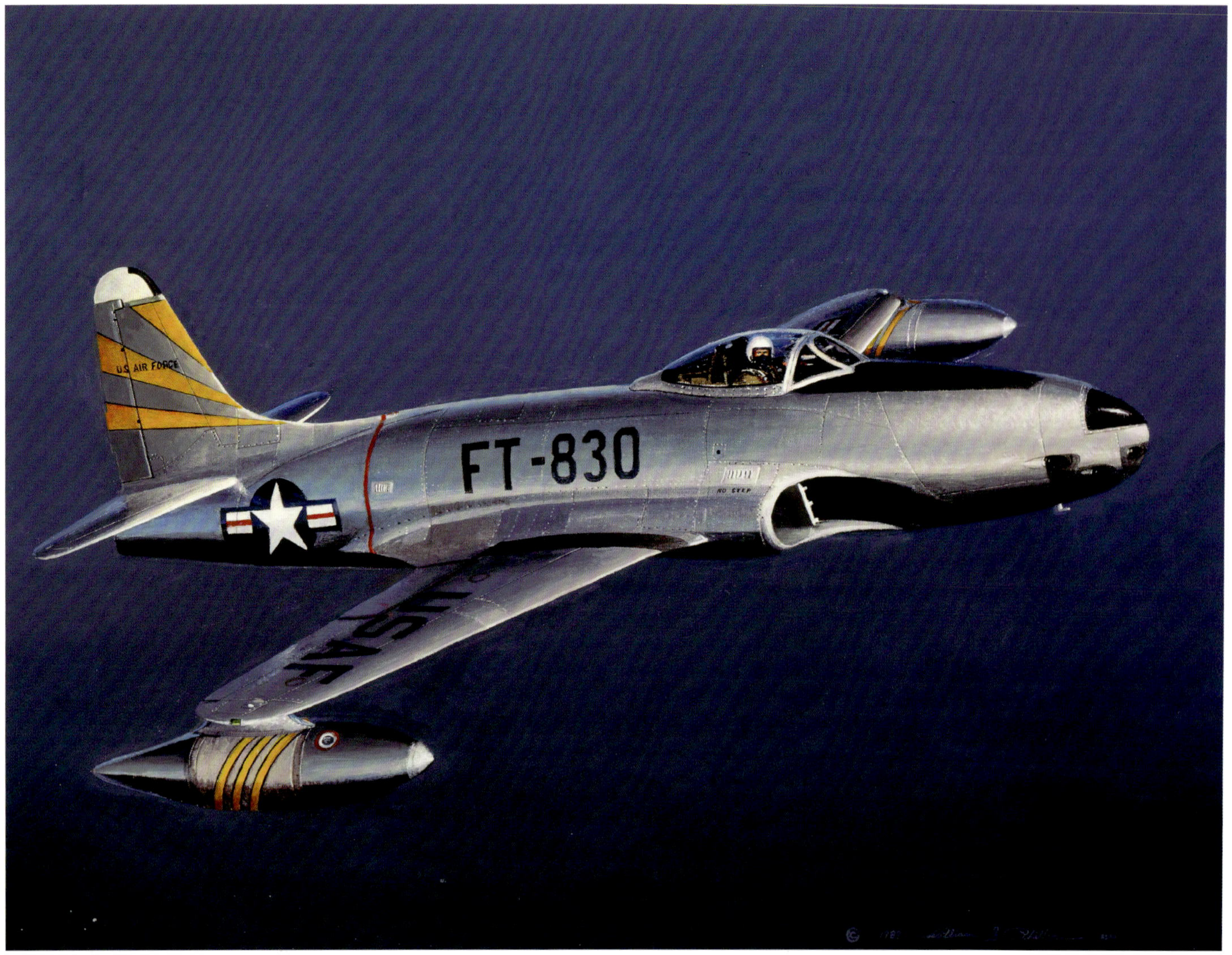


## Jet Pilot

This Lockheed F-80 Shooting Star (a cover illustration for a book by test pilot Tony LeVier and John Guenther) represents all crewmembers who risked lives during the Korean conflict. Representative of the 80th Fighter Bomber Squadron of the 8th Fighter Bomber Wing, the "Headhunters," the jet honors those who flew it from Itazuke Air Base in Japan and Kimpo Air Base in Korea. It was America's first jet fighter to enter aerial combat, although its straight wing rendered it less maneuverable than the swept-wing MiG 15 flown by Soviet pilots in the early days of the Korean conflict.

## Sunlit Sentinels

Reflecting the sun's warm glow, the crimson of the Grand Canyon's O'Neil Butte is repeated in the serpent striping the fuselage of Arizona Air National Guard's F-86A Sabre Jet out of Davis-Monthan Air Force Base. The base, named in honor of two World War I lieutenants, dates its inception as a military field to 1935. It became a training site for B-18 Bolos, B-24 Liberators, and B-29 Superfortresses and later was selected as a storage site for the preservation of U.S. military aircraft. In 1953, it welcomed the jet age when Strategic Air Command units converted to the B-47 Stratojet and the Air Defense Command arrived with North American Sabres. Early models were fitted with narrow-chord swept wings, a revolutionary development that allowed Sabres to achieve greater speed and maneuverability than the straight wing craft that preceded them.

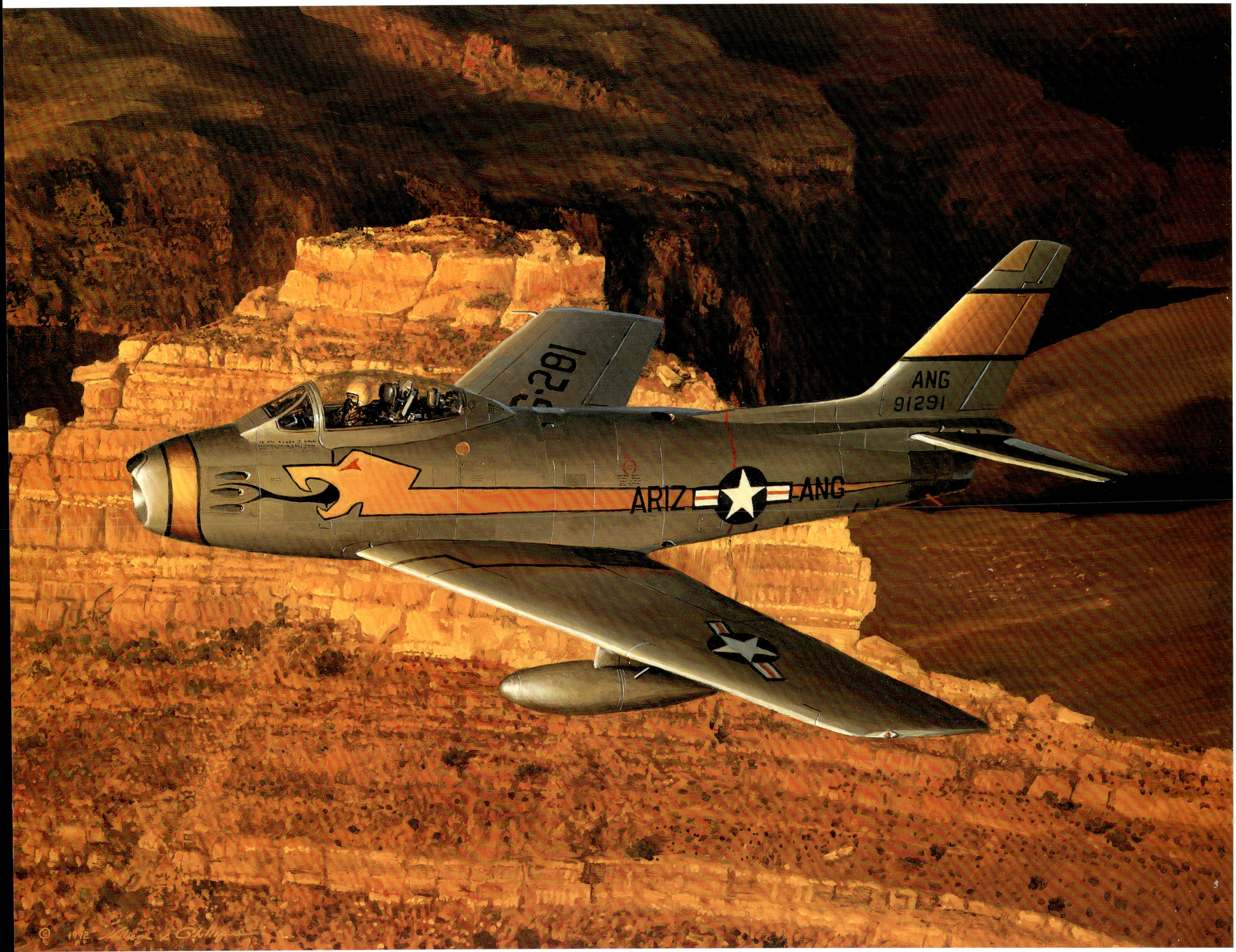
ANG
91291
ARIZ
ANG

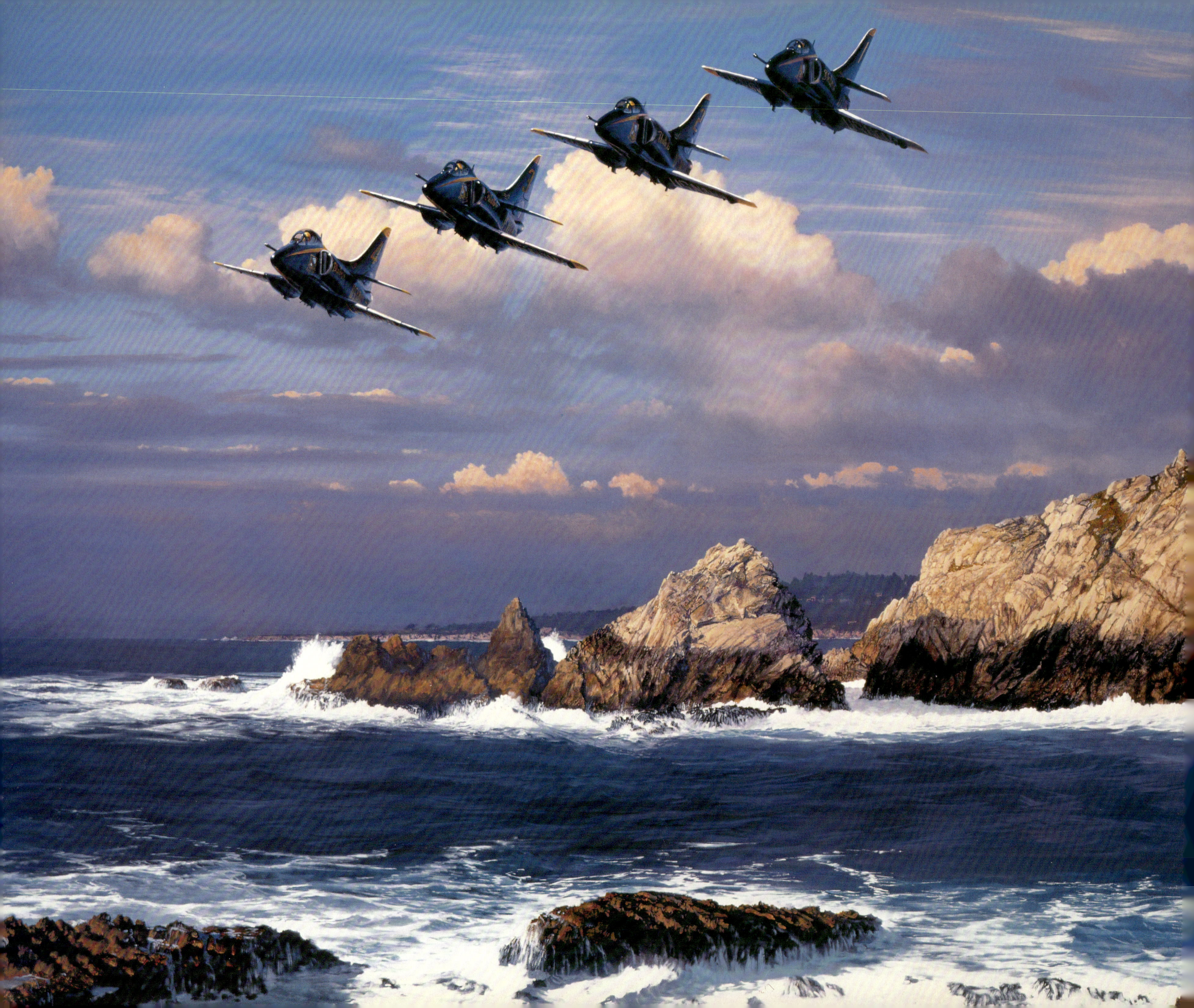

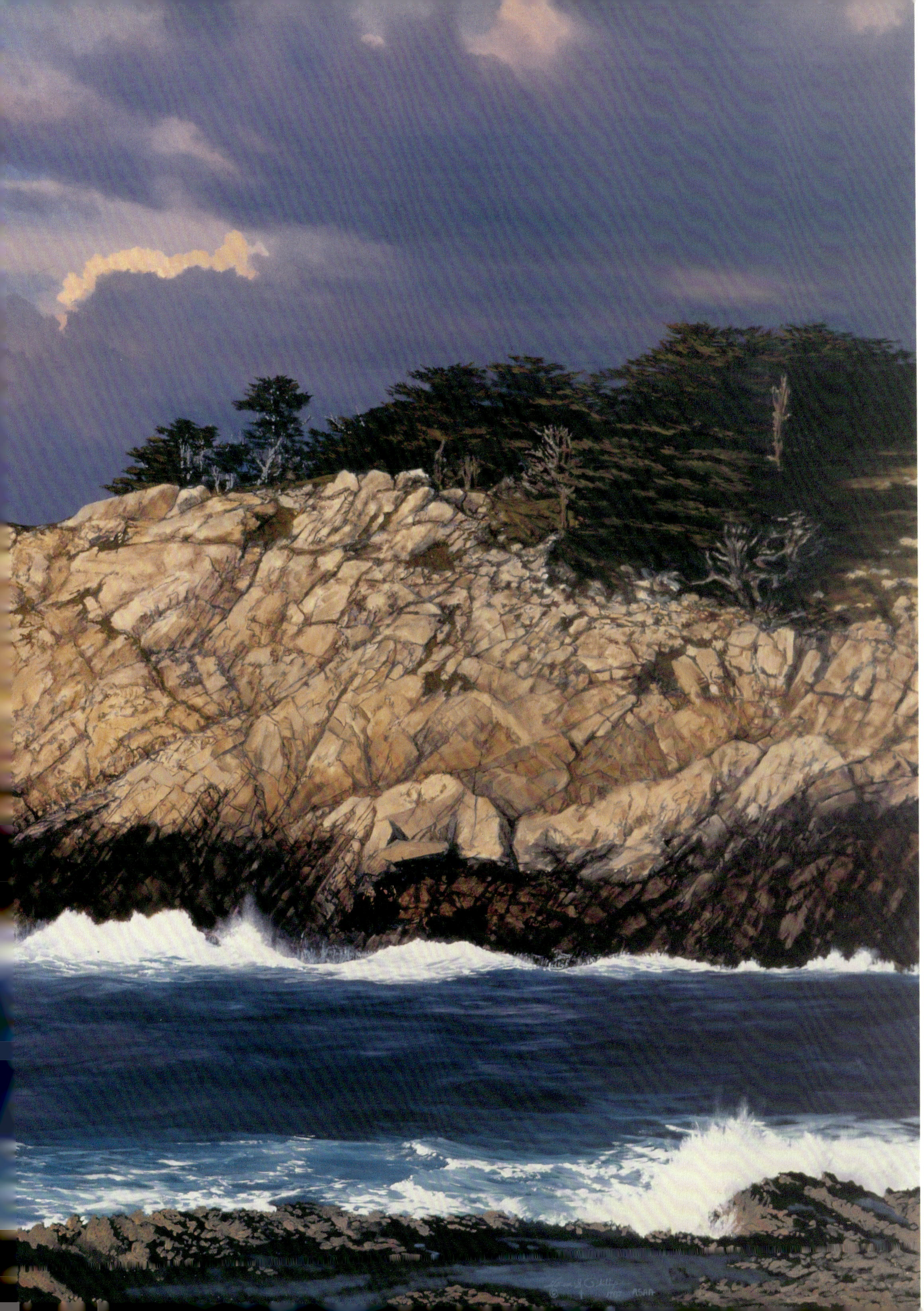

## Shore Birds at Point Lobos

Not equipped to participate in any amphibious maneuvers, the Blue Angels, the U.S. Navy's elite Flight Demonstration Team, primarily operated from land-based venues. Nonetheless, the demands of the Korean Conflict in 1950 made these "shore birds" the nucleus of Fighter Squadron 191 (VF- 191), that flew from the aircraft carrier USS *Princeton* (CV-37) as "Satan's Kittens."

In the mid-1940s, the Blues had been created by an order from Chief of Naval Operations Admiral Chester Nimitz, to exhibit the capabilities of naval aviation and the beauty of naval aerial flight. The Team, flying Grumman's F6F Hellcat, performed for the first time in June 1946. From the Hellcat, the Blues transitioned into a jet—Grumman's F9F-2 Panther. The Team reorganized and moved into the -5 version of the Panther, a faster jet, and moved from its Corpus Christi, Texas, field to today's locale, Sherman Field, Naval Air Station Pensacola, Florida.

Before taking to the A-4F Skyhawk II, which they flew from 1974 to 1986, the Team members flew Grumman's F11F-1 Tiger and the McDonnell Douglas F-4J Phantom II. In this scene, careening around the rugged, windswept promontory of Point Lobos, near Monterey, California, the foursome of McDonnell Douglas A-4F Skyhawk II shore birds evoke the wonder and awe felt by the artist for the creations of man, and for the creations of God.

Far from an aircraft portrait, the painting combines rugged landscape, compelling seascape, and dramatic skyscape, celebrating the combination of the three basic elements. Surrounded by the ocean, windswept trees, rocky cliffs, and billowing clouds, these four members of the Blue Angels sweep in tight formation, adding their high-pitched roar to the pounding of the waves. You can see them rushing in, but you won't hear them until they are overhead and passing behind you.

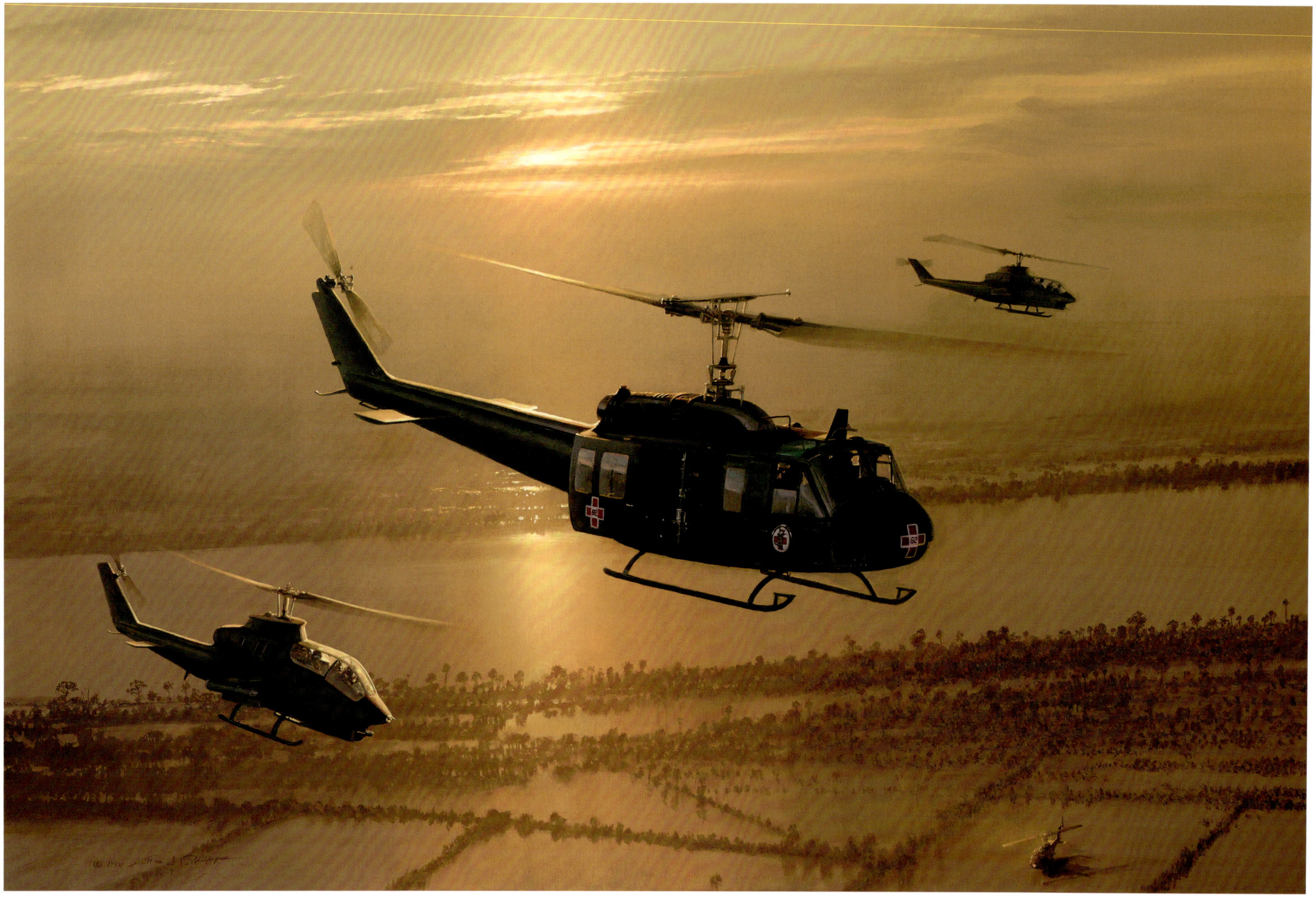

## DUSTOFF: Angels of Mercy

Of 500,000 DUSTOFF missions flown between 1962 and 1970, Bill Phillips has depicted one of CWO4 Michael J. Novosel's missions in the 82nd Medical Detachment. Circling repeatedly under heavy fire, Novosel rescued twenty-nine Vietnamese soldiers from certain death, despite having taken shrapnel in his hand, calf, and thigh. DUSTOFF, the radio call sign given to medevac helicopter ambulances in Vietnam, applied to pilots and medics who valued the lives of the wounded above their own. For his bravery, Novosel was presented the Medal of Honor. Phillips' depiction reflects his own experiences in Vietnam. He *knows* the rivers, the rice paddies, and the humid sunsets of Southeast Asia.

## Rolling on the River

*Rollin' on the River. . .* Weaving as if dancing to the rhythm of the classic chorus made famous by Rock and Roll Hall of Fame singer Tina Turner, these two rocket-firing Bell UH-1 Iroquois helicopters are in combat over South Vietnam's hazy Mekong Delta. Better known as the Huey, these rotary winged birds started augmenting the aircraft in Vietnam in 1963 and, prior to the end of the conflict, they numbered 5,000. In great demand and used extensively, the Huey served in command and control, in medical evacuation, in the transportation of personnel and materiel, in air assault and as gun ships.

UH-1D Hueys carried 20mm cannon, rocket launchers, grenade launchers, up to six anti-tank missiles and could be armed with door guns. The UH-1Ds had a longer fuselage than previous models, increased rotor diameter, increased range, more powerful Lycoming engines, and had larger cargo doors and twin cabin windows on both sides. Lift was provided by two-bladed, semi-rigid, all metal main rotors and two-bladed, rigid, all metal tail rotors. When serving in the U.S. Air Force, the artist had opportunities to fly in Hueys out of Tan Son Nhut Air Base outside of Saigon, now Ho Chi Minh City.

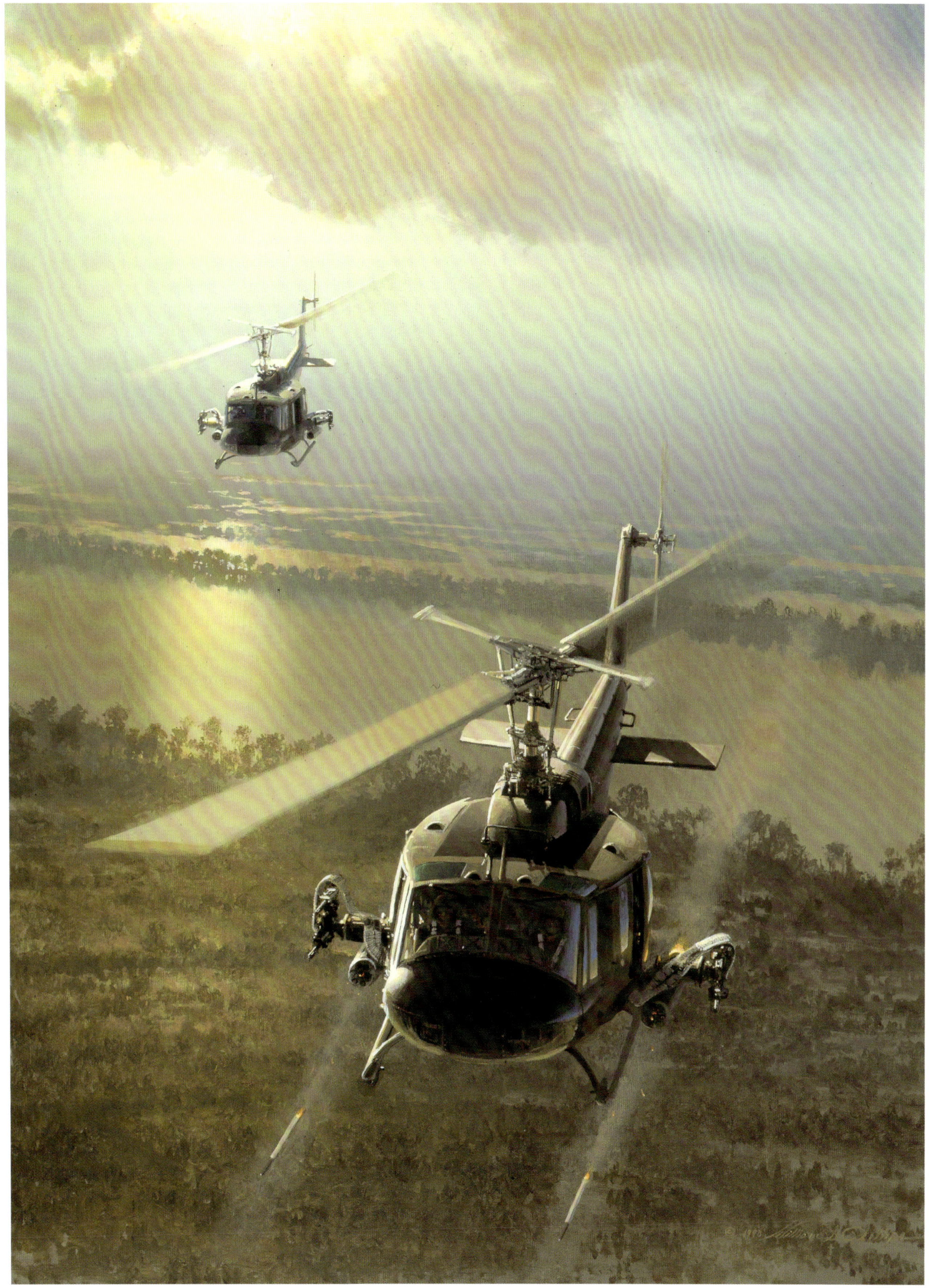

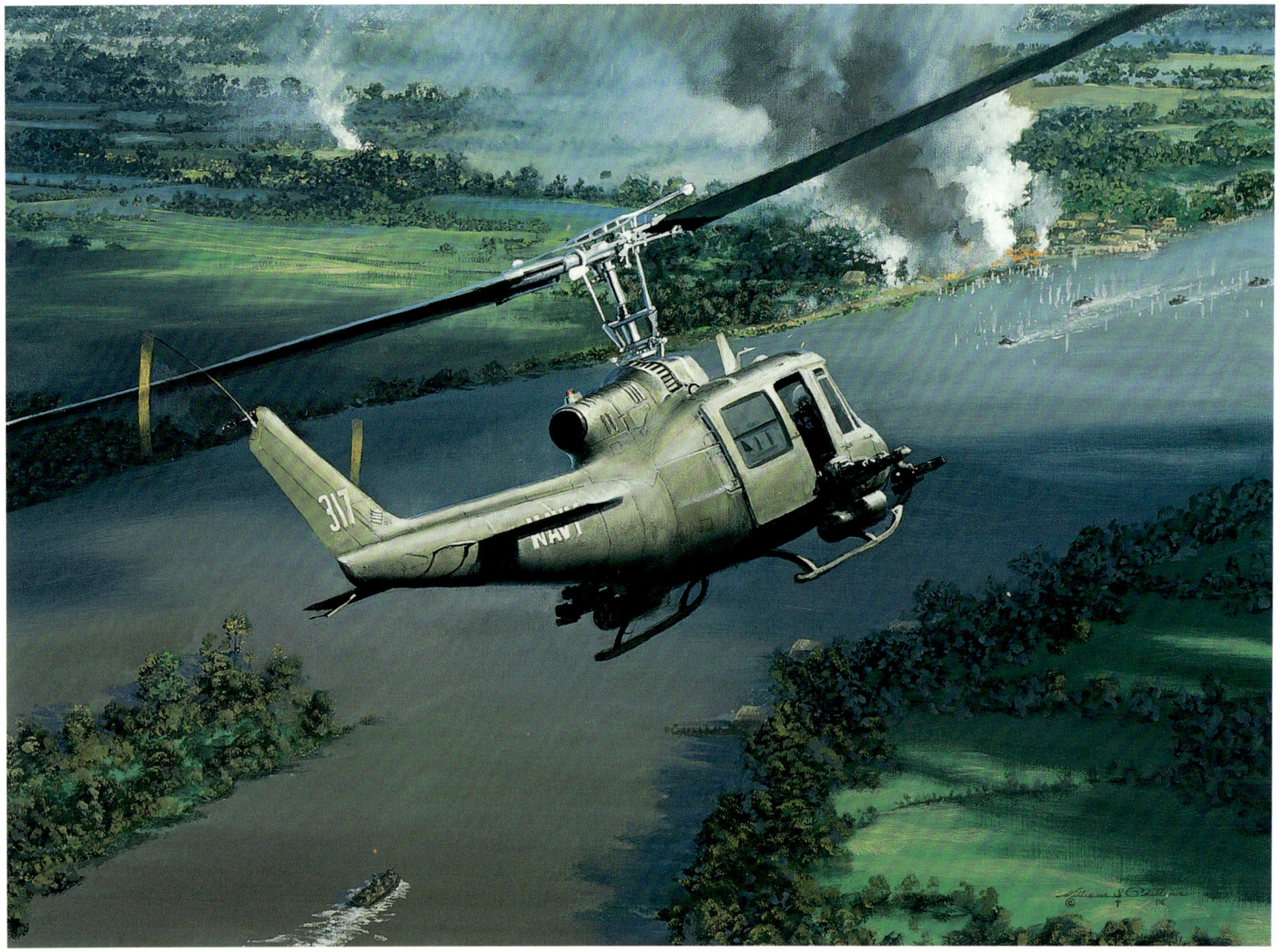

## Riverine Support

In Southeast Asia, Bell's UH-1 Hueys served in Medevac, air assault, command and control, transportation of personnel and materiel, and as gunships. The first armed Hueys, equipped with rockets and machine guns, began flying in offensive combat support in Vietnam in October 1962. Initially, gunships provided escorts for Army and Marine transport helicopters. Later, flying cover for riverine units (ships equipped for operating on rivers) Hueys were quick to strike with devastating firepower when called in by radio from river boats or ground patrols. In the painting, a Navy Huey weaves above a patrol boat on a river in Vietnam.

## Going in Hot

Grumman's A-6 Intruder served the U.S. Navy as its "tip of the spear" in combat ops for more than three decades. With a state of the art electronic system, the Intruder first became operational in February 1963 and was equipped with Digital Integrated Attack Navigation Equipment (DIANE). The A-6 was an all-weather, two seat, medium attack bomber with a solid-state weapons release system and track-and-search radar. Here, the fully loaded A-6 Intruder of VF-196 drops through leaden skies toward its target in North Vietnam. Its pilot, Stephen Coonts, then an A-6 driver, now a best-selling author, is fully engaged. He's going in hot—ordnance ready, speed adjusted, angle of attack right on, and depending upon DIANE to help the crew navigate toward release of their laser-guided weapons.

Naval Aviator Coonts completed flight training in Pensacola, Florida, in 1969. Transitioning into the A-6, he was assigned to VF-196, NAS Whidbey Island. Between 1971 and 1973, he deployed twice to fly combat from CVN-65, the USS *Enterprise*, the first U.S. nuclear aircraft carrier. Few knew the Intruder as well as Steve Coonts and with his best selling novel, *Flight of the Intruder*, he made it better known to all.

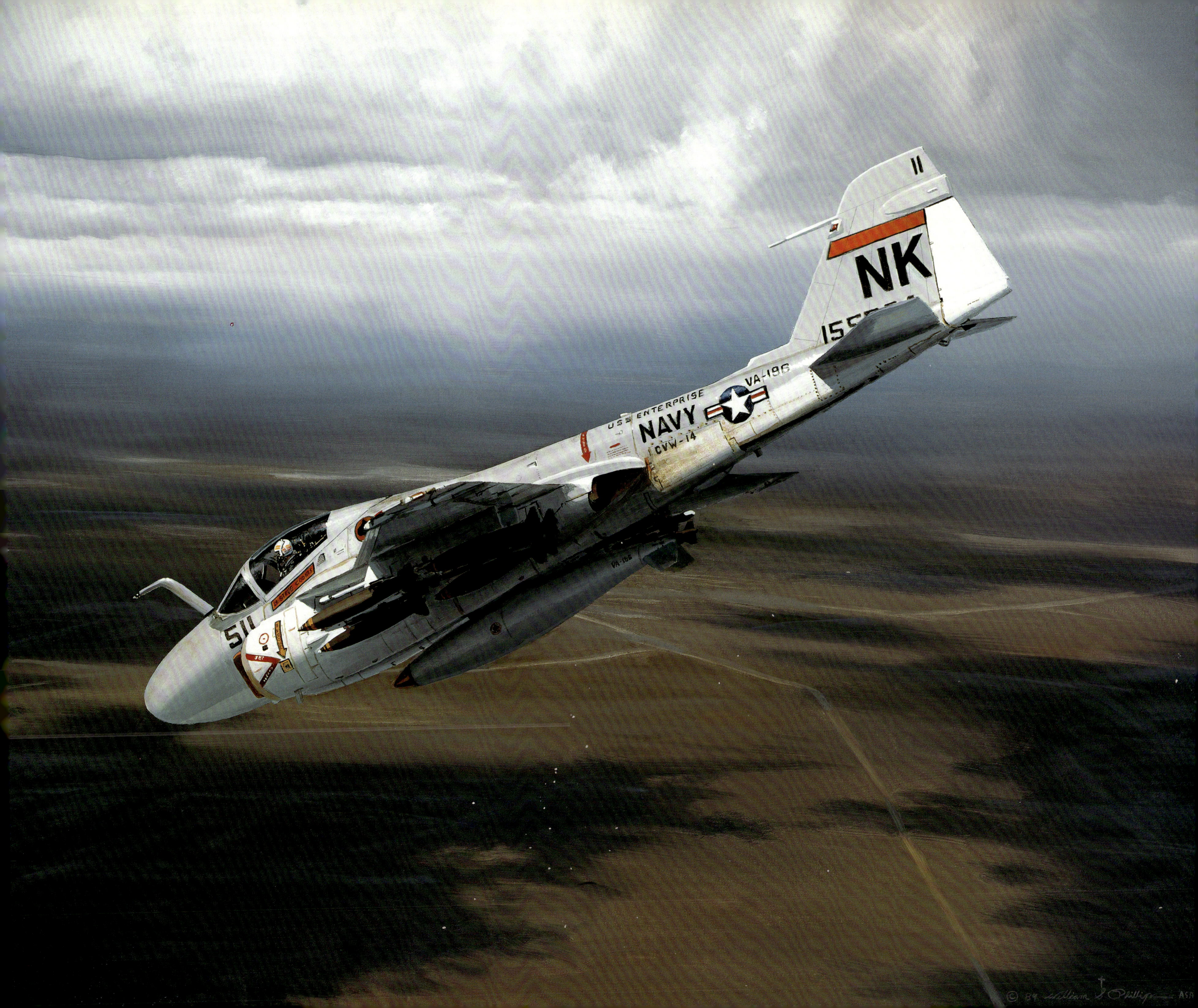
NK
USS ENTERPRISE
NAVY
VA-196
CVW-14

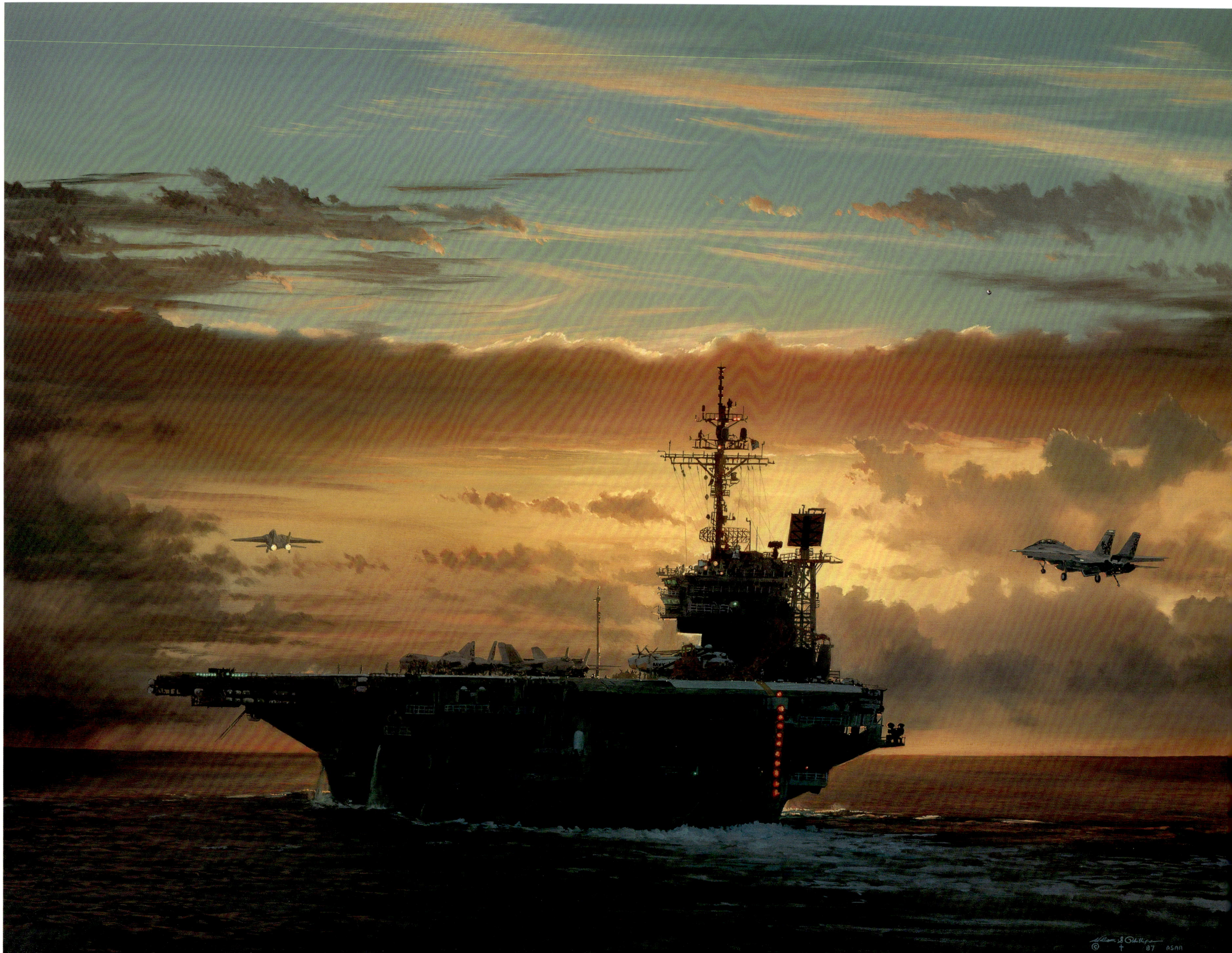

## Those Last Critical Moments

The opportunity to document the activities aboard the USS *Kitty Hawk* (CV 63) in 1984 left an indelible memory in the artist's heart and mind. It was the first carrier he visited, and his first chance to witness the challenges faced by naval personnel in every facet of shipboard operations. Flown in by helicopter, he immersed himself in the workings of the carrier, and research for subsequent paintings including photographs and video.

The *Kitty Hawk* was the first of the super carrier class, constructed by the New York Shipbuilding Corporation and commissioned at the Philadelphia Naval Shipyard in 1961. Bearing the image of the Wright Flyer as her logo, she was the second ship named for the location of Wilbur and Orville's success with controlled, powered flight. Deployed eight times in support of U.S. Forces in Vietnam, she earned the designation CV (multi-mission carrier) after a conversion in the early 1970s added antisubmarine warfare capability. A later, more complex overhaul configured *Kitty Hawk* to operate with the F-14 Tomcat and the S-3A Viking, which was its state in 1984, the year of Phillips' visit. The carrier traveled over 62,000 miles when deploying as the flagship for Battle Group Bravo and plying the North Arabian Sea. The following year, she repeated the deployment and her crews logged more than 18,000 flight hours and 7,300 arrested landings.

The painting sets a mood and exudes the tensions of night flight ops for the Tomcat pilots—those last critical moments. Sweeping toward the deck, tailhook extended, aligning with the lights, taking into consideration the speed of the ship, the speed of the wind and the speed of the aircraft, the pilot is also concerned with the movement of the black storm clouds off the bow, the lowering visibility of impending sunset and bringing the 38-million-dollar variable-geometry-winged craft safely home to *Kitty Hawk*.

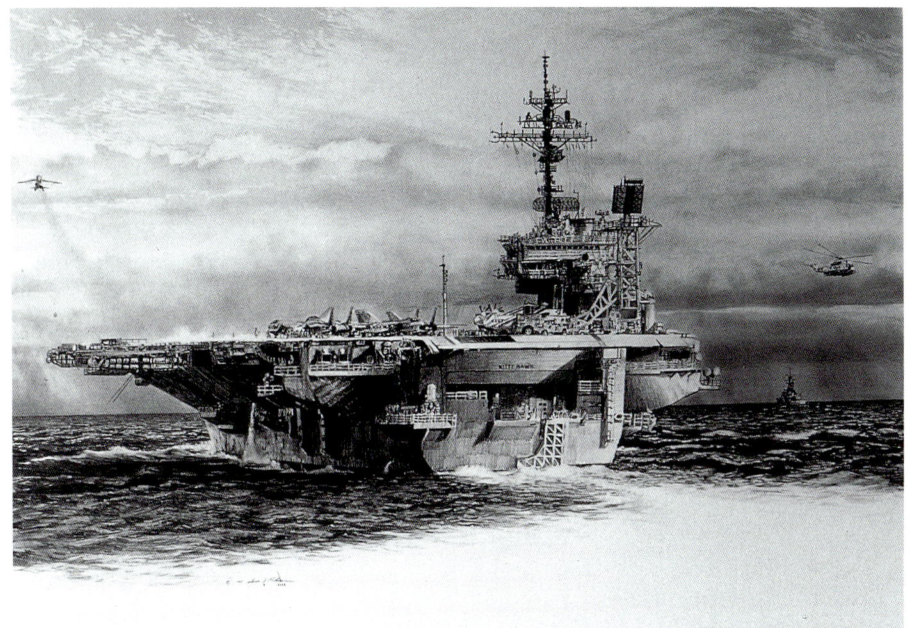

Artist's sketch of the USS *Kitty Hawk*.

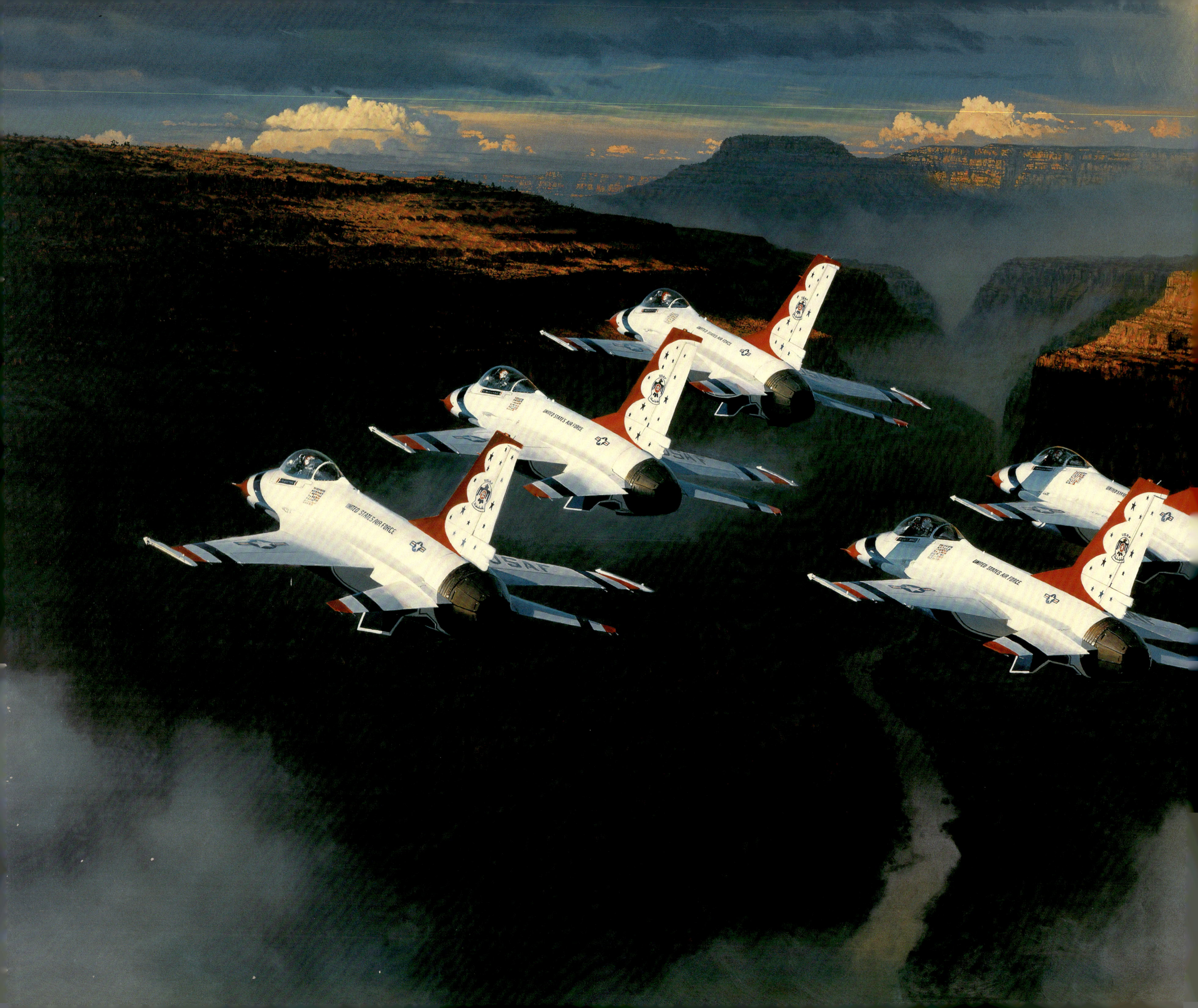

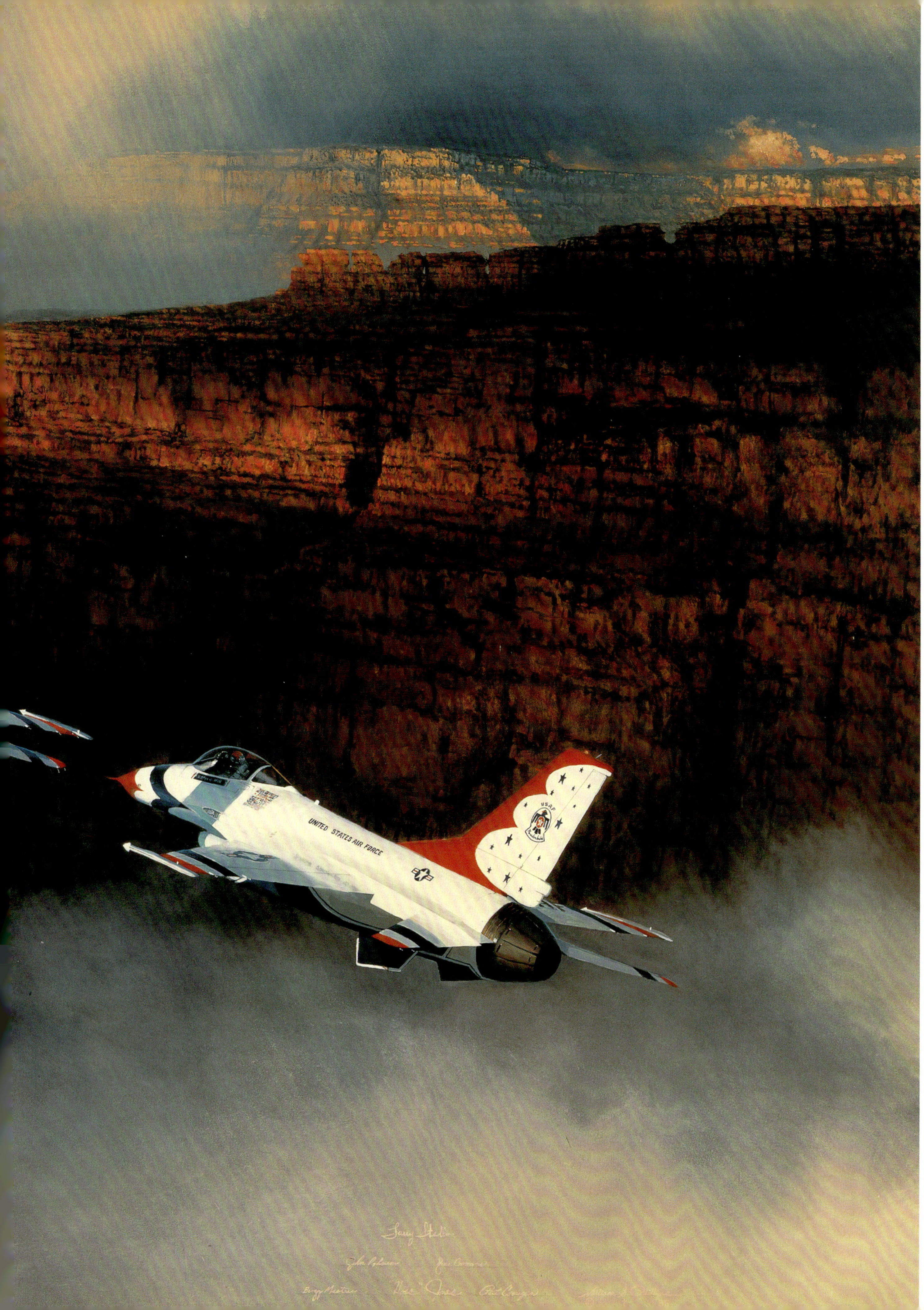

## Thunder in the Canyon

We hear the dominant roar of freedom's voice as the U.S. Air Force Demonstration Squadron, the Thunderbirds, pulls tightly into its famed delta formation and slices into turbulent air above Arizona's Grand Canyon. We listen for an echoing growl of thunder as a mythical Thor reigns throughout the sky, leaving distant vestiges of a passing storm. Our eyes are drawn throughout the painting, from the Colorado River far below to the crimson, sun-tipped heights, from the misty clouds hovering over the gorge to the tops of the cumulus build-ups reflecting a welcoming sunlight. We are invited to share the artist's two loves—landscape and aviation.

Phillips has researched his subjects well. He has flown over, and hiked throughout the majestic canyon, familiarizing himself with its dramatic buttes, mesas, and the variegated beauty of its eroded walls. He has studied the strata and knows the geology. He has flown with the Thunderbirds, the name honoring the Native American legendary and masterful bird, and with tactical F-16 squadron pilots.

The Thunderbirds, resplendent in red, white and blue, represent the U.S. Air Force and the United States of America and invoke feelings of pride and patriotism in the millions of people who are treated to their aerial displays. The elite team, which performed its inaugural flight in 1953, was equipped initially with the F-84G aircraft. Thirty years after rising from the Southwestern desert at Luke Air Force Base, the team selected the sleek F-16. Since 1983, this multi-task fighter has benefited the Squadron and all who fly her in protection of the boundaries of twenty-four nations.

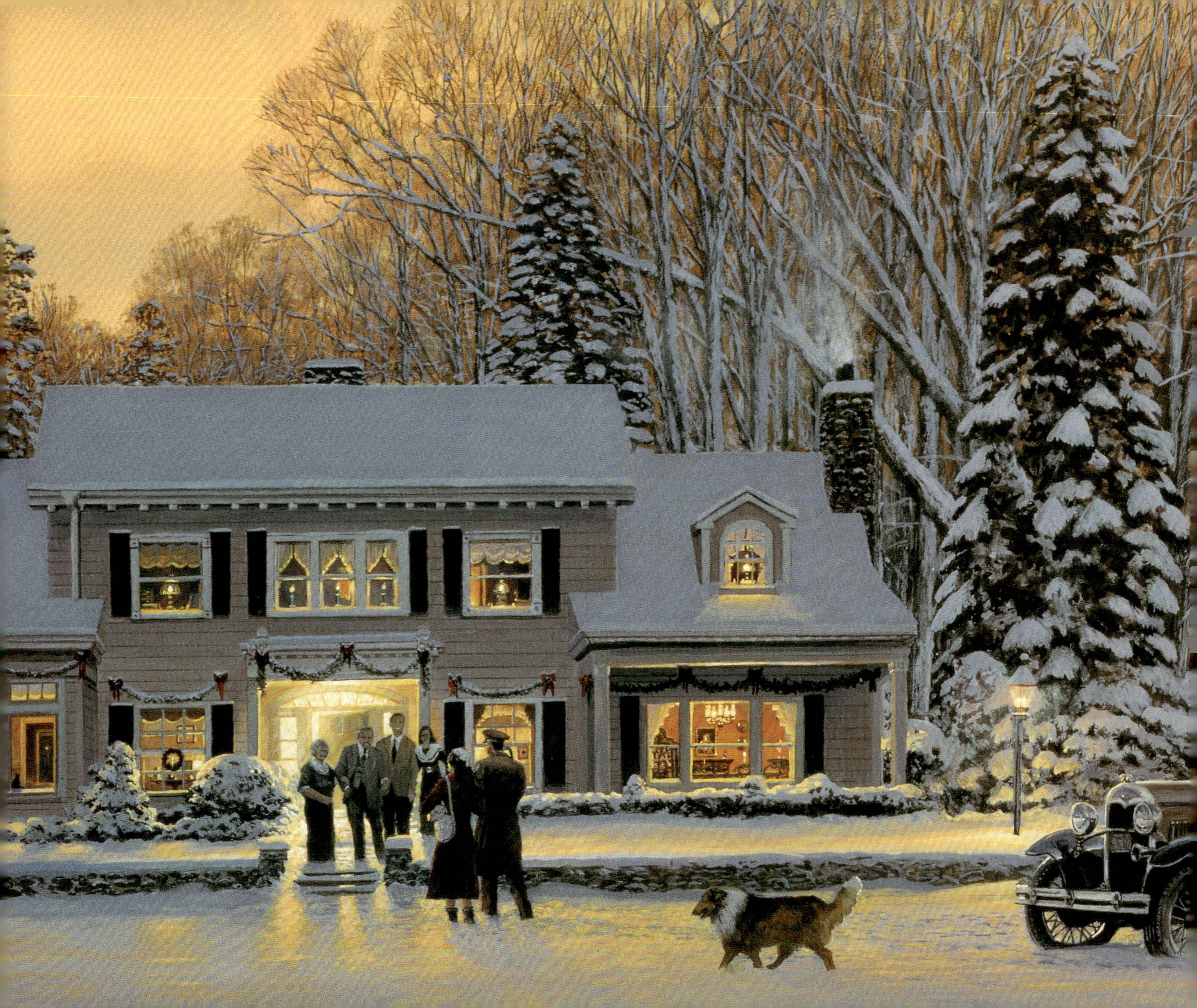

# MOTIVATION

## ON WINGS AND A PRAYER

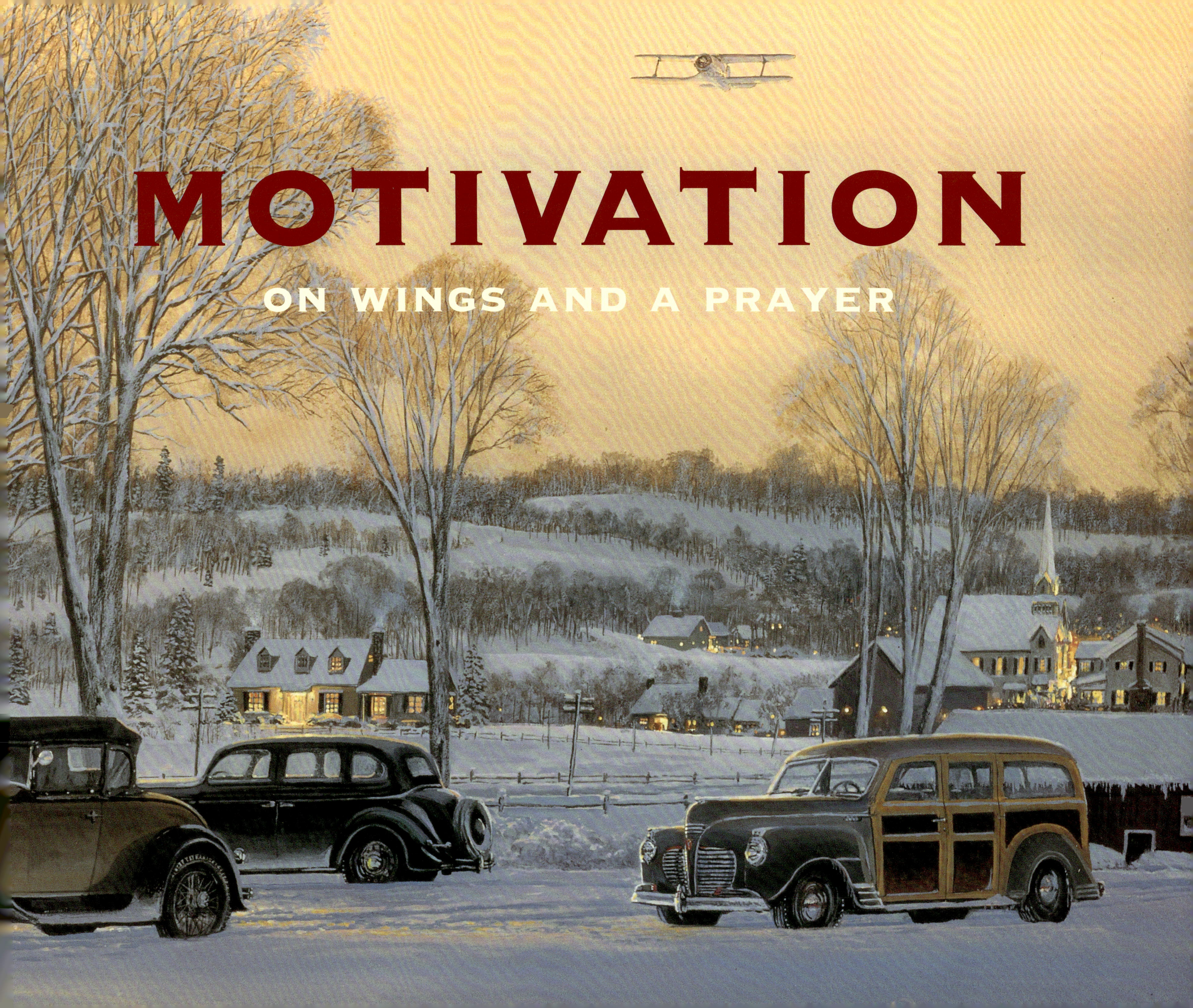

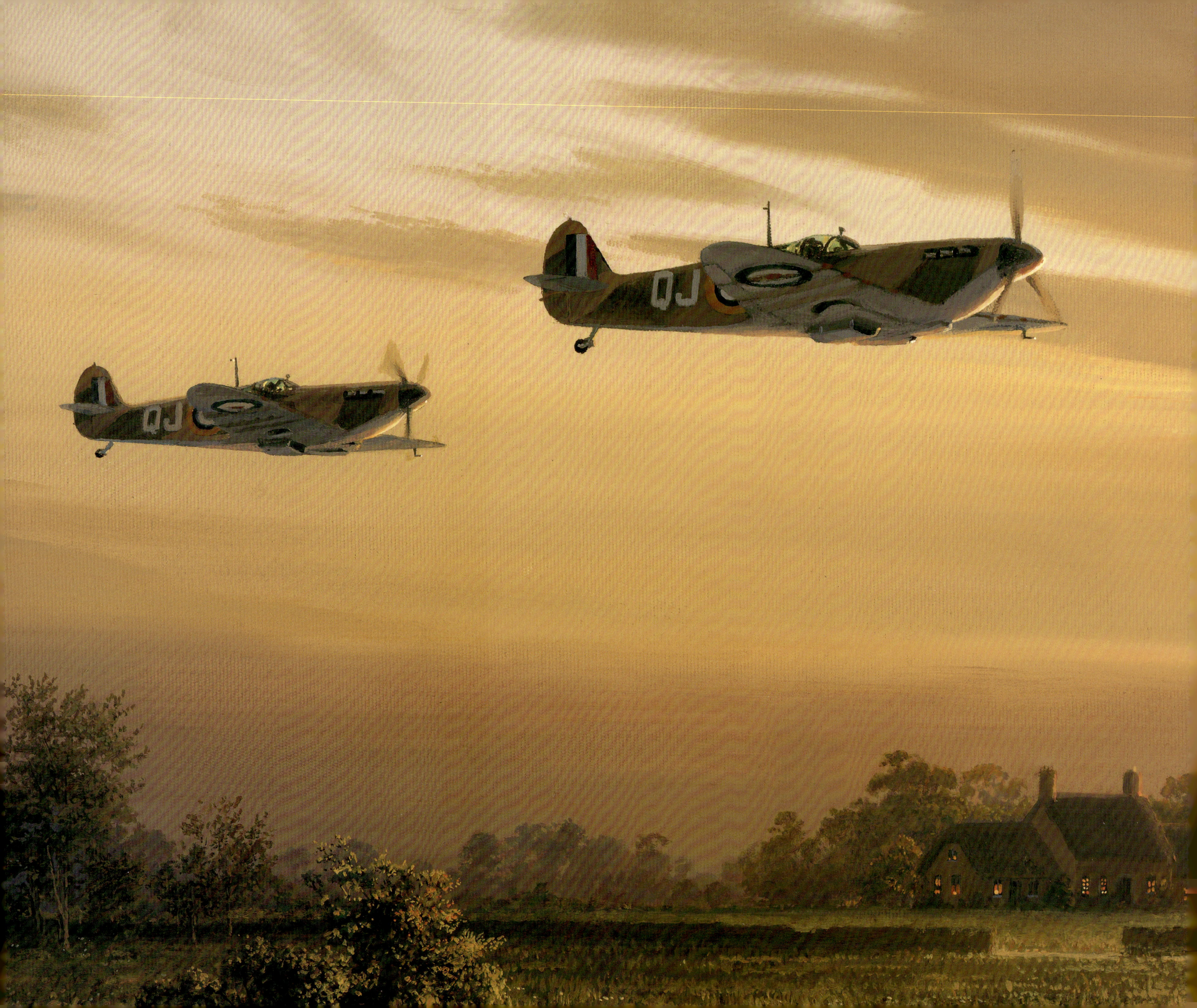
QJ
QJ

*The artist must train not only his eye, but also his soul.*

Wassily Kandinsky

Bill Phillips' life as an artist is rooted in some of his earliest experiences. On his first day of Catholic school, the first-grader bee-lined for a window seat knowing airplanes would be flying overhead. Invariably, his concentration drifted and airplane sketches decorated his papers. By second grade, the Sisters knew that "Billy Phillips should never be seated by a window." Sister Mary Regina, who still remains in contact with Bill, appreciated that she often had to deal with an artistic passion that refused to be stilled. Thanks to her, and other mentors in high school, Bill eventually received a "B" in algebra and was named to the honor roll. His confidence soared. "It was as though a huge weight had been lifted," he said.

Despite being discouraged to draw in class, art became a constant in his life. "I loved drawing," he said, "but instead of pursuing it, I retreated to it. Art became a refuge." Bill was a senior at Silverton High before taking his first art course, but encouragement from Mrs. Asbury, an art teacher who recognized his potential, accelerated him from beginning levels through intermediate. By the end of the year, seventeen-year-old Bill was working on commercial accounts for grocery stores, men's stores, and local newspapers. Like a sunbeam shimmering on water, the ephemeral idea of art as a potential career became conscious thought. Bill stashed the dream in a corner of his mind.

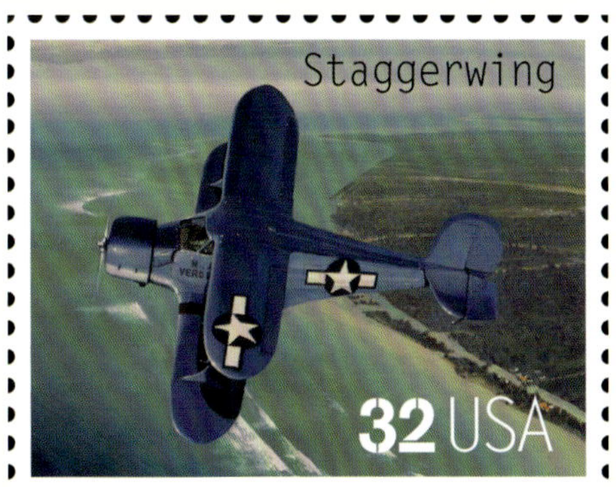


**Beechcraft Staggerwing**
U.S. Postal Service, Classic American Aircraft series, 1997

To market his Beech 17 Staggerwing, Walter Beech entered it in the 1936 Bendix Transcontinental Race. Winning the prestigious trophy and $10,000 prize, the innovative biplane, its upper wing offset behind the lower—a negative stagger—took aviation by storm.

Before his senior year, Bill applied to spend the summer as a wild land fire fighter in Oregon. He joined an elite group, the Buck Springs Crew, which imposed a demanding, disciplined, and Spartan existence. The fire fighters lived in tents and rose at five in the morning for calisthenics, followed by a run with full equipment in their backpacks. Calisthenics preceded every meal and, by nightfall, the men dropped into their beds, exhausted. This initial foray into fire fighting taught him about himself and honed his stamina and his strength. He learned to accept challenges and, more importantly, to see them through.

When there were no fires, the Buck Springs Crew trained. When fires broke out, "We charged into areas that were virtually inaccessible and couldn't be reached conventionally," Bill explained. "Using shovels, hoes, axes, and chainsaws in almost impenetrable terrain, I discovered I could survive some of the hardest, dirtiest work I'd ever done in my life. It was tough! The hours were long and hard; the demands were physically and psychologically stressful. Yet, when a fire was brought under control, the feeling of accomplishment was phenomenal."

In 1963, immediately after high school graduation, Bill enlisted in the United States Air Force and became a Security Policeman. Sent to Minot Air Force Base, North Dakota, and assigned to missile escort duty, he traveled throughout the Plains state and discovered a study in contrasts to his mountainous Oregon. In Oregon's Willamette Valley, summer thunderstorms were few and because of the mountain ranges, they rarely roiled eastward in huge convective build-ups like those that rage across the prairies. In North Dakota, Bill was fascinated with the deliberate movement of the majestic, classic thunderheads, their tops sheared into anvil shapes by strong winds, their bases dark with slashing rain and their movement visible for miles. Especially unforgettable were the flashes of lightning that flicked like rattlers' tongues from the huge clouds or glowed brilliantly within, bathing the entire cloud with golden light. In Oregon, winter's generally warm, moisture-laden clouds are forced up windward slopes to loose feathery snowflakes and deep powder on the mountain peaks. In North Dakota, winter brought frigid blizzards that blasted out of Canada, driving horizontal, icy snow across the prairie.

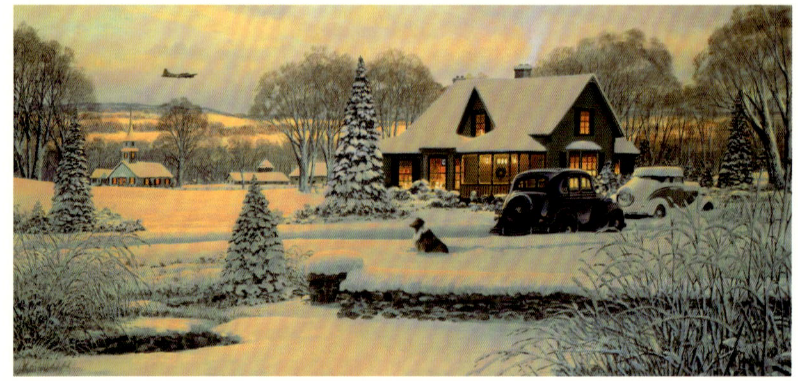

**If Only In My Dreams**
Part of the artist's "Homefront" series of paintings, a B-17 in the sky and a star in the window to honor the family member in service remind us that the war goes on. The loved ones wait, the collie maintains a vigil, and far away servicemen and women dream of being home for the holidays.

Occasionally, Bill flew aboard the UH-19 Missile Site Transport helicopter. His mobility gave him excellent opportunities to etch scenes into his mind – scenes that became the catalyst for artistic expression. Art increasingly became more than a retreat or a refuge. Bill began to hand-paint tee shirts and sell them to his friends. He pocketed the first payment for his art – a harbinger of greater things to come.

When his mother suffered a heart attack, Bill received a compassionate transfer to Portland, Oregon, to serve with Air Defense Command's 377th Combat Support Squadron. Then, in 1965, with optimism and a yearning to be where the action was, to be part of the forces allied against Communist North Vietnam, and for the opportunity to see Southeast Asia, he volunteered and was assigned to the 6250th Combat Support Group, Tan Son Nhut Air Base in Saigon, Vietnam.

A mortar attack became a pivotal experience in his life. "It was an isolated incident," he remembered. "There was no bravery involved in what I faced. Yet, my name might have been etched on Washington, DC's Vietnam Veterans Memorial Wall just doing what I did every night I was there. On April 13, 1966, I was guarding a B-57 when the first mortar shells struck nearby. I took cover behind some sandbags within five feet of where I'd been standing. We were getting pounded; my ears were ringing, dust and debris showered down. I prayed, 'God, please don't let them hit me. Please don't let anything hit me, God.' At the start of the attack, I'd believed myself to be a confirmed atheist. But, when the attack was over, I went back to the B-57 and dug the tail of a mortar shell out of the ground where I'd been standing. If I had merely dropped to the ground, I would have been killed. I've often reflected on the thin line that separated my having become an artist, from having been listed on that Wall."

When his military commitment was over, Bill decided to pursue a degree in criminology with the intent of becoming a lawyer. He earned college funds by working as a security policeman at the Hollywood Burbank Airport where he and his dad had lunched. There, he was surrounded by the airplanes he was motivated to draw. When he had saved enough to return to school, he drove north and enrolled in Southern Oregon College in Ashland. He became a student fire fighter with the Ashland Fire Department. There was no pay involved, but the job provided him with a place to live.

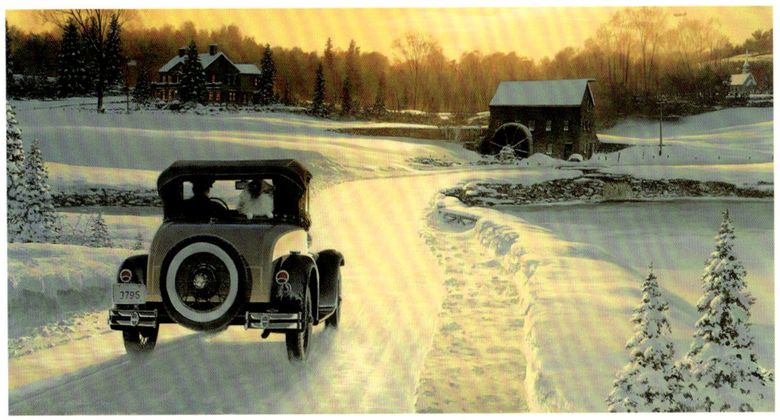

**A Christmas Leave, When Dreams Come True**

The collie's master is home, if only for a while, in a happy reunion. The car chugs past pristine snow banks. In the distance, the DC-3, known as a C-47 in its military use, is a passenger airliner and symbolizes a hopeful start to a world at peace.

Not long after arriving in Ashland, Bill met a beautiful young woman named Kristi Bittle and asked her for a date. Kristi was initially cool to the suggestion and, despite himself, he said, "I don't want to *marry* you; I just want to take you out for dinner." Two years later, Bill joined the fire department and he and Kristi were married. They have celebrated more than thirty years of marriage. Kristi has been at the center of his life. This was forcefully brought home to him the day he received Southern Oregon University's Alumni of the Year Award in 1997. When he glanced down and found Kristi in the audience, he immediately choked up. He told her later, "I looked out into the audience and found you, realizing that you are always there. You do so much of the work and you never get any of the glory. I was overcome."

Bill graduated from college and decided that he could not, and would not, go deeper into debt by attending an art school. He began what has evolved into a life-long study of art, in keeping with Winston Churchill's motivating words, "Continuous effort—not strength or intelligence—is the key to unlocking our potential." No longer was art his refuge. It moved to a prominent place in Bill's daily life.

Taking advantage of museums, art magazines, and the Southern Oregon University library's vast collection of art books, he pored over the works of Thomas Cole and Asher Durand. He discovered the Luminists and examined Frederic Church's paintings. He studied landscape masters Albert Bierstadt, Thomas Moran, Carl Evers, Eric Sloane, John Stobart, Robert Grant "R.G." Smith, and Wilson Hurley. He said, "I studied Evers' techniques in depicting water, R.G. Smith's skill in creating clouds and the atmosphere and perspective of Wilson Hurley." Study took tremendous concentration, discipline and desire. Art became his motivation, the cord that wove the tapestry of his life.

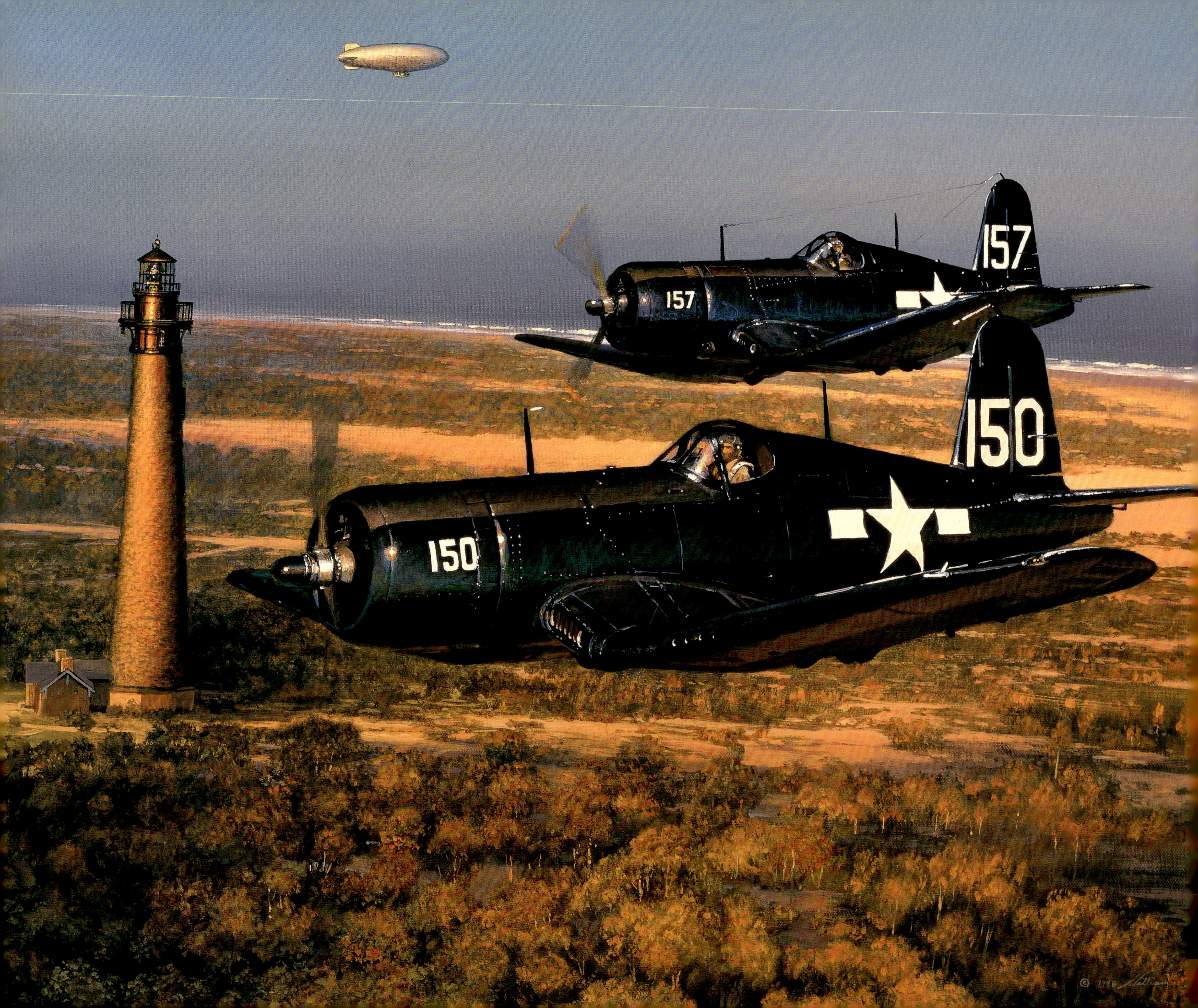

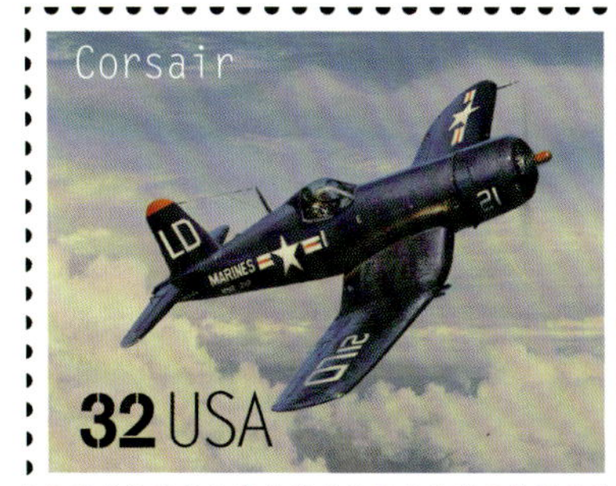


## Corsairs: Inbound at the Light

In 1941, as war raged in Europe, the United States assessed its defensive capabilities. Clearly, the most imminent danger lay beneath the cover of ocean waves in the stealthy form of torpedo-bearing ships that had been imagined and sketched by Leonardo da Vinci and actually built in 1885 by an English engineer, George Garret.

The German submarine threat, begun during World War I, was heightened when Germany's announced that its U-boats (*Unterseeboot*) would engage in unrestricted submarine warfare. The U.S. turned to blimps for submarine surveillance. Since airships could hover, perform slow flight for extended periods, and be equipped and armed against the threat, Congress authorized construction of facilities to house and support forty-eight airships. Weeksville Naval Air Station (the second station built after Lakehurst, New Jersey) was built in North Carolina close to the large U.S. Naval presence in Norfolk, Virginia.

A Navy K-ship from Weeksville is shown here, patrolling the vulnerable Outer Banks, its crew in search of the elusive wake of an up-periscope. Two U.S. Navy Vought F4U Corsairs, on a training mission designed to prepare them for combat, skim in close formation over the sand. They cruise past the 162-foot, historic Currituck Beach Lighthouse, built in 1875, the last of four beacons built and equally spaced between Cape Henry, Virginia, and Cape Hatteras, North Carolina. The string of beacon lights was designed to warn offshore mariners of the dangers of running aground in fog, storm, or darkness. More than 500 wrecks clutter that particular portion of the Atlantic floor and are sometimes referred to as "The Ghost Fleet of the Outer Banks."

For the Corsair fighter pilots, a training mission could be a lark. But here, the patrolling blimp and the Currituck Beach Lighthouse are vivid reminders of the war in which they will fight and the very real threat of an enemy that, for the moment, remains unseen.

## Corsair

U.S. Postal Service, Classic American Aircraft series, 1997

Built by the Chance Vought Aircraft Corporation of Stratford, Connecticut, this F4U Corsair—here piloted by a U.S. Marine during the Korean Conflict—represents more than 12,000 built by Chance Vought. The F4U, with its long fuselage, huge radial engine, and inverted gull-wings, enjoyed outstanding longevity. It was first flown in 1938 and continued its service into the 1950s.

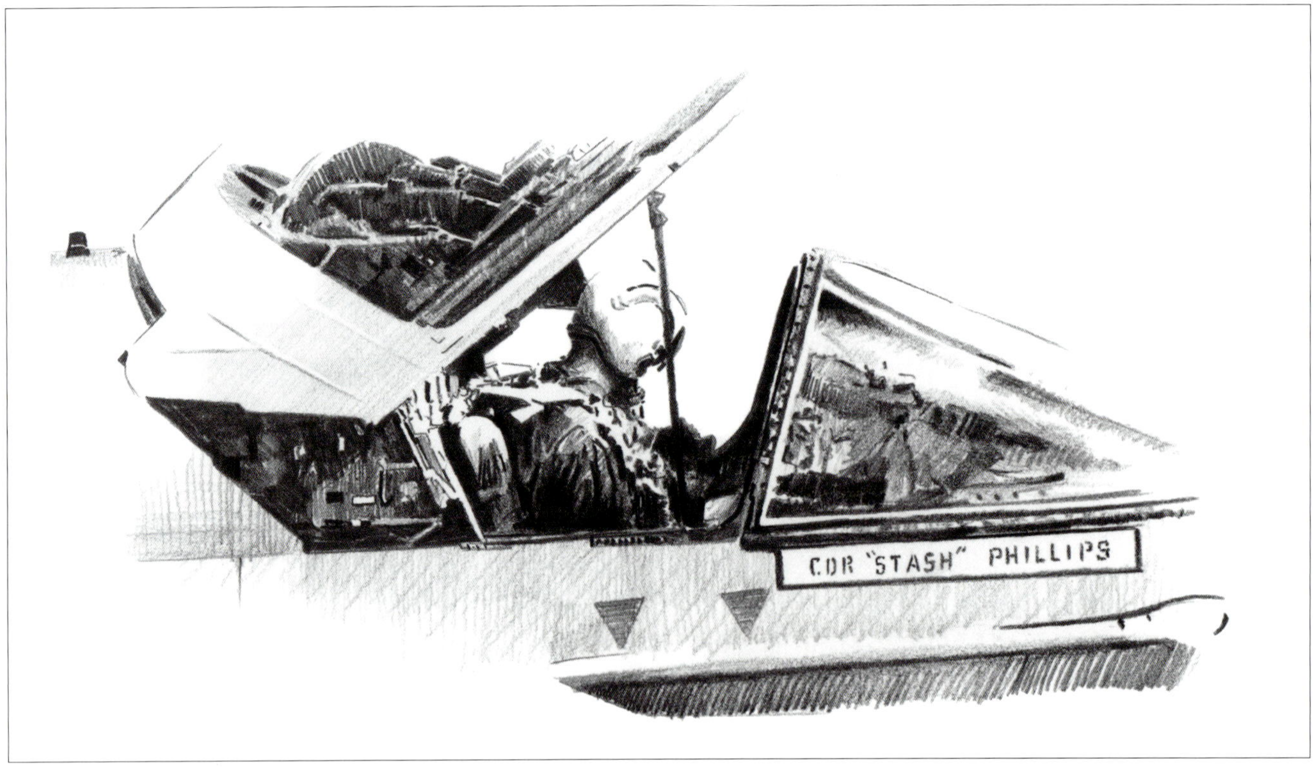


The A-7 Corsair II was introduced in 1965 and underwent a number of modifications during its service. Now retired, the A-7 Corsair II flew over 120,000 combat sorties including those in Vietnam, Libya, and Desert Storm.

## Those Clouds Won't Help You Now

Many brave acts during World War II created heroes of ordinary mortals and the U.S. Marine's Marion E. Carl was such a man. In a Wildcat during the 1942 Battle of Midway, outnumbered by swarms of enemy aircraft, Carl was one of two, out of twenty-eight, in VMF-221 to survive. After Midway, he was assigned to VMF-223 as one of the first U.S. pilots in the "Cactus Air Force," a name for all Allied pilots out of Guadalcanal that grew from its code name. He became an ace in August 1942. He survived a bailout from his damaged Wildcat in September and, throughout the fall of 1942, he and his squadron faced rigorous daily combat. His commitment up, he returned to the States in October and the following summer became Commanding Officer of VMF-223 and returned for his second combat tour. That December his squadron played a key role in the skies over Rabaul, New Britain Island, New Guinea.

The painting depicts Marion Carl in a F4U-1 Corsair during his 17th victory over a Japanese Ki-61 "Tony" that dove out of the clouds into Carl's sights. In a tragic epilogue, the man who became Major General Marion E. Carl, more than triple ace, an outstanding test pilot, honored with two Navy Crosses, three Distinguished Flying Crosses, and thirteen Air Medals, was gunned down at the age of eighty-three while trying to protect his wife from an intruder who'd broken into their Oregon home. This giant of a man will not be forgotten.

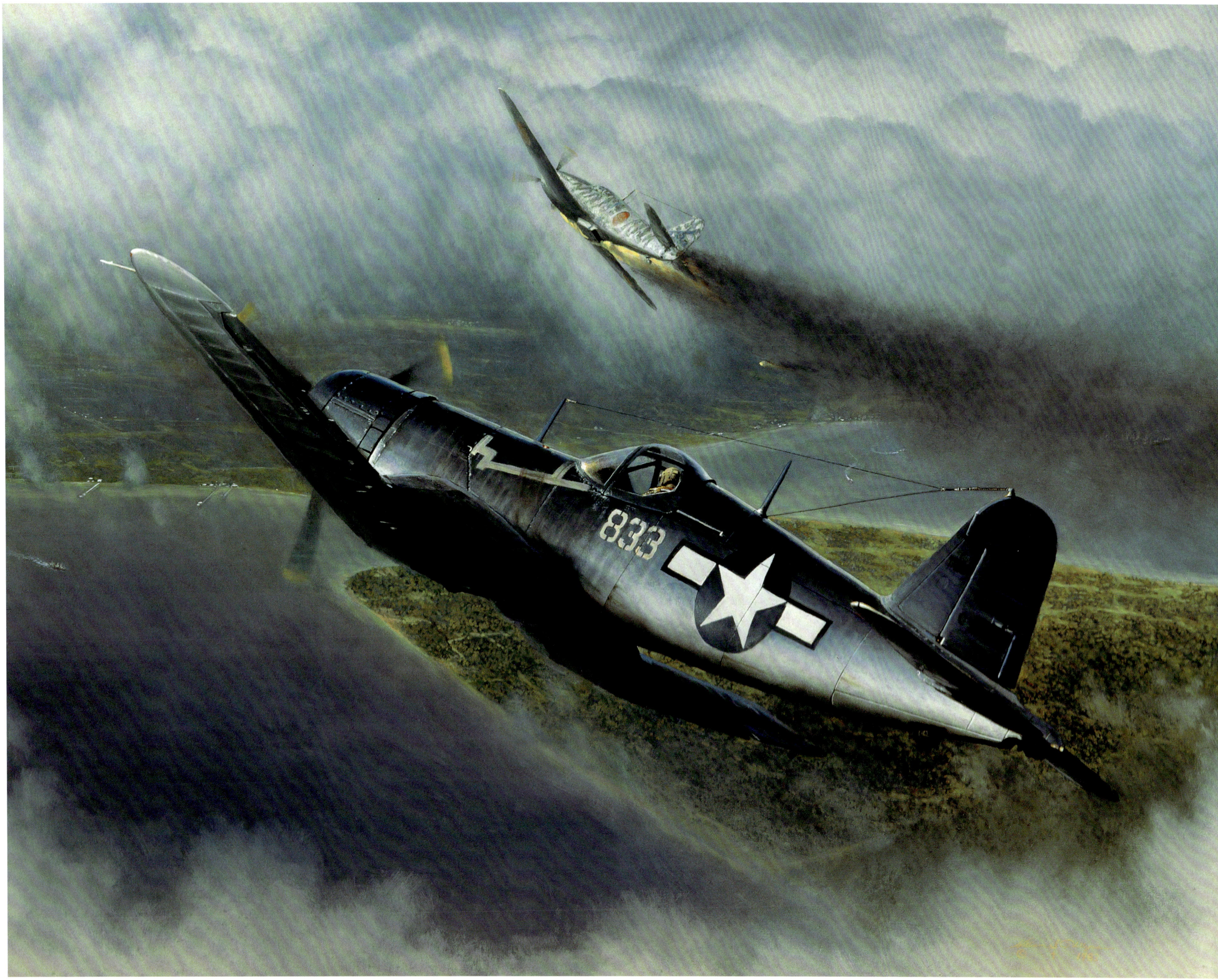
833

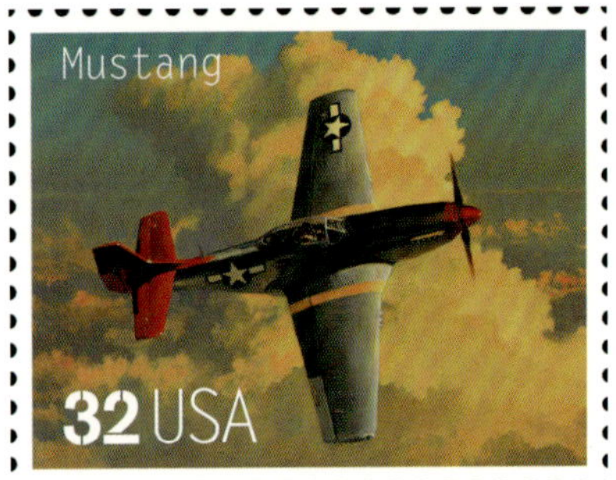


## P-51 Mustang

U.S. Postal Service, Classic American Aircraft series, 1997

The North American P-51 Mustang became the leading long-range escort fighter of World War II, responsible for saving many bombers and their men. When bomber crews saw arriving escort fighters that bore the Red Tails of the famed Tuskegee Airmen, they rejoiced. The Tuskegee Airmen could honestly boast that, when flying cover, they never lost a bomber to enemy fighter action.

## A Bandit Goes Down

Clarence E. "Bud" Anderson learned to drive before he could reach the pedals. He slammed into the garage and his father told him to get back behind the wheel and try to improve. Assertive and persistent, Bud learned to fly before the outbreak of World War II and, after the Japanese attacks in the Pacific, he volunteered for aviation cadet training. He was with the 363rd Fighter Squadron, 357th Fighter Group when it became the first 8th Air Force unit to fly the P-51 Mustang. Anderson fit the plane like a glove.

From his first victory on March 8, 1944, he battled Me 109s, FW 190s and, in July 1944, completed his first combat tour having scored 12.25 victories. Bud Anderson went on to become leading ace of the 363rd FS. In the painting at right, created as cover art for Anderson's biography, *To Fly and Fight,* Anderson is shown the day prior to the June 6, 1944 onslaught at Normandy. He was flying escort for bombers bound for Germany. Intercepted by Messerschmitts, Bud strafed his combatant and the German dove for the deck. As Bud closed in, the German pilot made a fatal mistake. Inching forward on his stick, he impacted the ground at full power and exploded.



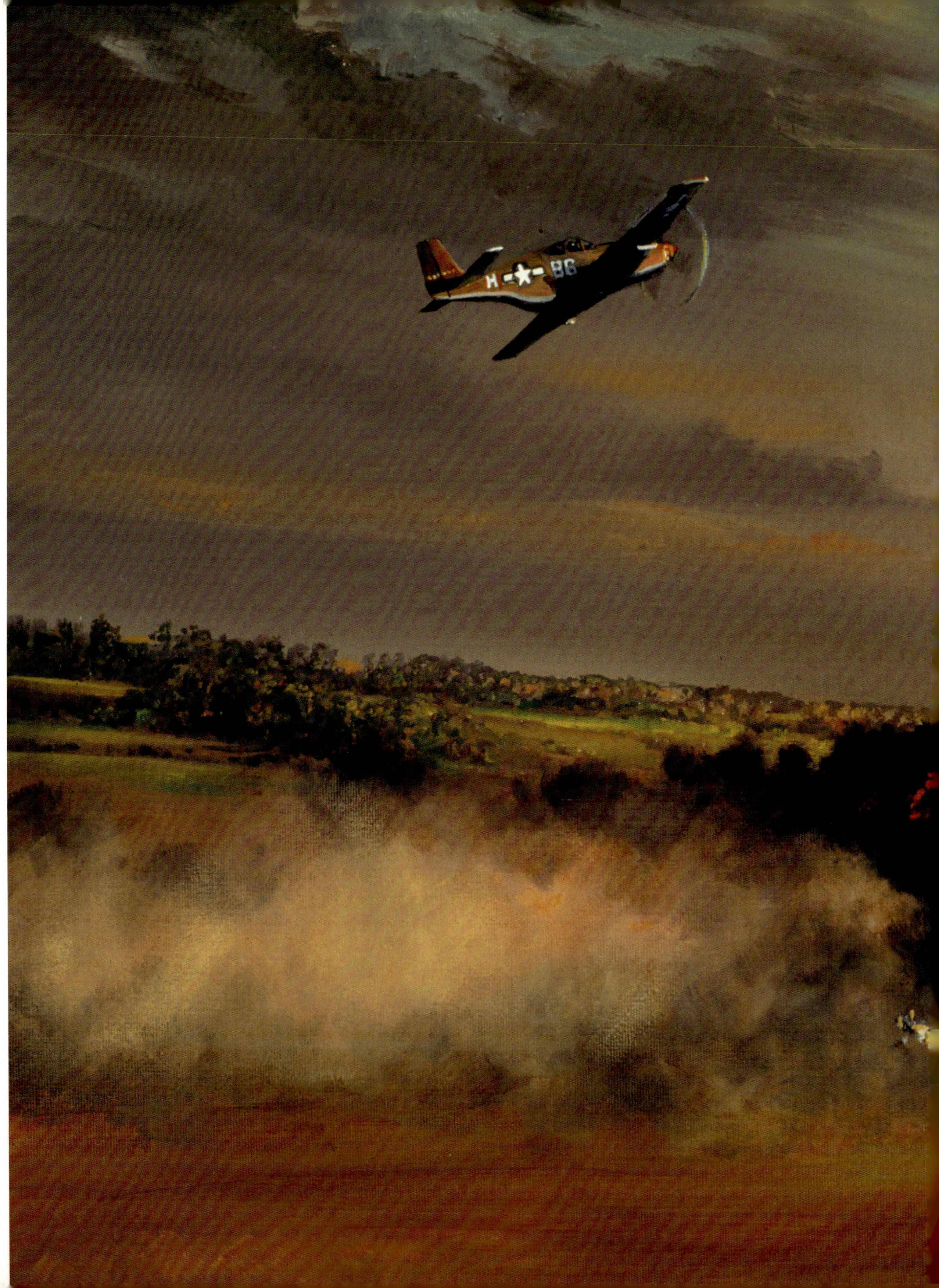

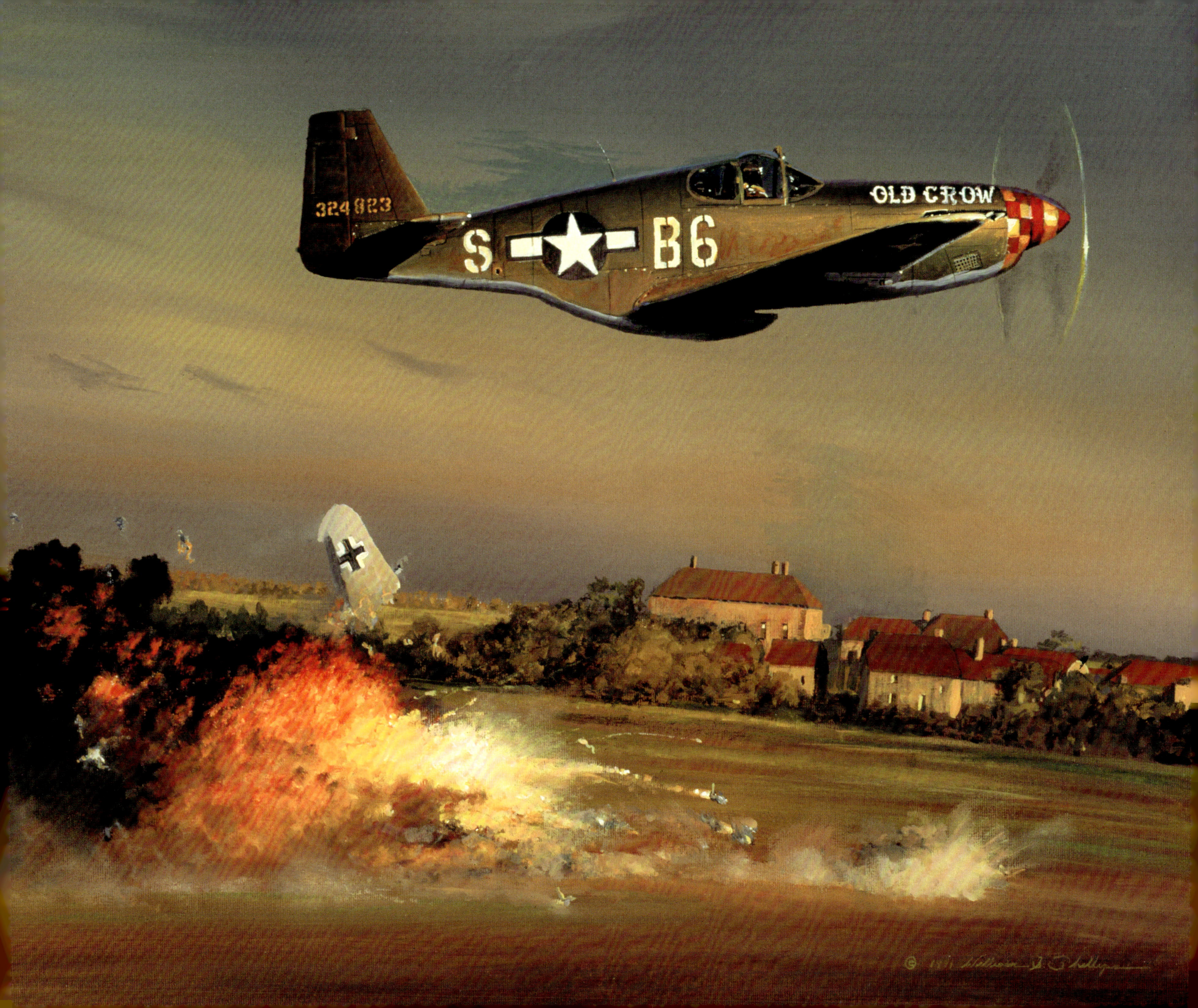
324823
S B6
OLD CROW

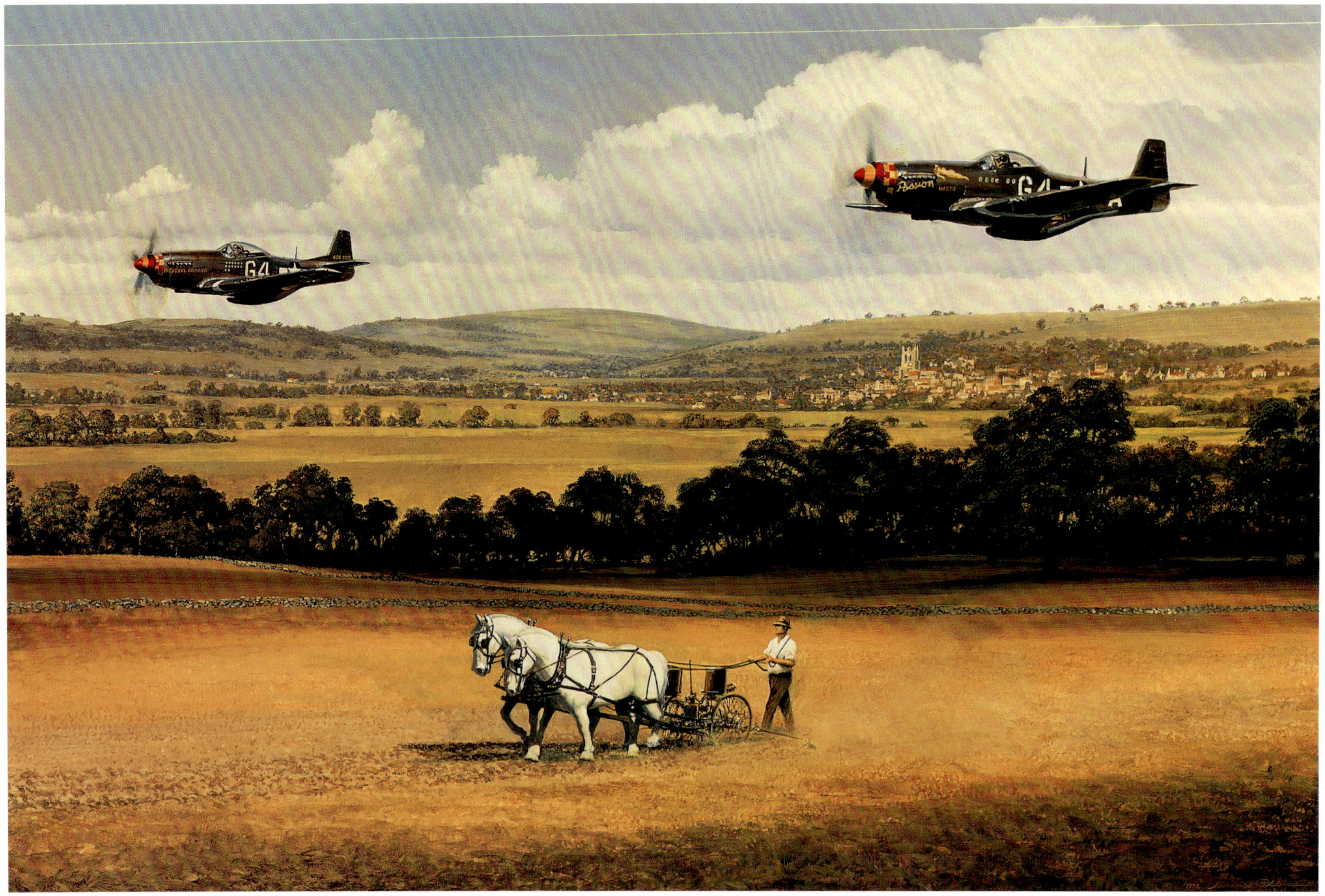

## Old Hat

In early 1940, the British asked North American Aviation to build a plane to British specifications that became the original Mustang. Powered by the Allison V-1710 engine, it was delivered to England, flight tested by a Brit, Ronald Harker. Modified with the Rolls Royce Merlin engine, it was piloted by Yanks from 1943 to 1945.

The 357th Fighter Group, 8th Air Force, transitioned to P-51s in January 1944 and, here, out of Leiston, two Mustangs roar past plow horses guided by a farmer. The horses are wearing blinders but, as it is now later in the spring of 1944, the horses and the farmer are accustomed to the Mustangs. Their presence, though appreciated, is "old hat."

## On Wings and a Prayer

In this evocative work, the artist invites us to share the emotions of a young woman who is so moved by the beautiful Supermarine Spitfires passing overhead that she stops to watch their departure. Might it be that her brother, or her lover, or a school friend is at the controls of one of the planes? Might she have been bicycling to church to pray for the safe return of all? Is it September 3, 1939, and her country has just announced its declaration of war? Or, is it July 10, 1940, and the Battle of Britain has just begun? Perhaps, like citizens the world over, she cherishes her love of country and is paying homage to those who put themselves in harm's way. Perhaps, too, the golden glow of morning symbolizes hope for a better day.

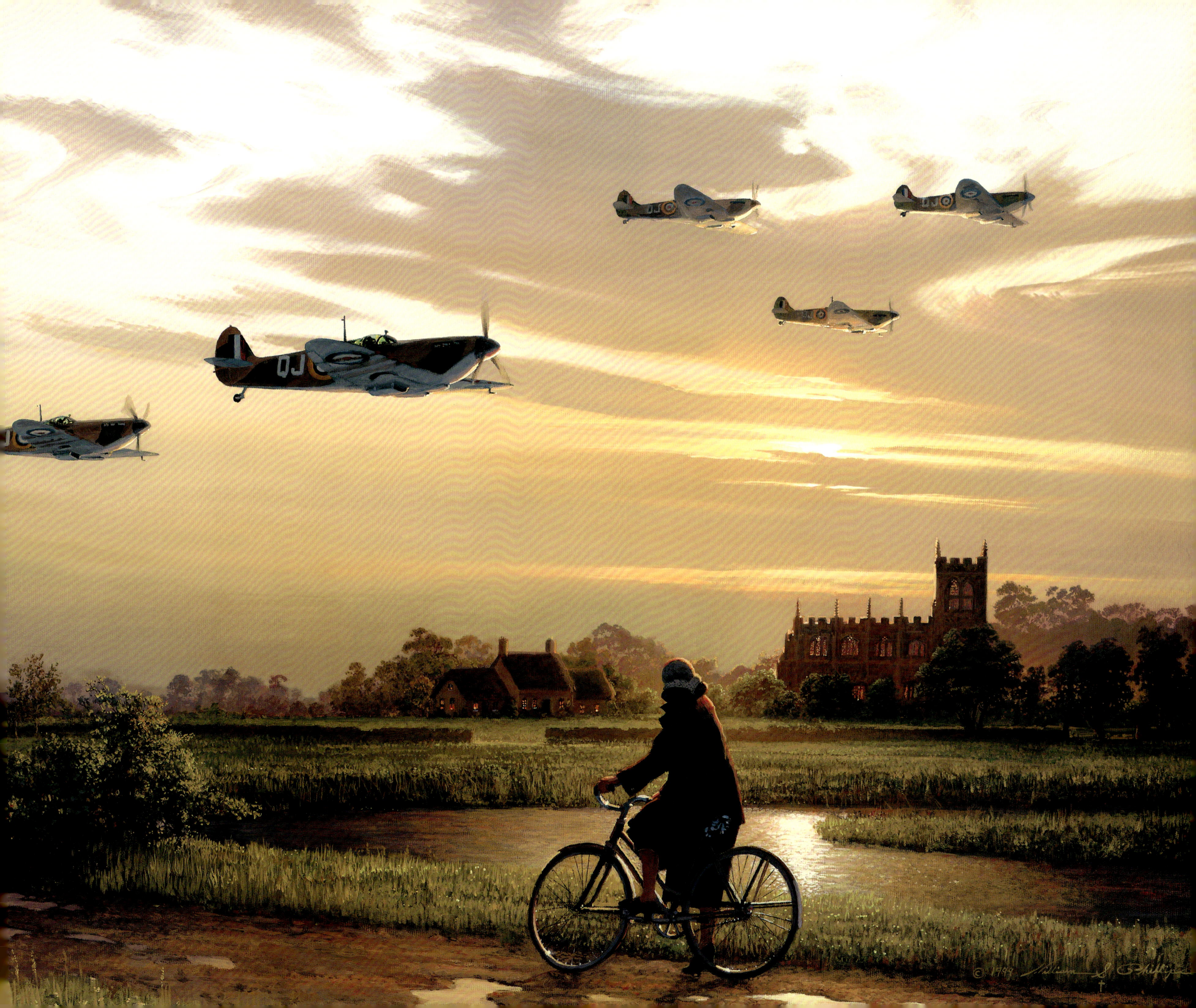

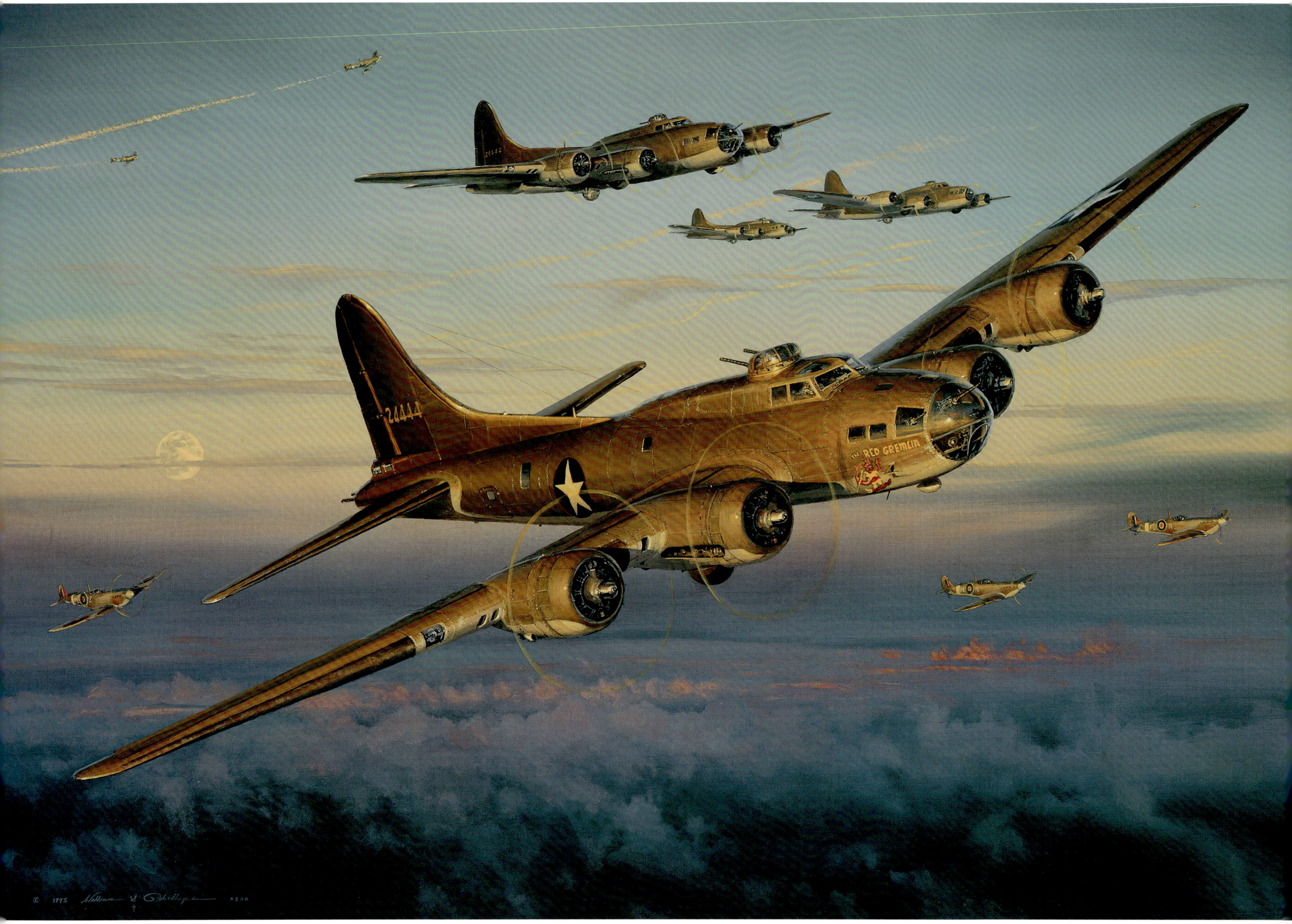
24444
THE RED GREMLIN
© 1995
ASAA

## Return of the Red Gremlin

Some pilots and their aircraft are so indelibly connected that they cannot be separated. General Paul W. Tibbets, who celebrated his ninetieth birthday on February 22, 2005, was linked to his B-29, *Enola Gay*. Yet, in the summer of 1942, then Major Tibbets piloted a B-17, *Red Gremlin*, on an escort mission—a turn-around for a bomber accustomed to seeking escort from "Little Friends." Beginning on June 15, 1942, Operation Bolero involved eighty P-38 twin-engine fighters, fifty-two C-47 transports and forty-nine B-17 Flying Fortresses. The plan was designed to move a significant number of fighters to England, where their numbers were sorely needed in the war against Fascism, and to avoid the growing successes of German submarine warfare.

Rather than ship the P-38s, it was determined that they could be flown across the North Atlantic. They had the range capability, but lacked the necessary navigational equipment. A B-17, like *Red Gremlin*, would shepherd two to four P-38s along the difficult distances, stopping in Labrador and Iceland. In the seven months of Operation Bolero, 366 heavy bombers, 150 medium bombers, and 185 P-38s were delivered to England.

It was during this massive logistical mission that a squadron of B-17s and P-38s were forced to land on the Greenland Ice Cap. The aircraft remained there and were eventually buried under 265 feet of ice. The P-38, *Glacier Girl*, trapped for decades, was arduously located and exhumed for restoration in 1992. In an incredible postscript to the 1942 mission, *Glacier Girl* was flown to share in the celebration of General Tibbets' ninetieth birthday. Also on hand were a few members of his crew: Orville Splitt, Dick Wiley, Charlie Peach, Red Horning, and Dutch VanKirk.

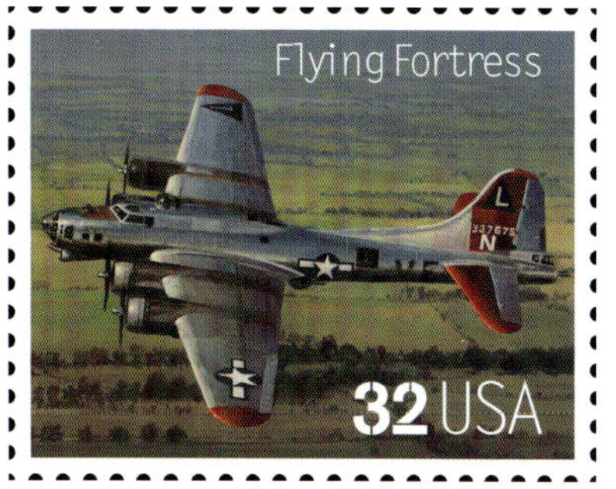


## Flying Fortress

U.S. Postal Service, Classic American Aircraft series, 1997

A backlit Boeing B-17G banks gracefully as it departs its home station on another daylight bomb run against the enemy. This variant—the G model—introduced a "chin" turret that afforded greater protection against attacking Luftwaffe fighters, and represented sixty-eight percent of the more than 12,000 Flying Fortresses built at Boeing, Douglas, and Lockheed factories.

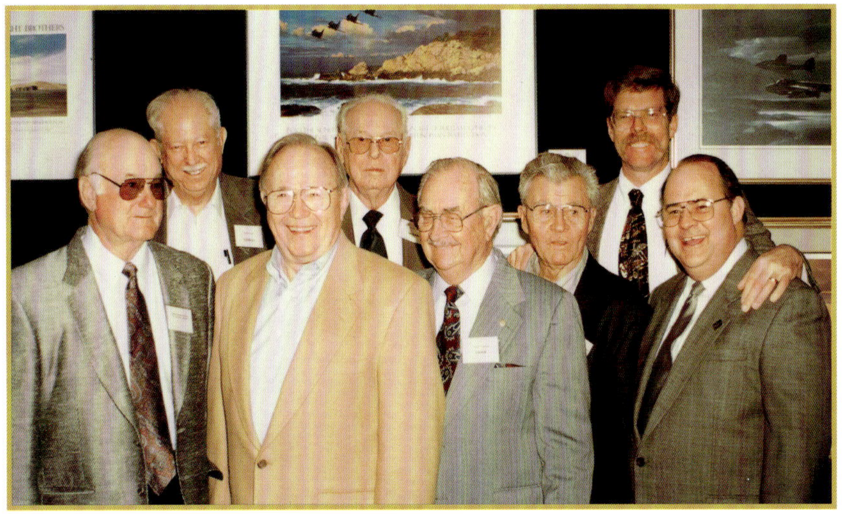

From left to right bombardier Thomas W. Ferebee, ball turret gunner Zackie T. Gowan, Jr., navigator Theodore J. Van Kirk, flight engineer/top turret gunner Herman Haag, assistant crew chief Charles H. Peach, pilot Paul W. Tibbets, artist Bill Phillips, and military historian George E. Hicks.

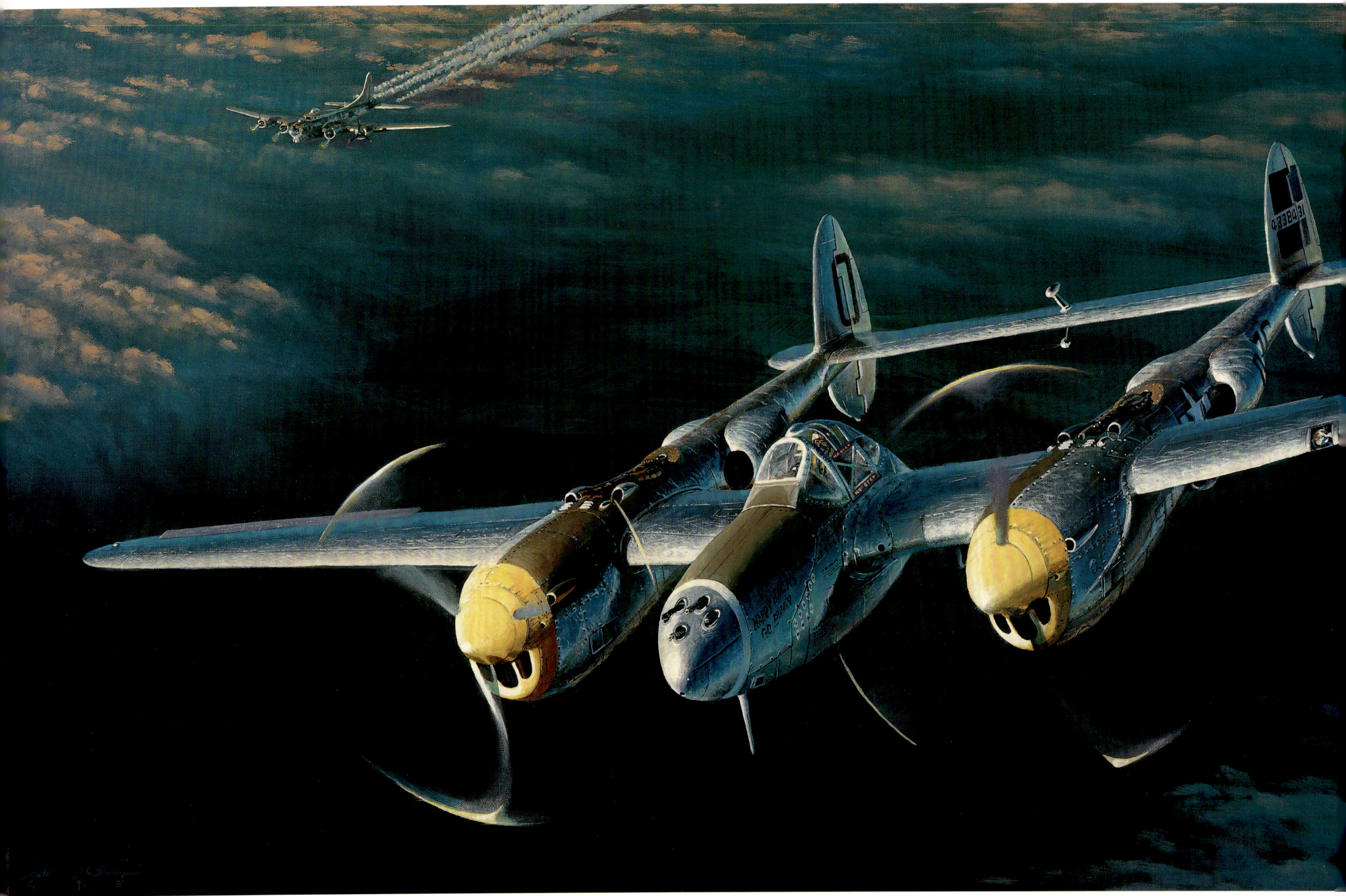

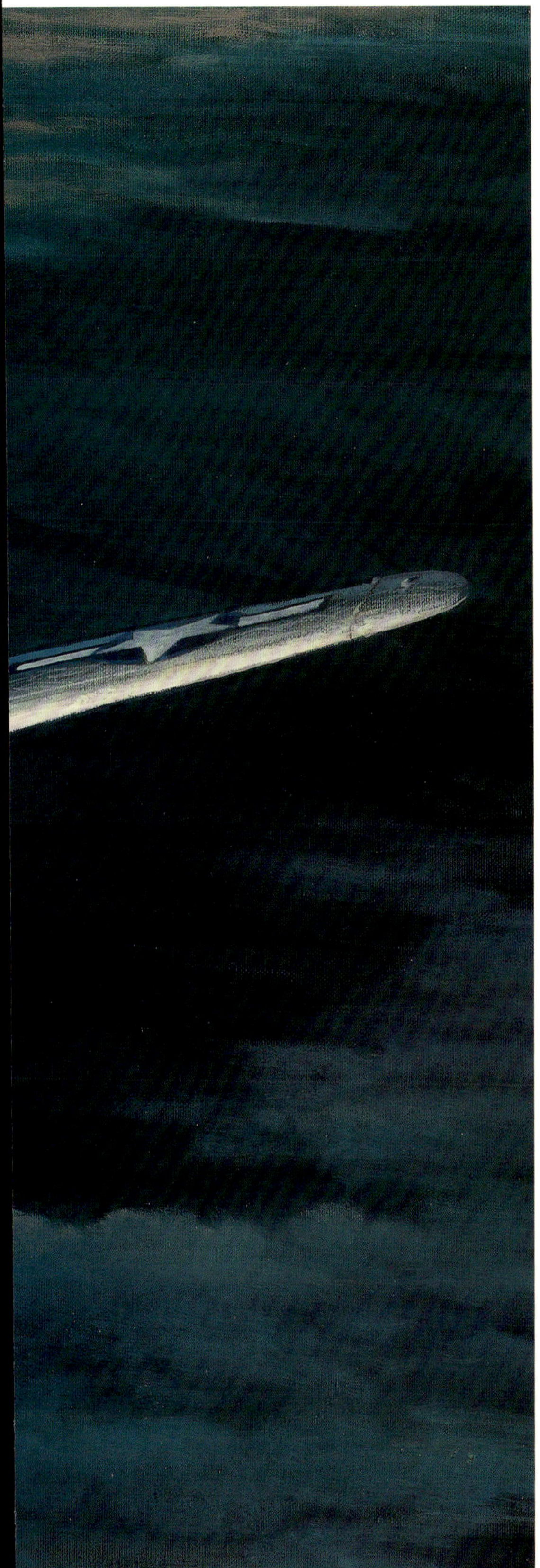

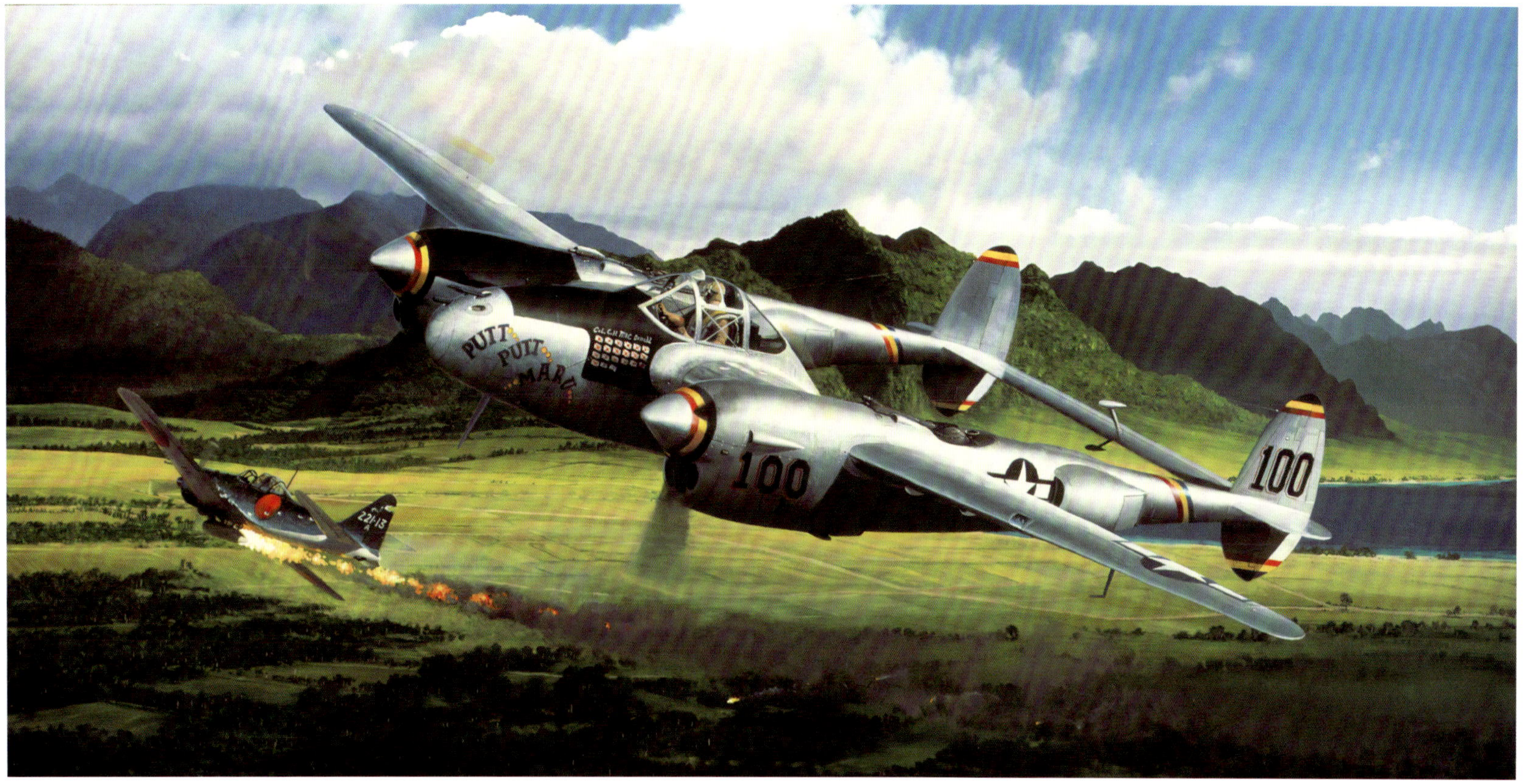


## Top Cover for the Straggler

Jack M. Ilfrey served as a member of Texas A&M's Corps and took flight training in its Civilian Pilot Training Program. Joining the Army Air Corps in April 1941, Ilfrey checked out in the P-38 and a year later participated in Operation Bolero. He ferried his P-38 from Maine to England, joined the 94th Fighter Squadron, 1st Fighter Group in Lincolnshire, and was quickly swept into the war. That November, his Group flew a non-stop 1,500 mile flight to Algeria. Shortly after takeoff, Ilfrey's plane lost one of its 150-gallon auxiliary tanks. Low on fuel and opting to land in neutral Portugal, he barely escaped after learning they planned to intern all combatants and aircraft. As his airplane's name, *Happy Jack's Go Buggy*, suggests, Ilfrey was a boisterous sort. He became an ace in the skies over Tunisia and was later named C.O. of the 79th Fighter Squadron, 20th Fighter Group in England. Serving two combat tours, he completed 142 missions and was awarded the Silver Star, Distinguished Flying Cross with five Oak Leaf Clusters, and the Air Medal with 13 OLCs. Here, as top cover for a straggling B-17, Happy Jack reverses the role offered him in Operation Bolero.

## Lightning Response

P-38 Ace Colonel Charles MacDonald survived the December 7, 1941 Japanese attack on Hawaii on and joined the 475th Fighter Group, 5th Air Force. He became Group Commander in November 1943 and, among hundreds of flight hours, flew with Charles Lindbergh in a flight that was designed as non-combat. They were both surprised by a Japanese fighter and, in a head-on rush, Lindbergh pulled up and the enemy crashed.

MacDonald was punitively removed from combat for endangering a U.S. national hero, but returned in time to participate in the liberation of the Philippines. Ironically, MacDonald downed this Japanese Mitsubishi JM2 "Jack" on December 7, 1944. It was said of Mac that he was one of those rare people who was able to lead, command, teach, and inspire at the same time.

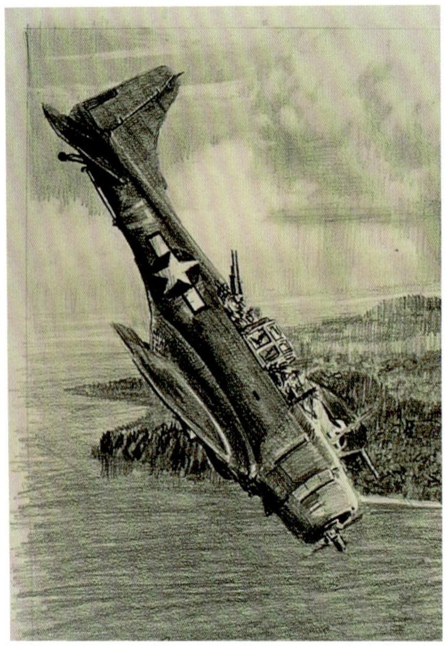

SBD-5 Dauntless

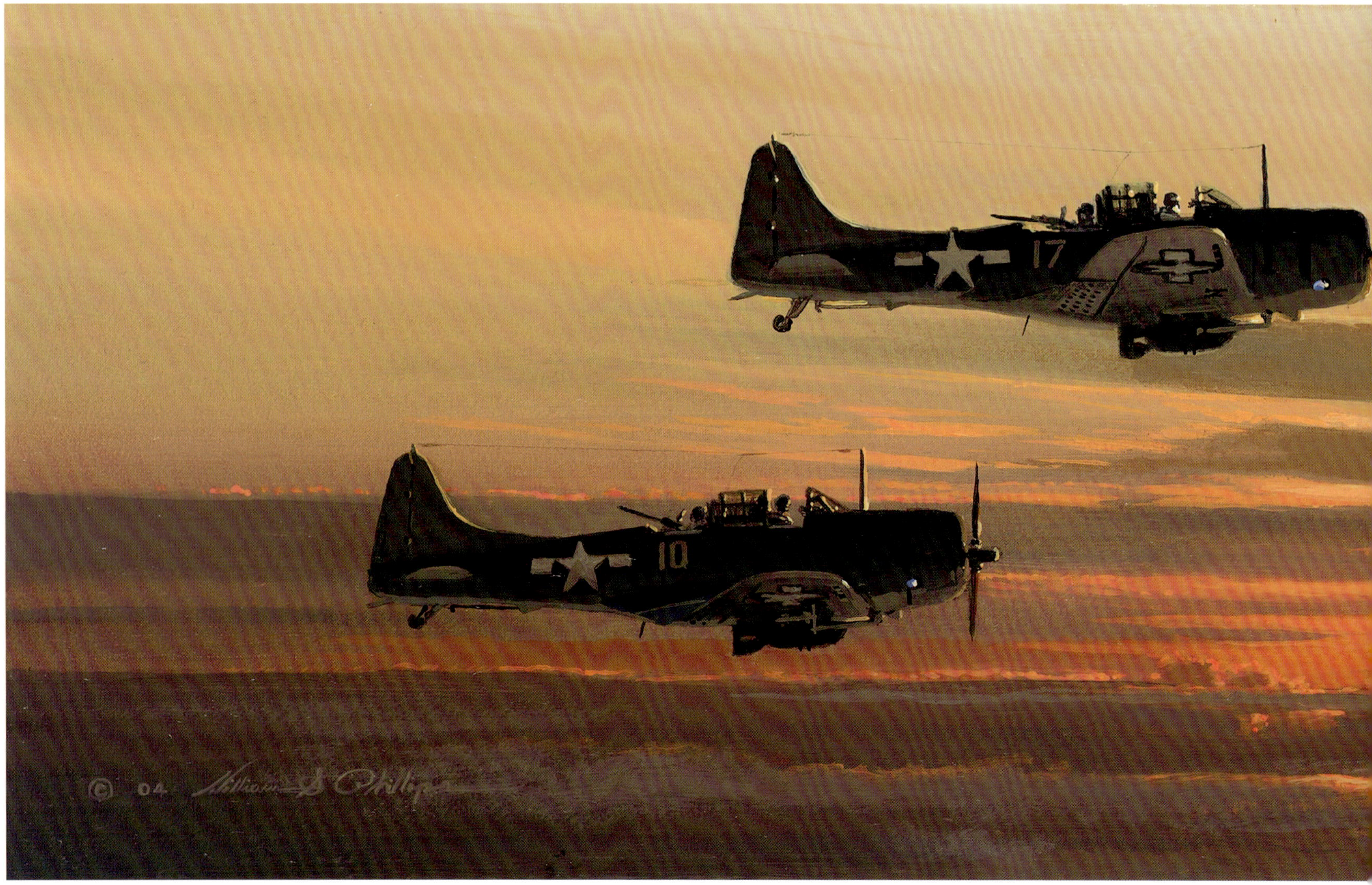

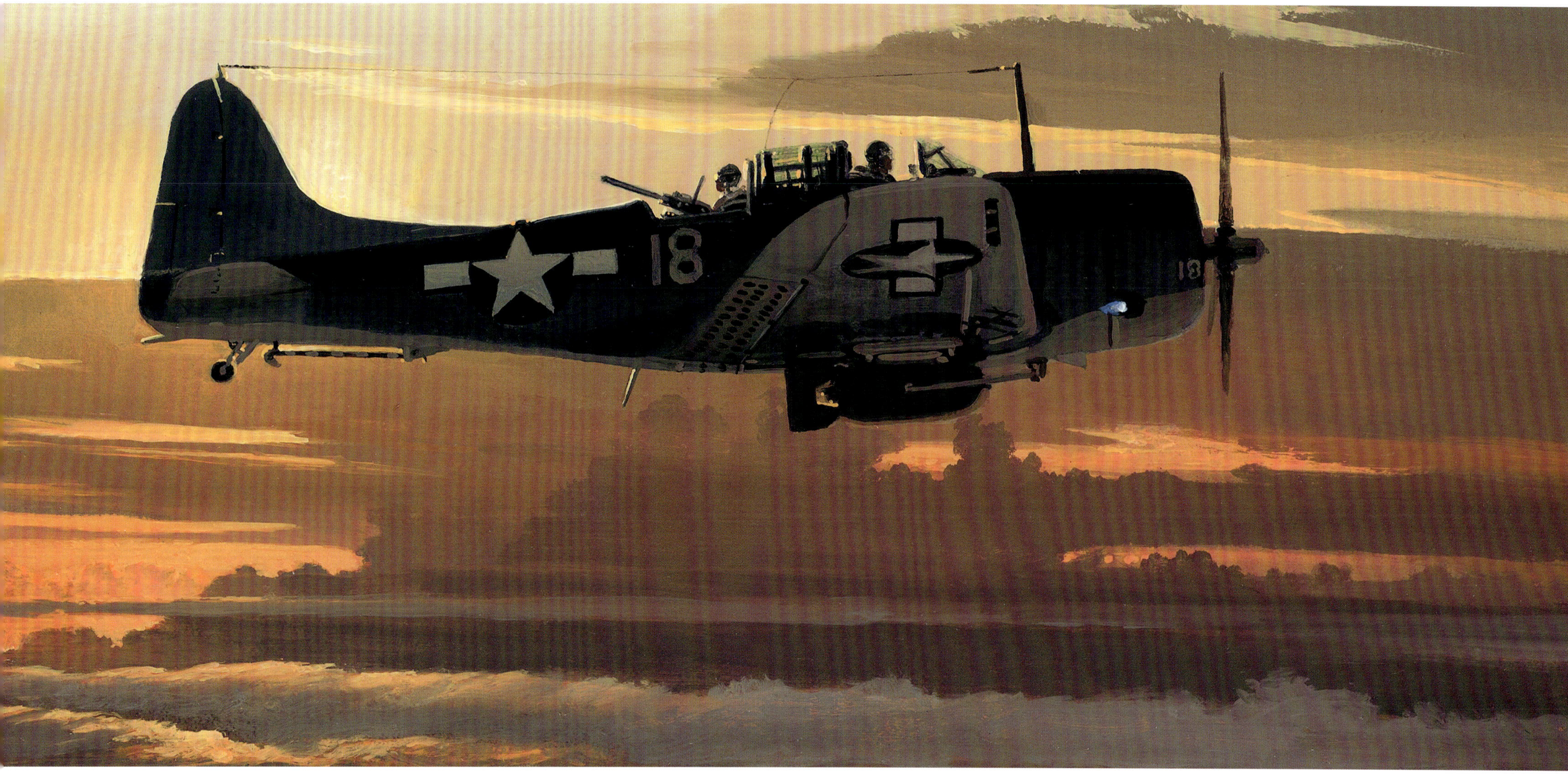



## Dauntless: Rising to the Sun

Symbolism gives pause for thought as one considers this painting. The fearless, the undaunted, saddled into their SBD-5s, are once again up before the sun and cruising toward its light, after climbing from the deck of their carrier. It is early 1944, the year all combatants came to believe would be the year in which World War II would be won or lost. From their position high above the endless waters of the Pacific Ocean, these pilots sought the enemy that bore the sun as its image. The SBD-5 dive-bomber—the husky, durable, two-place, low-wing Navy scout bomber—was the heroine of the Battle of Midway. Arriving on the disastrous scene that began over Midway Island, a squadron of SBDs attacked all four Japanese carriers and other ships, turning the tide of that battle and the outcome of the war.

The SBD-5 ranged the vast Pacific from over the Coral Sea, to the final victory flight over the USS *Missouri* in Tokyo Bay. One of the last squadrons equipped with SBDs was VB-10 off the carrier USS *Enterprise* CV-6, the most decorated ship of the Second World War. The Dauntless was considered, by those who manned her, "slow, but deadly," yet she flew on, ending the war with the distinction of being the bomber that sunk more enemy ships than any other U.S. aircraft.

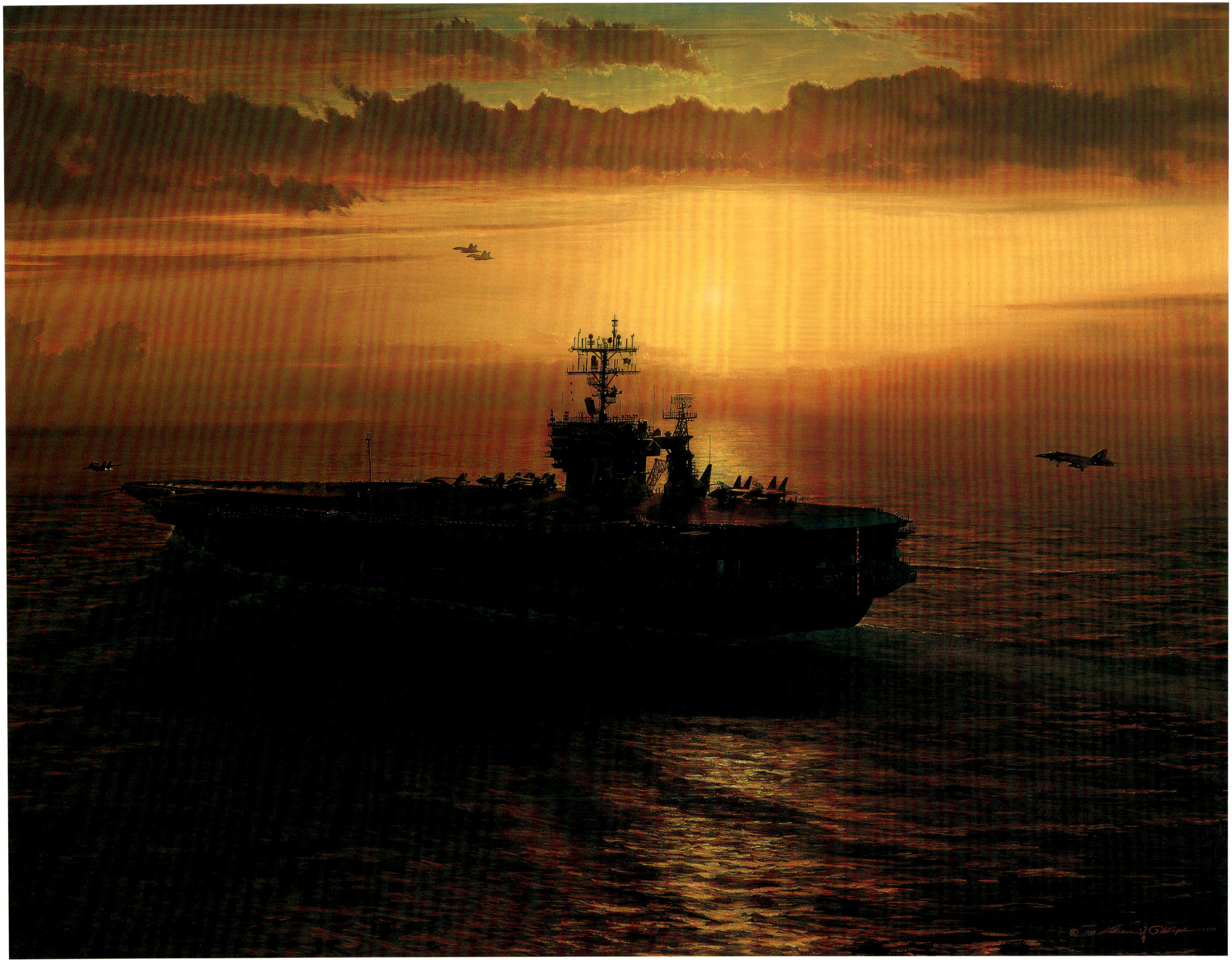

## Threading the Eye of the Needle

It has been said that landing a jet on the deck of a moving aircraft carrier is like "threading the eye of a needle." Even on a super carrier like this Nimitz class, it requires timing, judgment, concentration, coordination, a responsive hand on the stick, and utmost respect for the craft, the ship and all crewmembers involved. As a U.S. Navy combat artist, Bill was deployed to the Mediterranean and the Persian Gulf and conducted firsthand research on the subject.

Located here on the watery expanse of the Mediterranean, this CVN-73 *George Washington* played host to arriving and departing jet fighters. *George Washington* has a proud series of accomplishments to its credit. She played a starring role on her maiden deployment in 1994 as a backdrop for the impressive ceremonies celebrating the 50th anniversary of D-Day, and has made a total of three Mediterranean cruises since then. During its second cruise, *George Washington* was host to an historic meeting of the Joint Military Commission that brought together military leaders of the former combatants in Bosnia/Herzegovina. It provided a vital peacekeeping role during Operation Decisive Endeavor in Bosnia/Herzegovina and participated in enforcing the "no fly" zone over Southern Iraq. In 2004, *George Washington* plied the waters of the region traveling more than 51,000 nautical miles and remaining at sea for six months. That voyage represented its sixth deployment to the area, prior to returning to Norfolk, Virginia in July.

By September, 2004, *George Washington's* sailors returned to the sea to perform carrier qualifications, conduct training drills, and hone skills necessary to maintain readiness for subsequent cruises. Also in 2004, equipped with the Navy's newest fighter, Strike Fighter Squadron (VFA) 122—the Flying Eagles—was welcomed aboard *George Washington* with its fleet of F/A-18 E/F Super Hornets and represented the first time those fighters had been launched and recovered on the carrier. As one of the largest warships in the world, the *George Washington* is powered by two nuclear reactors, carries 85 aircraft, a ship's company of 3,200, and an air wing of 2,480.

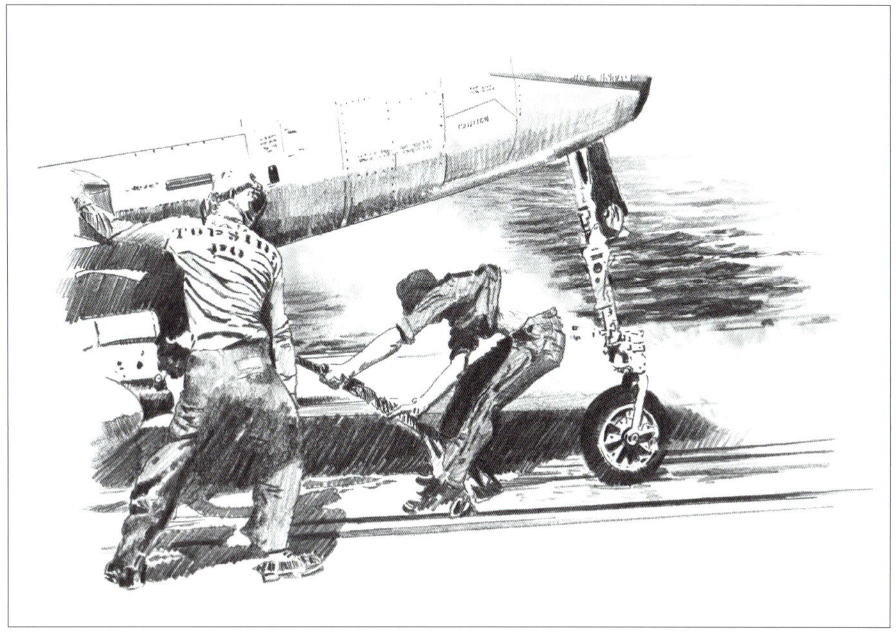

Deck handlers secure an FJ-4 Fury to the catapult of its carrier prior to launch. The men and women who work on the deck of aircraft carriers have one of the most dangerous and demanding jobs in the U.S. Navy.

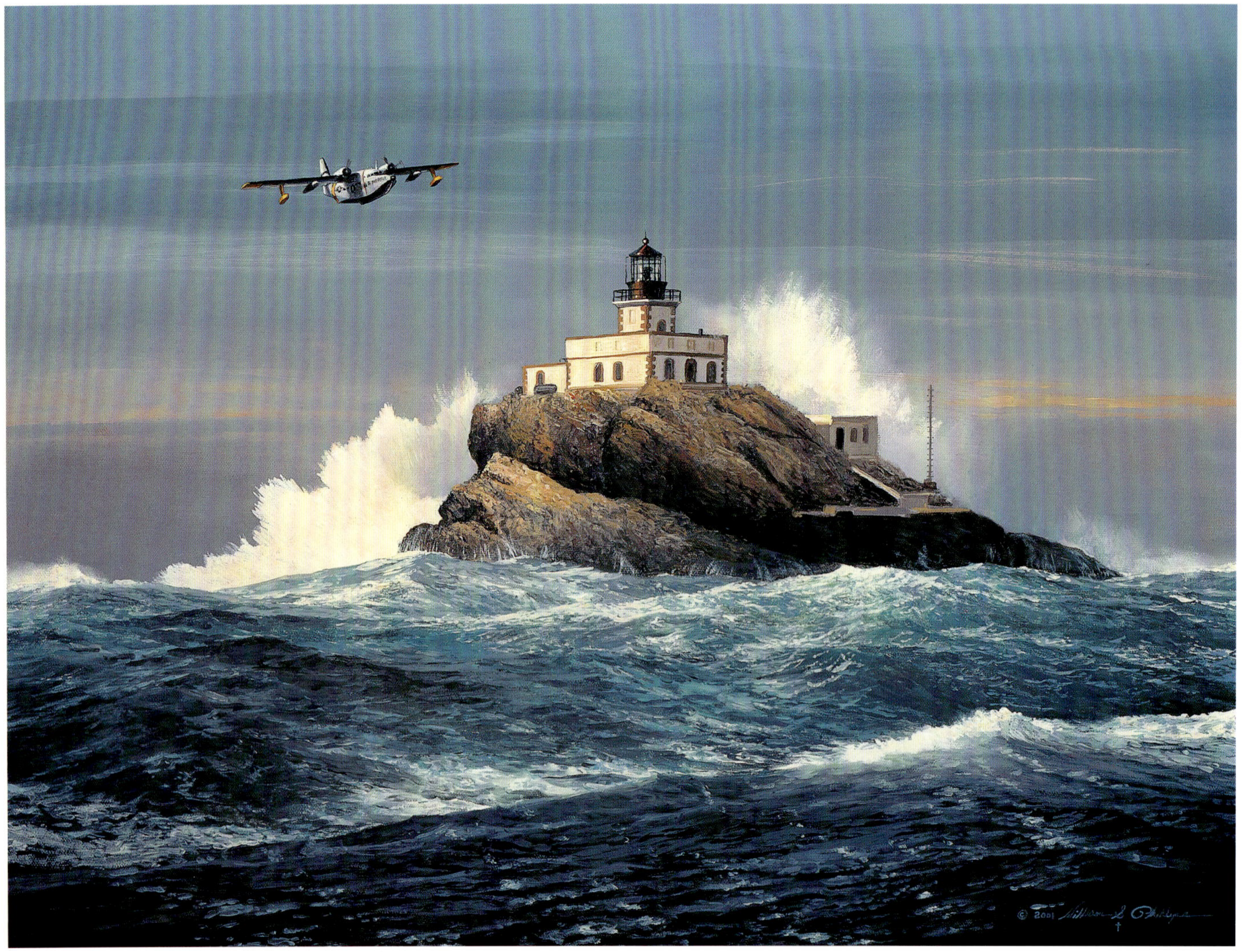

## Homecoming at Terrible Tilly

The noble lighthouse descends from its beginnings on the island of Pharos at the Nile Delta, Egypt, which protected mariners for 1,600 years. The beacon shown above graces Tillamook Rock, Oregon. Although "Terrible Tilly" guarded sailors from the dangers of running aground or being crushed against the rocky Oregon Coast for a far shorter period than the Pharos Lighthouse, Tilly remains a unique landmark and a reminder of the efforts of lighthouse keepers to protect those at sea.

Sharing the role of protector in the painting is the U.S. Air Force Grumman SA-16 Albatross. Like lighthouse keepers, search and rescue crews play significant roles in insuring safety to those in peril. Albatross aircraft, first built in 1947, served during the Korean and Vietnam conflicts and were in service with the U.S. Coast Guard, Air Force, and Navy. Artist Bill Phillips flew as an observer on Albatross missions when he was stationed with the Air Force in Portland, Oregon.

## CAPing the Tico

As bomber pilots of World War II counted on "little friends" to orbit above and protect them from enemy fighters, so aircraft carrier groups depend upon fighters to fly CAP (Combat Air Patrol) as escorts. Fighters establish aerial patterns and patrol, prepared to detect, intercept and destroy. CAP could divide fighters between high- and low-altitudes to shorten response time if attacked and a Barrier or BARCAP positioned fighters between the carrier group and the direction from which attack was most likely. In the painting, Crusaders climb to CAP altitude above CVS-14 USS *Ticonderoga*, an aircraft carrier that served from 1944 to 1973, from World War II through the Vietnam conflict. Piloting one fighter was James B. Stockdale, famed thirty-seven-year U.S. Navy veteran and recipient of the Medal of Honor. Commander Stockdale, brutalized as a prisoner of war in Hanoi, has written eloquently on how man can maintain dignity in the face of adversity.

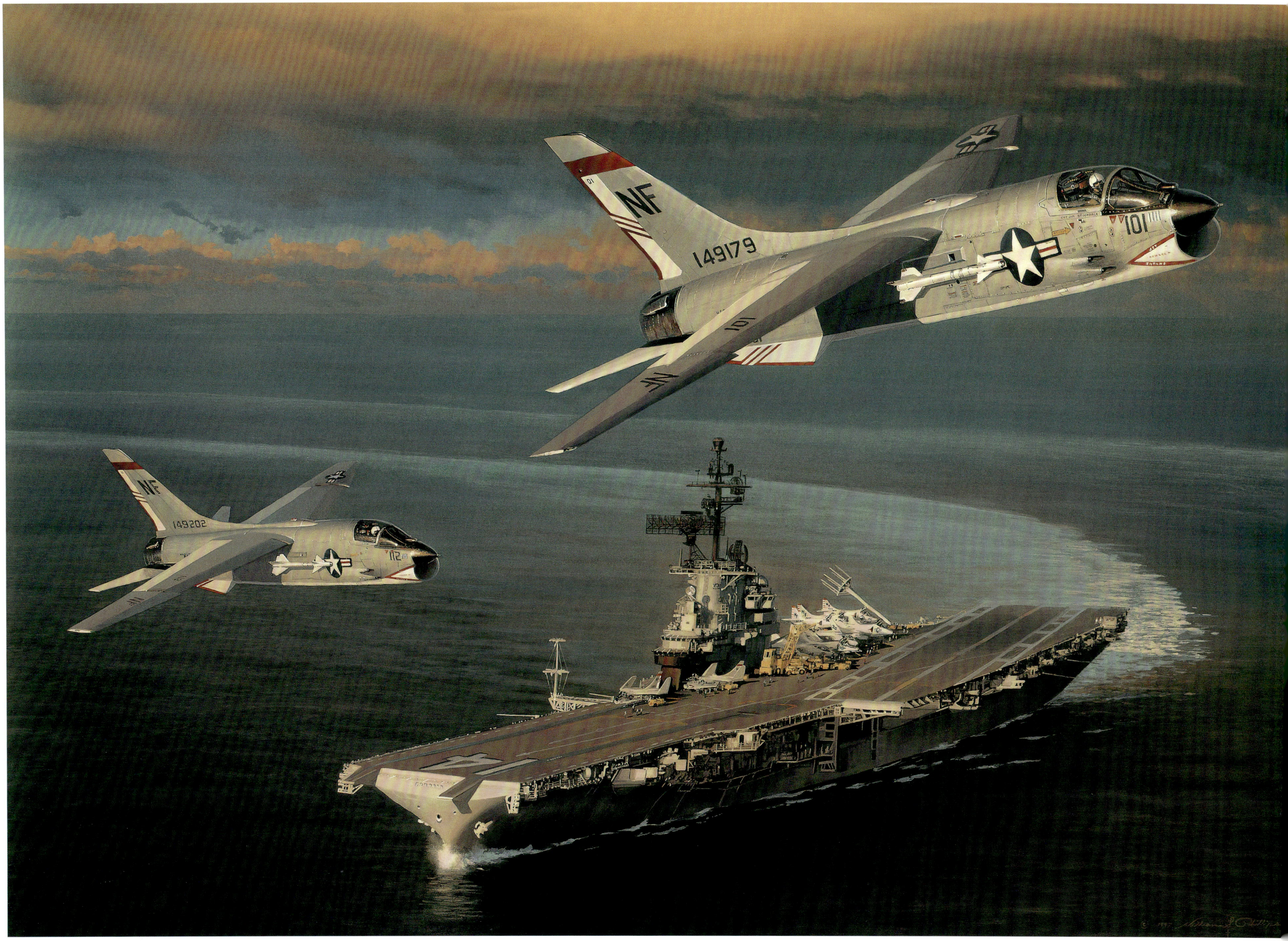

## Into the Throne Room of God

Carrying the logo of the U.S. Navy's Fighting Squadron Thirty-Two, which was created in 1945, this F-14 Tomcat of the VF-32 Swordsmen thrusts itself into the heavens with a reverberating roar that echoes the surrounding cumulonimbus maelstrom. The Swordsmen made the first Atlantic Fleet F-14 deployment in 1975. In the 1990s, they participated in the skies over Iraq and Bosnia-Herzogovina in Operations Desert Shield, Desert Storm, Southern Watch, and Deny Flight. By 1997, they were committed to support of Operations Southern Watch and Deliberate Forge and combat operations against Iraqi targets in Operation Desert Fox.

Having sailed with the maiden voyage of the USS *Harry S. Truman* in 2000, the Swordsmen's next deployment was in 2002 in support of Operations Noble Eagle, Northern Watch, and Iraqi Freedom, flying reconnaissance missions and liberation efforts. Theirs is a proud tradition and captured here is one moment in one mission into which the pilot is thrust into the mystery and the grandeur of God's work.

After shooting through the clouds' ominous darkness and turbulence into the highest reaches of the storm, the pilot suddenly charges into a sanctuary of brilliant warm and golden light. Like the strange and compelling stillness in the eye of a hurricane, the Tomcat has entered a corridor within the convective giant that offers a place of serenity within the surrounding inky walls of torrential rain and bone-jarring turmoil. With awe and respect, pilots privileged to glimpse such majesty welcome the invitation to enter "into the throne room of God."

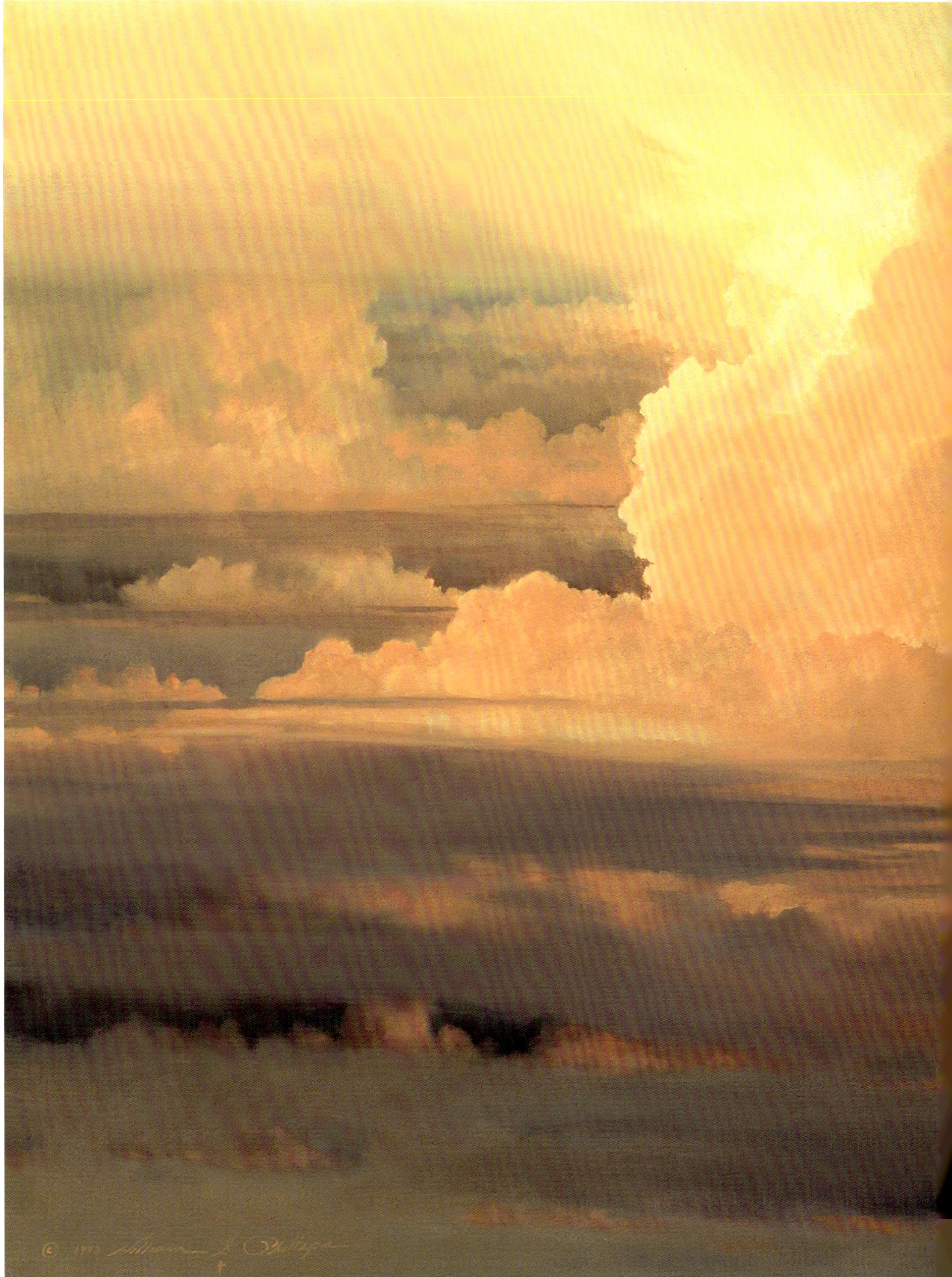

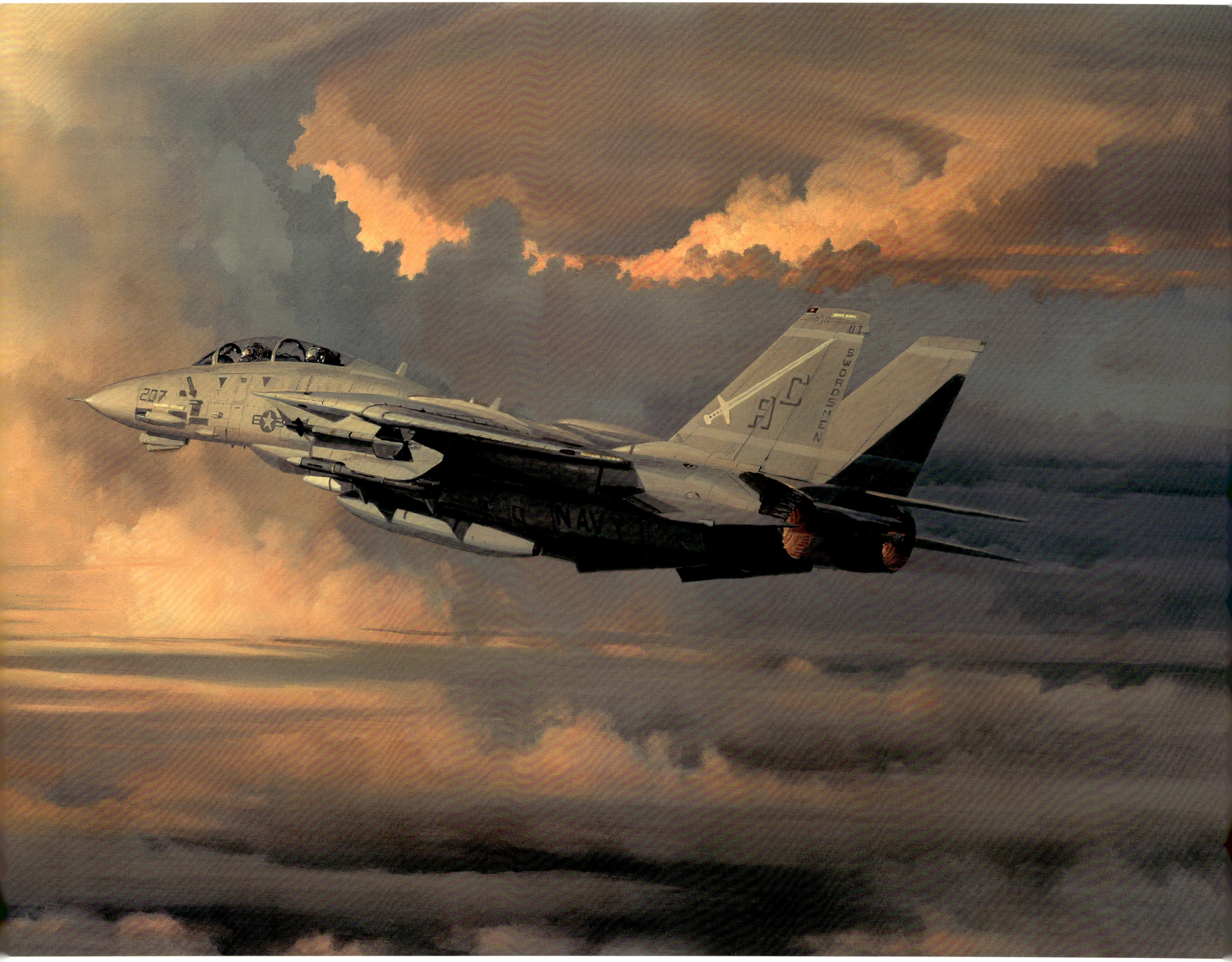
207
SWORDSMEN

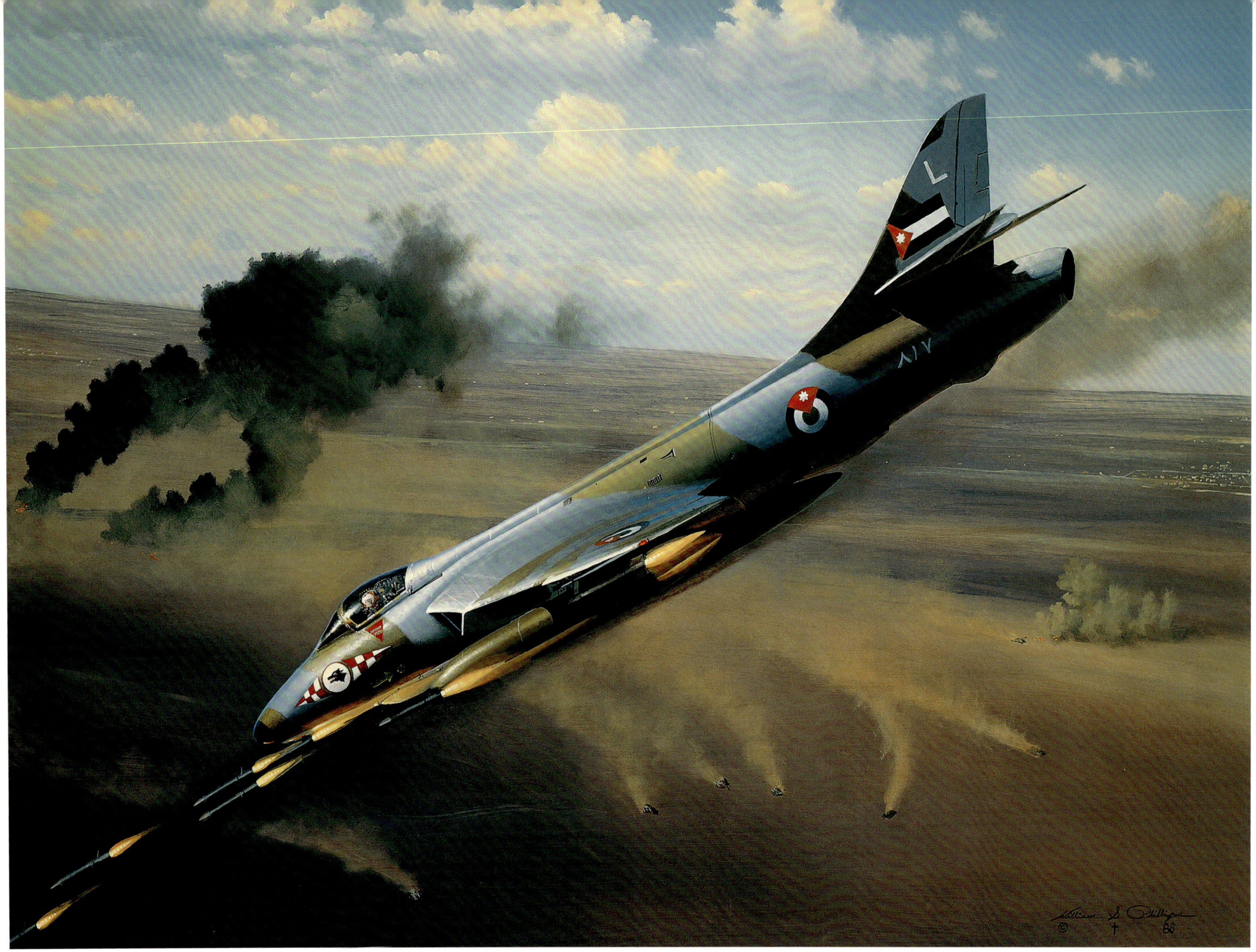

## Hunter on the Attack

As Britain evacuated the air stations at Amman and Mafraq, King Hussein sought to expand his Royal Jordanian Air Force (RJAF) with replacements for the Vampires that had served well. At the end of the revolution in Iraq in 1958, the RJAF received its first Hawker Hunter aircraft for the formation of Number 1 Fighter Squadron. This aerial attack on maneuvering ground elements brings out the best from the Hawker Hunter, marked with the Wolf's Head within the red and white checkerboard of the squadron.

Prized for its ease of handling and reliability, the Hunter was thought by many to be the finest transonic fighter. In the first Jordanian air combat, December 1964, four Hunters scrambled to battle Israeli Mirage IIIs. Known to the RJAF as the Battle of the Dead Sea, one Mirage was downed and three were damaged. With its 8,000 fpm climb and 710 mph top speed, the Hunter's combat success was assured.

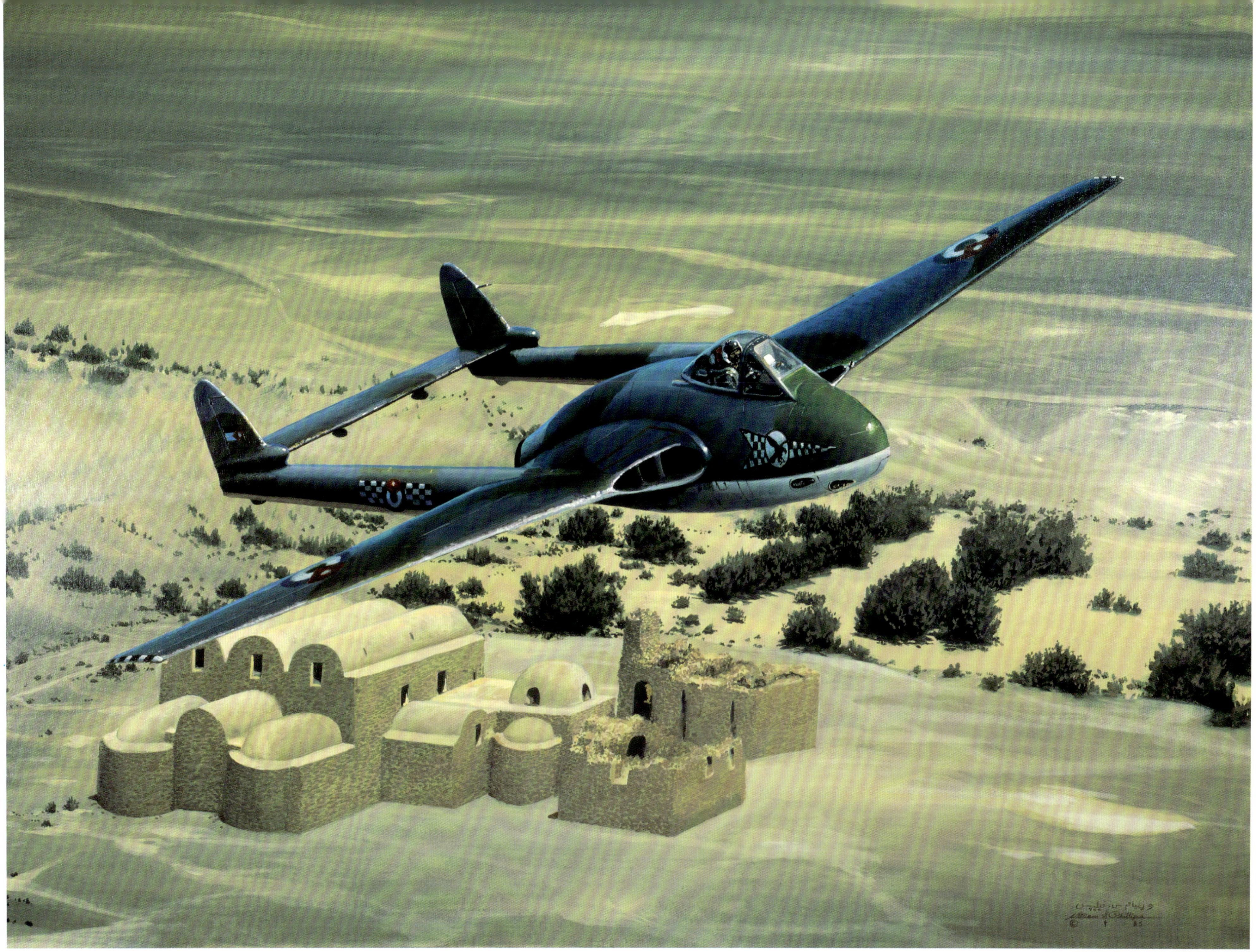

## Sentinel of the Desert

Britain provided air support when Jordan became a separate country in 1920 and, under King Hussein bin Talal, Jordan established air operations utilizing RAF Amman and RAF Mafraq, seeking to create a small air force. By 1955, the Arab Legion Air Force was renamed Royal Jordanian Air Force (RJAF) when Britain donated nine Vampire jets. This deHaviland Vampire FB.Mk 9, a symbol of modernity over the ancient abandoned castle, bears the RJAF insignia and an eagle on black and white checkerboard represents Number 2 Squadron of the RJAF. The Vampire FB.Mk 9 was an air-conditioned version of the FB.Mk 5 that had been optimized for ground attack with a strengthened and shortened wing. The Mk 9 carried four 20-mm cannon in the nose and could carry eight 60-pound rockets, two 1,000 pound bombs or two expendable fuel tanks under the wing. Having proven itself as a capable jet fighter, approximately 1,900 single-seat Vampires were built from 1945 to December 1953 when production ceased.

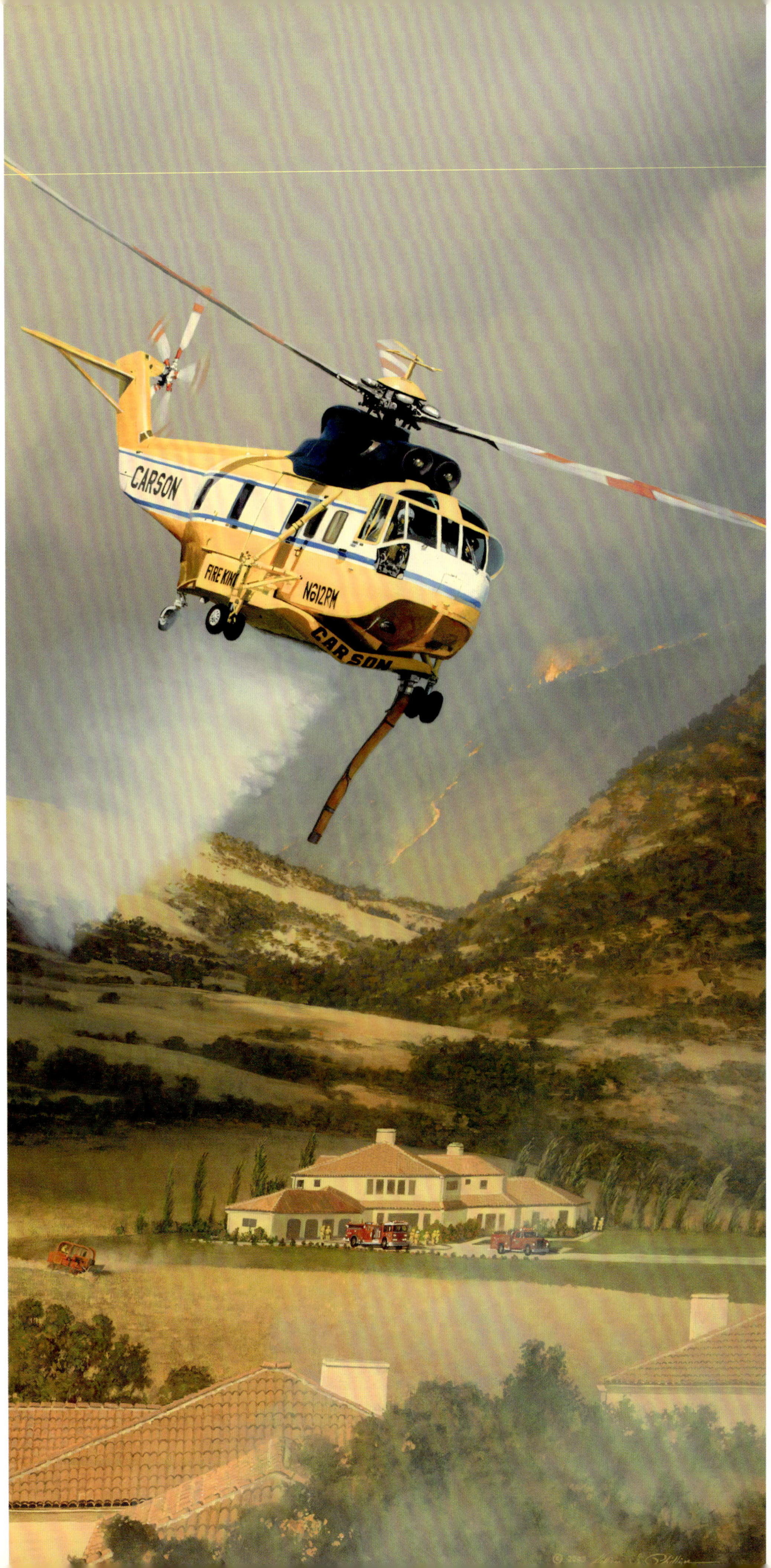


## Hitting the Spot

When Igor Sikorsky perfected the first successful helicopter design, he had a vision of a multitude of uses for his innovative craft. His first production model appeared in 1945 and in 1961 Sikorsky's manufacturing company created the S-61. Just four years prior, the U.S. Navy had awarded Sikorsky a development contract to produce an amphibious anti-submarine warfare helicopter. The S-61 Sea King first flew in 1959; production deliveries started in 1961 and Sea King was followed quickly by a civil version that increased the payload of freight or passengers.

In 1963, Frank Carson, another pioneer in rotary wing flight, put the S-61 aerial workhorse to its initial industrial use. Carson Helicopters started in 1958 and developed external lift capabilities that had not yet been explored. Today Carson Helicopters, Inc. operates eleven Sikorsky Super S-61s that are capable of air lifting external cargo loads up to 11,000 pounds as well as internal cargo and passengers. The company uses the Super S-61 for transporting equipment and personnel, building, lifting, rescuing, logging, and, as shown in this dramatic Southern California landscape, fire fighting. In the midst of a blazing inferno consuming the brush-covered hills and threatening a housing development, Carson's crewmembers release water that has been sucked up through the hose dangling next to the gear. The crew is "hitting the spot" to specifically concentrate water near an imperiled home.

## Last Chance

(opposite)

Although the Alaskan bush pilot generally flies by the seat of the pants and pushes through weather conditions that would keep a Canada goose on the ground, this one is keeping one eye on the weather and one on crucial decisions about weight and balance. If the deHavilland Beaver is loaded too heavily in the tail, the takeoff might turn into a mushing taxi across the water and no departure at all. The pilot is trying to be patient with his passengers, but weather waits for no man. If those silky clouds slipping down over the tops of the peaks meet the fog gathering across the land, the pass may well become completely obscured in zero-zero visibility. It's the last chance for takeoff and, while the adults come to grips with the baggage, one last chance for the son to catch that wily trout that he's been after.

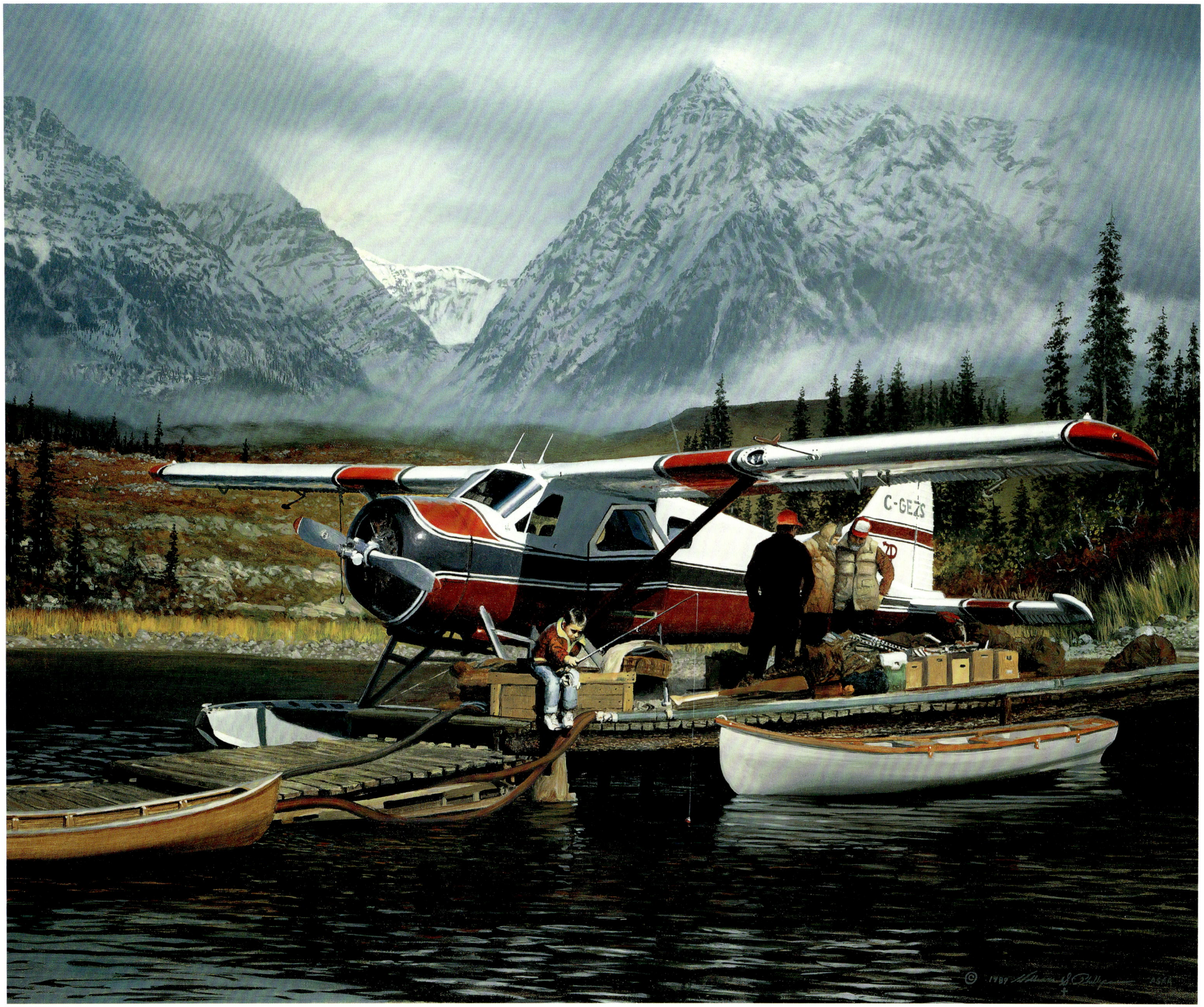
C-GEZS

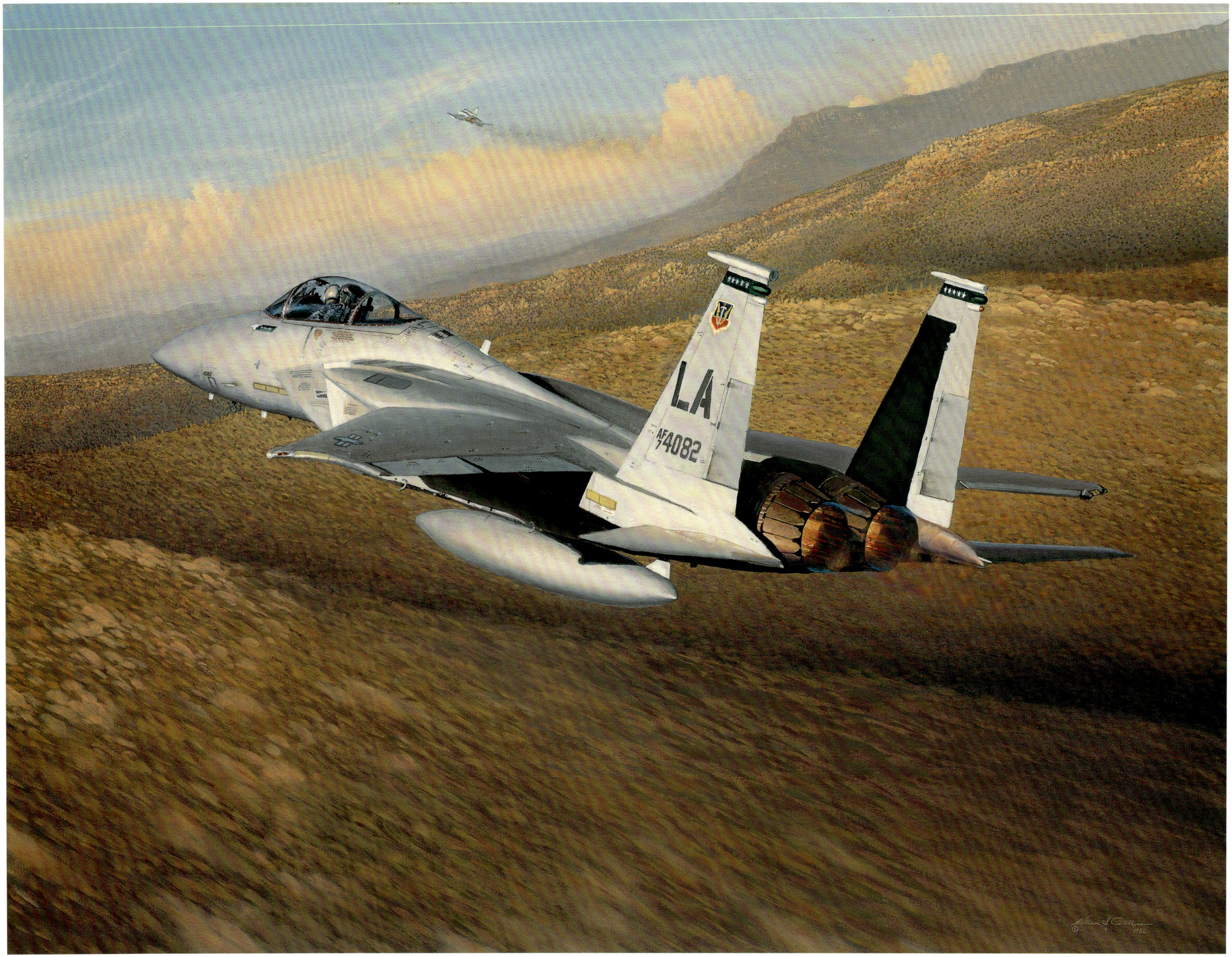

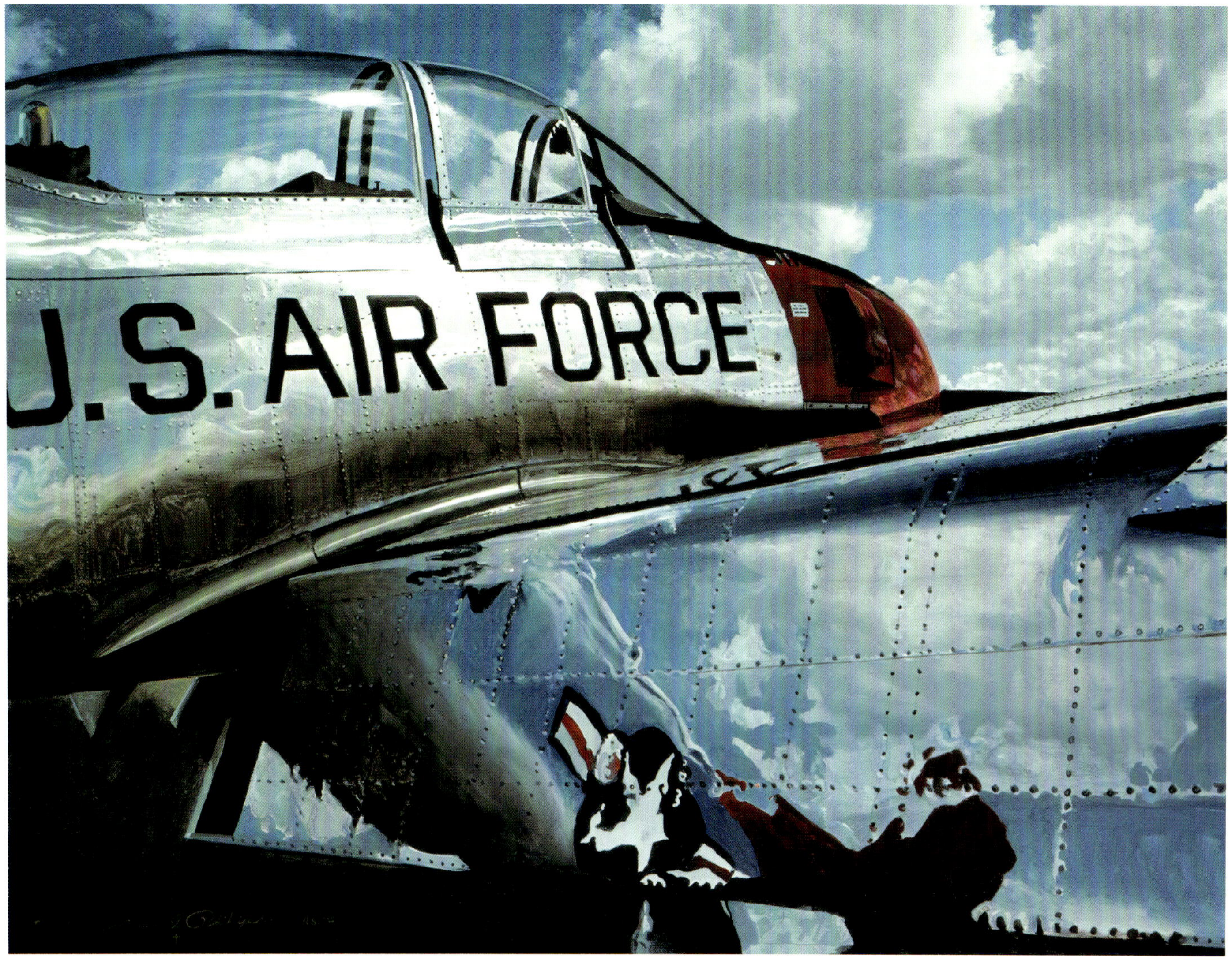


## Advantage Eagle

Air superiority requires responsive agility. Speed, sought since Barney Oldfield raced an airplane around an auto track in 1912, isn't a complete answer. Maneuverability is key. And when specifications were issued for new fighter aircraft in 1965, performance was the goal. The F-15 Eagle, advanced, agile and fast, was the result.

Luke Air Force Base, the 58th Tactical Fighter Training Wing and its famous "Triple Nickel" 555th Tactical Fighter Training Squadron were the first recipients of the F-15. From the first delivery on November 14, 1974, the Eagle proved equal to its nickname. With its better than one-to-one thrust ratio, the fighter can accelerate when flying vertically. One of its pilots said, "The Eagle can out-climb, out-maneuver and out-accelerate any fighter threat in existence or on the horizon."

Currently the top-of-the-line fighter of the U.S. Air Force, the F-15 is capable of Mach 2.5, accelerating in vertical climb and, with an aux tank, flying 1,300 to 1,500 miles unrefueled. Bill Phillips flew from Luke AFB, with the wide expanses of the Sonoran Desert below, and, at 500 knots, he saw the ground blur and felt the surge of energy. In his translation of that momentum to canvas, you can feel the Eagle's speed and power.

## Air Force Reflections

This T-28 Trojan came from the same drawing boards as the B-25 Mitchell bomber and P-51 Mustang, North American Aviation, a manufacturer of outstanding aircraft. In an effort to capture the lucrative training market at the end of World War II, North American designers developed a high performance trainer to follow their successful T-6/SNJ Texan. The T-28 Trojan, powered with the Wright R-1300 engine (the A Model) served the Air Force, and a "B" with the Wright Cyclone R-1820 was built for the Navy. By 1955, a C-model with the tailhook necessary for aircraft carrier landings was introduced. Here, flaps down, aluminum gleaming, the polished T-28A reflects the sky and itself.

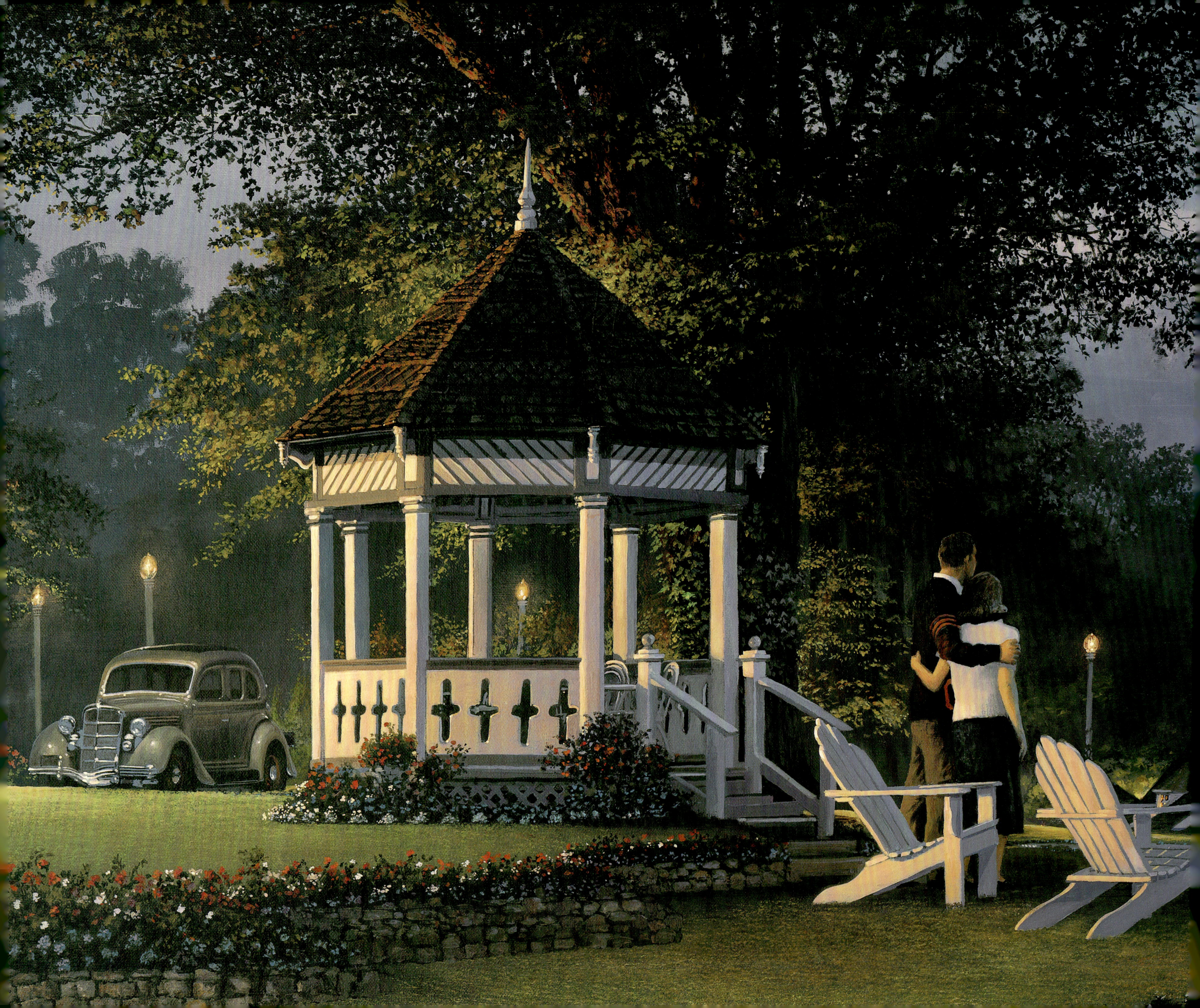

# VISION

## INTO THE THRONE ROOM OF GOD

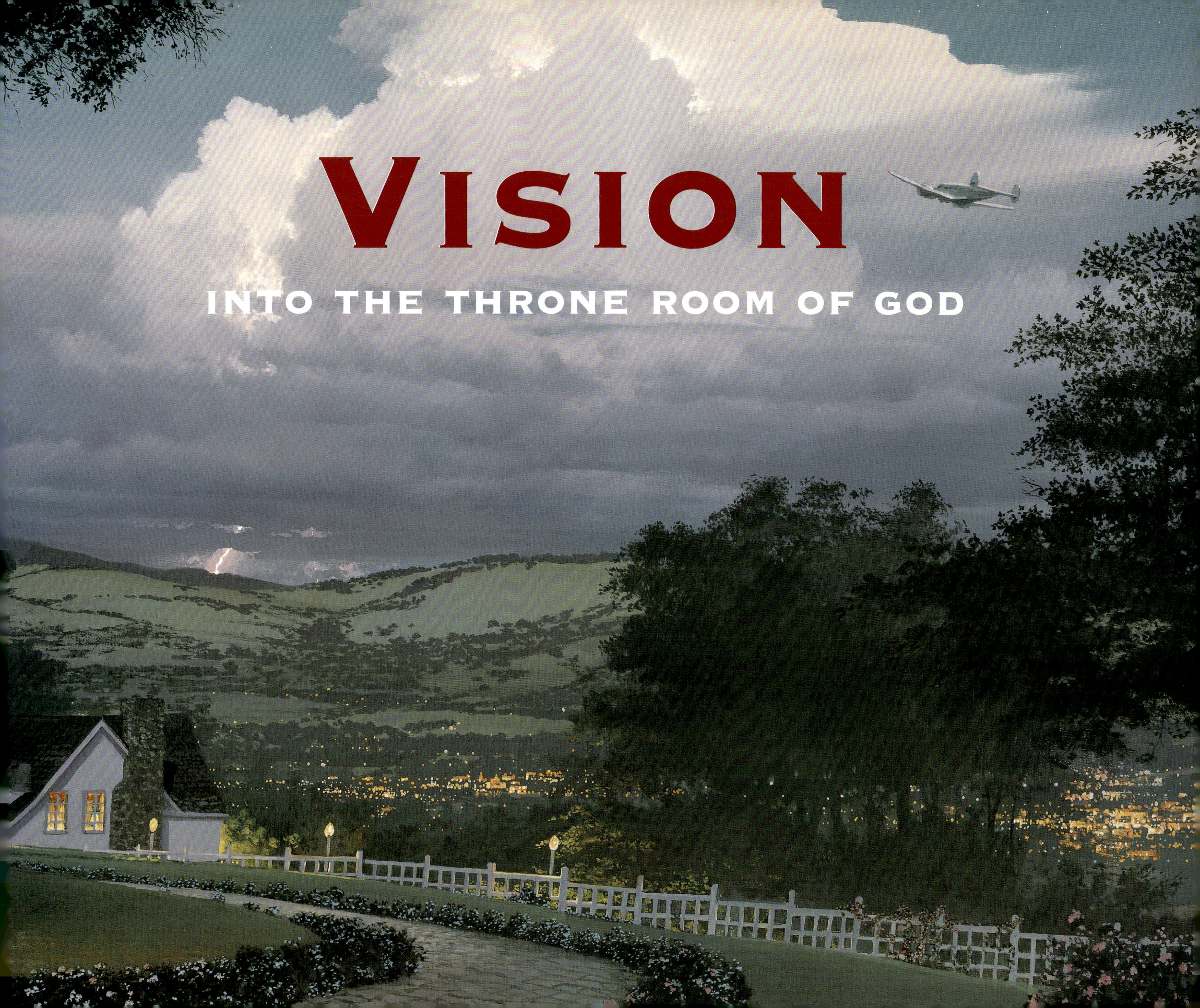

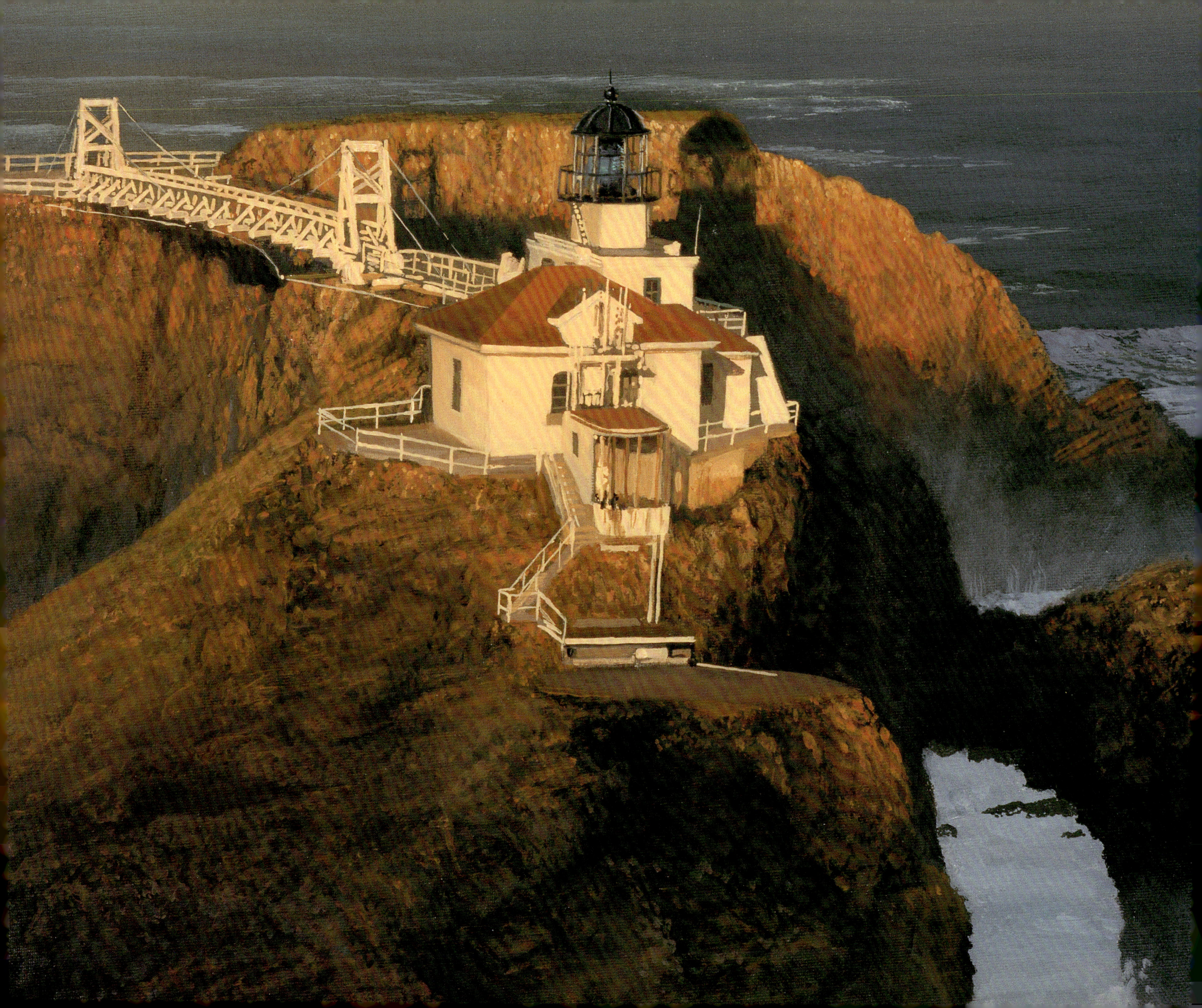

*What is life but the angle of vision? A man is measured by the angle at which he looks at objects. What is life but what a man is thinking all day? This is his fate and his employer. Knowing is the measure of the man. By how much we know, so much we are.*

RALPH WALDO EMERSON

Some aspiring artists may believe that other artists have merely picked up magical paintbrushes and mastered them. "However, there are no shortcuts," insists Bill. "I was attracted to art, yet I spent years believing I could *never* become an artist. With only a few exceptions, those around me believed that the pursuit of art demonstrated a lack of maturity and no real appreciation of life's responsibilities. It would be far better, I was told, to pursue a lucrative and intellectually challenging profession and to pursue it with vigor. I chose a course of study toward a career goal for which I had no passion—a direction I thought to be socially responsible, but not personally rewarding. I closed in on a bachelor's degree in criminal justice and looked toward a degree in law. Just prior to graduation and a scant three months prior to entering law school, I sold four small paintings at a local airport. With the check for $100 in hand, my wife and I made the decision for art that led to a paradigm shift that forever changed our lives.

"For years the vision of an art career had flickered as a small flame in my soul; yet, like others who have stepped out of the box, my decision placed me at the brink of a great adventure. This journey not only carried me forward artistically, but, unknown to me at the time, also spiritually. An instinctive trust began to emerge, surpassing any sentimental, momentary impulse."

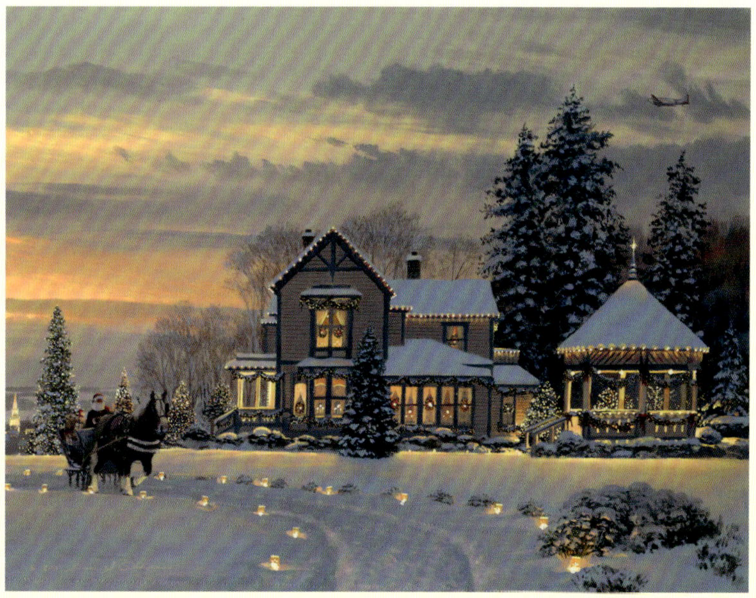

**CHRISTMAS EVE AT THE WINCHESTER INN**
The Christmas Eve celebration at Phillips Bay's historic inn is filled with pinpoints of light and a very welcome sleigh bound visitor.

Bill decided *not* to attend law school or to reenroll as an art student in college. "It would have plunged me further into debt," he explained, "and, as the thrust of most art courses in colleges and universities was toward abstract modernism, it held no appeal for me. My passion is for representational art." Inherent in a program of self-directed study is persistent, resolute discipline. Furthermore, self-directed study requires making a commitment to a vision, a goal. "In my case," Bill admitted, "the vision was to master the principles that have served fine artists for thousands of years. Since I was especially drawn to great landscapes, I made a study of the rules and methods established by great artists to accurately depict images of nature or of contemporary life."

He studied, he analyzed, he drew, he painted; he applied techniques new to him and called upon his own creativity in his approach to landscapes, skyscapes, and other subjects that intrigued him. He devoted himself to intense study and, he added, "I have not stopped nor will I ever cease to learn."

One of Bill's fellow firemen proved to be a catalyst for another pivotal change in his life. In 1975, a firehouse discussion involving philosophy and religion caused him to take a long and critical look at his own worldview and its hypocrisy. "I was raised a Catholic," he explained, "so I was convinced that I knew all there was to know about God and Christ. (I studied Latin, was dragged to Mass by my mother, and served as an altar boy). What more, I reasoned, could I possibly need to know about religion? After all, God had let my dad die. I turned my back on Him in high school and, as time passed, was not inclined to change. I delighted instead in postmodernist thought.

"In the firehouse discussion, I rambled on, praising the merits of Jean-Paul Sartre and the presumption that Christianity and the Bible were irrelevant in the modern, enlightened world. I was challenged by a fire fighter named Al Jenkins who asked simply, 'Have you ever read the Bible with an open mind?' At the time, I dismissed his question and resumed the debate. That night, however, I faced the truth that I did *not* have an open mind toward Christ or God. I began a search every bit as intense as my quest for artistic perfection. I temporarily put aside my study of art and pored over the Bible and any literature about Christianity, God and comparative religion that I could find.

"I had always believed in the importance of scientific verification; but I began to question my preconceived ideas. Many of my earlier arguments were dispelled by the writings of C. S. Lewis and other 'Christian intellectuals,' (a term I had shortly before dismissed as an oxymoron) and, as I gave myself to honest exploration, God began to prove His existence in ways that were indisputable—far beyond mere coincidence. My skepticism didn't disappear immediately, despite the stirrings of spiritual change. Initially in my quest to know God, I realized I had no clear vision at all. In my arrogance, I challenged Him. If He was unable to show Himself to me as all-loving and the ultimate form of goodness, then I proposed that I might be convinced if the antithesis were proven. I was, for the most part, self-assured and felt that my own perseverance could overcome almost any obstacle.

"Within a matter of weeks, I fell into a deep depression; I was overwhelmed by a terrifying darkness. Not long after, some Bible tracts were brought to the fire station while I was on duty. One article featured salient points I had never considered about Scripture and Faith and the other focused on suicidal depression. The two articles affected me deeply. For the first time in weeks the oppression began to lift and my skepticism abated. I felt a strong spiritual calling.

"When I reached my lowest point, feeling drained and unable to control my circumstances, I recognized that I had endured a cleansing, an erasure of doubt and I gave my life to Christ. I offered my

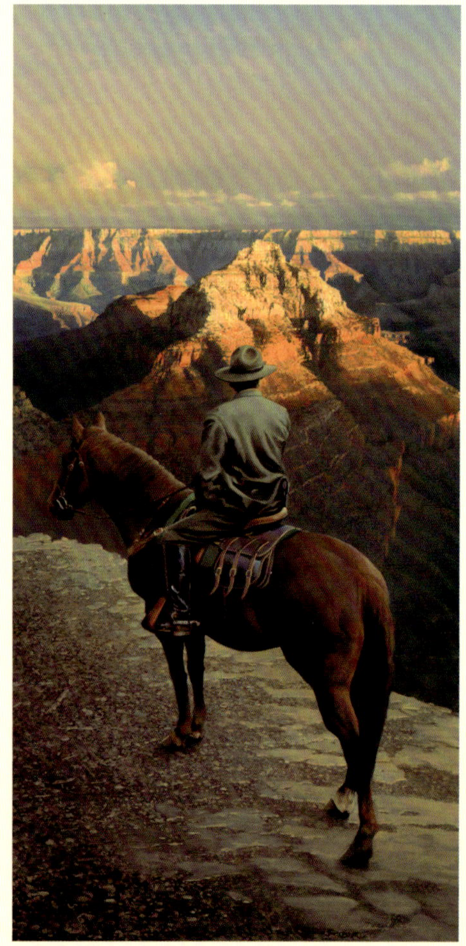

**REFLECTION**
Rangers of the North Kaibab Ranger District in the Grand Canyon maintain a close watch on those who might venture beyond their capabilities, inadequately prepared (and a majority of the 250 rescued annually from the Grand Canyon are young, healthy males). The North Kaibab Trail begins near the North Rim's Grand Canyon Lodge where Phillips was Artist in Residence in 2004.

talent to God by saying, 'I have always wanted to be an artist but, God, if You have another direction for me, I am ready to follow whatever course You set.' I was prepared to give everything to God. I was through living for myself. My conversion was powerful and mysterious; it was humbling and now it makes great sense to me. It has infused my vision as an artist."

By the 1970s, Bill had explored a wide variety of subjects for his art. He couldn't escape his fascination with flight and the vastness of the sky and, as an avid outdoorsman, his exploration of the deserts of the American southwest influenced his artistic direction. As his talent began to attract attention, he longed to excel as an artist and believed aviation art could succeed in fine art galleries. Idealism and reality, however, were on a collision course. Galleries rejected his aviation paintings, while praising his handling of his subjects. One gallery owner said, "You are a fine artist, but paintings of airplanes will never sell. They are too illustrative and commercial. What else do you paint?"

Bill compromised and painted upland and wetland game birds, and mountain landscapes. His work interested dealer Bruce Watkins of Soaring Wings Gallery in Salem, Oregon and Bill was invited to exhibit at the gallery's opening gala. The artist placed an open portfolio of aviation art beside his other paintings. "My bird paintings sold well," said Bill, "but, the real thrill came when a car dealer commissioned a painting of a Ford Phaeton automobile and a Ford Trimotor aircraft. That aviation commission broke the $1,000 mark among paintings that ranged between $150 and $300. I later sat next to the gallery owner at dinner and, somewhere between the green salad and the main course, Bruce asked, 'About those airplane paintings, do you think we could hang a few in the gallery?'"

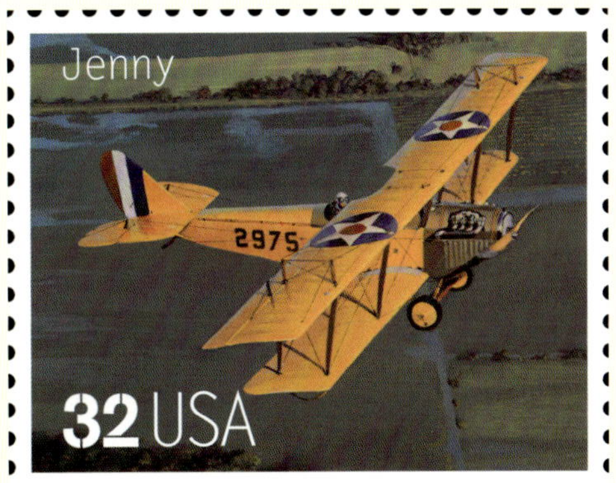


**CURTISS JN JENNY**
U.S. Postal Service, Classic American Aircraft series, 1997

Produced by Glenn Curtiss, the JN series of aircraft introduced flight to many. England's Royal Flying Corps ordered the JN2. In 1916, the Jenny served against Pancho Villa with Pershing's expeditionary force. The most widely used trainer for World War I, it became a favorite of barnstormers.

Bill had positioned himself for the success that followed. Once his aviation pieces were on the gallery walls, buyers entered a fine art gallery and associated aviation art as a fine art genre. "My wildlife art was a leader; my aviation art started to generate interest. As the aviation sales grew, other galleries expressed interest." In 1994 the United States Postal Service awarded him the contract to produce a twenty-stamp sheet entitled *Classic American Aircraft*. "I worked with two very gifted men, Phil Jordan, the stamp designer, and Terry McCaffry, the art director," Bill explained. "The aircraft were already selected and, although tight standards for accuracy were imposed, both men allowed me the freedom to be creative and to press forward with my vision for each stamp. Millions of these stamps were produced and the *Classic American Aircraft* series became the second best-selling stamp sheet in 1997. In 2004, Bill completed artwork for his second sheet of stamps.

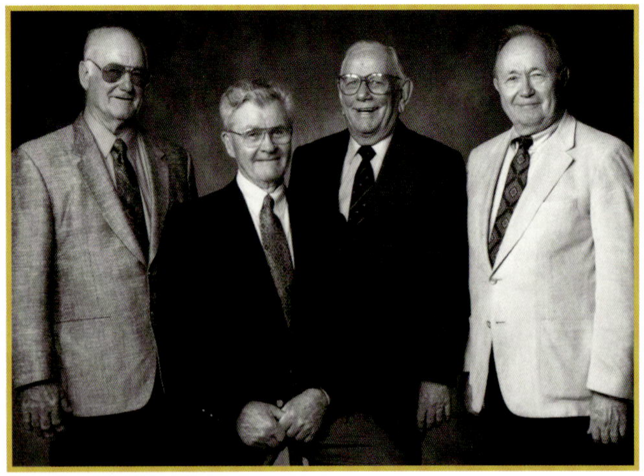

(l to r) Members of the *Enola Gay* and *Bock's Car* crews: *Enola Gay's* Colonel Thomas W. Ferebee (USAF-Ret.), bombadier; Brigadier General Paul W. Tibbets (USAF-Ret.), pilot; *Bock's Car's* co-pilot 1st. Lt. Charles Albury; and *Enola Gay's* Captain Theodore J. "Dutch" Van Kirk, navigator.

## The Beginning of the End

Crewmembers with Colonel Paul Tibbets completed several missions against Nazis in Germany before participating in the invasion of North Africa. A third, dramatic change occurred when Tibbets was named leader of the 509th Composite Group. He and an elite crew transitioned from the B-17 Flying Fortress to the largest aircraft flown during World War II, the B-29 Superfortress. The transition was cloaked in secrecy. The aircraft itself was designed to specifications calling for a bomber capable of flying 5,000 miles, at 400 mph and carrying 20,000 pounds in bombs. The requirements were extraordinary in the late 1930s. In 1934, Boeing engineers designed the Model 299 that became the XB-15 and the B-17. By 1936, Boeing engineers worked on an up-dated XB-15, Models 330 through the 334A. It wasn't until 1939 that Model 341 came along with an improved wing.

By then war drums were sounding across the Atlantic. Soon after, analysis of enemy bombers was possible and reaction to the findings pressed Boeing to follow the B-17 with a strengthened bomber that would be long-range and equipped with more, and higher caliber, machine guns and self-sealing fuel tanks. Sharing Billy Mitchell's vision of airpower's destiny, General H. H. "Hap" Arnold exhibited a bulldog's tenacity to implement the bomber. The United States Army Accessions Command (USAAC) was so desperate for this aircraft to fulfill the specs that the XB-29 was ordered in massive quantity before the prototype flew. It boasted innovations—an advanced gunnery system and pressurization—but it had difficulties from the start. Development was arduous. The first planes arrived in India in April 1944, and missions were plagued with engine failures, aborted takeoffs, and poor results. Logistics was a nightmare. Nevertheless, by March, 9-10, 1945, 302 B-29s flew the first raid against Tokyo. By April 12, Japan's aircraft engine industry was virtually gone. Surrender was no option for Japan, but *Enola Gay* and *Bock's Car* forced the issue. The Boeing B-29 Superfortress was a major contributor to Japan's defeat.

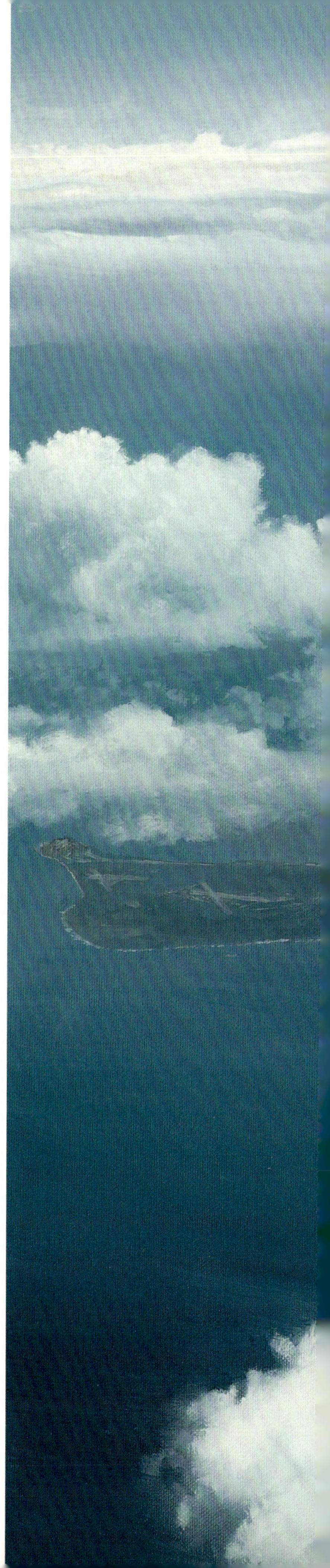

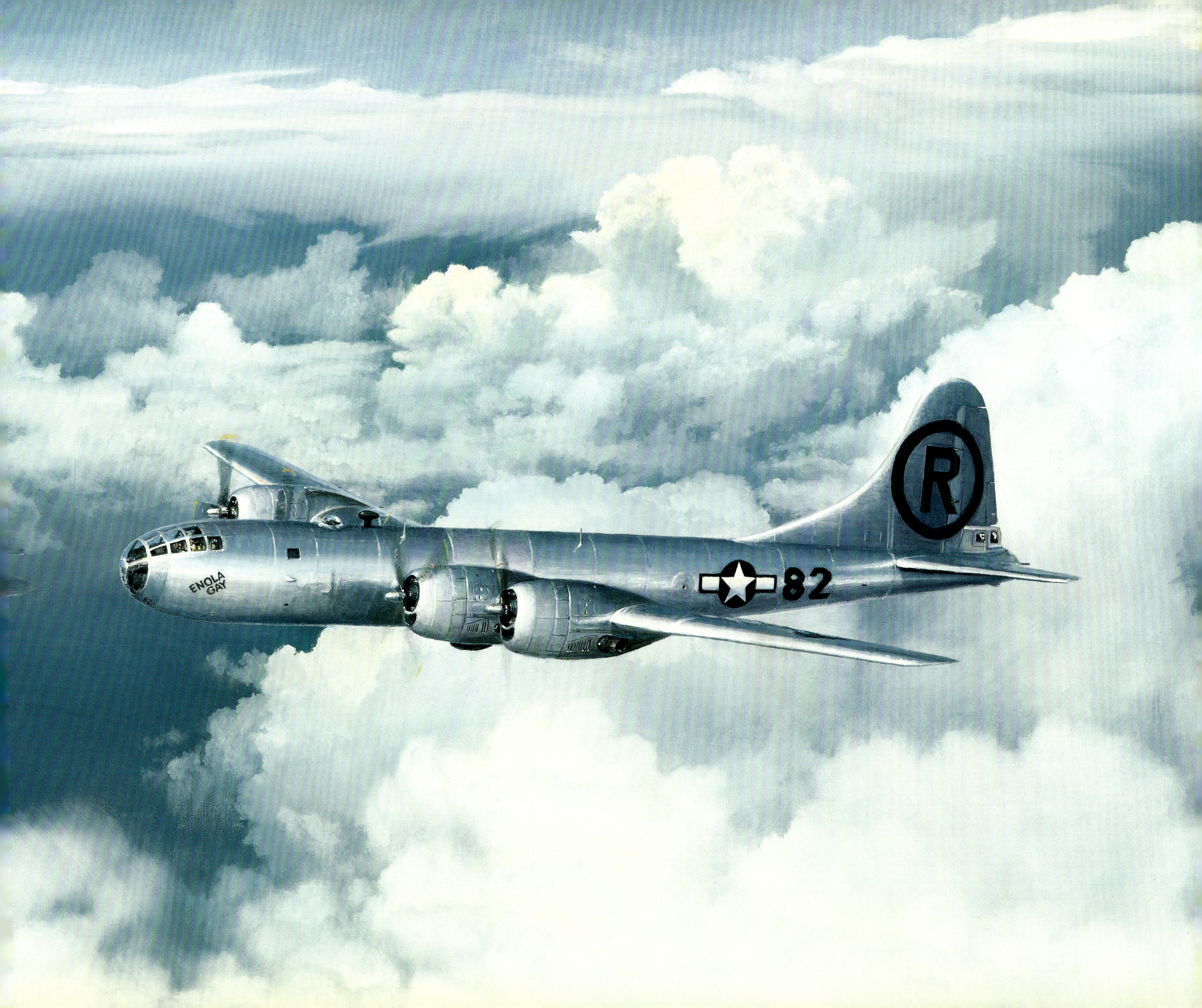
R
ENOLA GAY
82

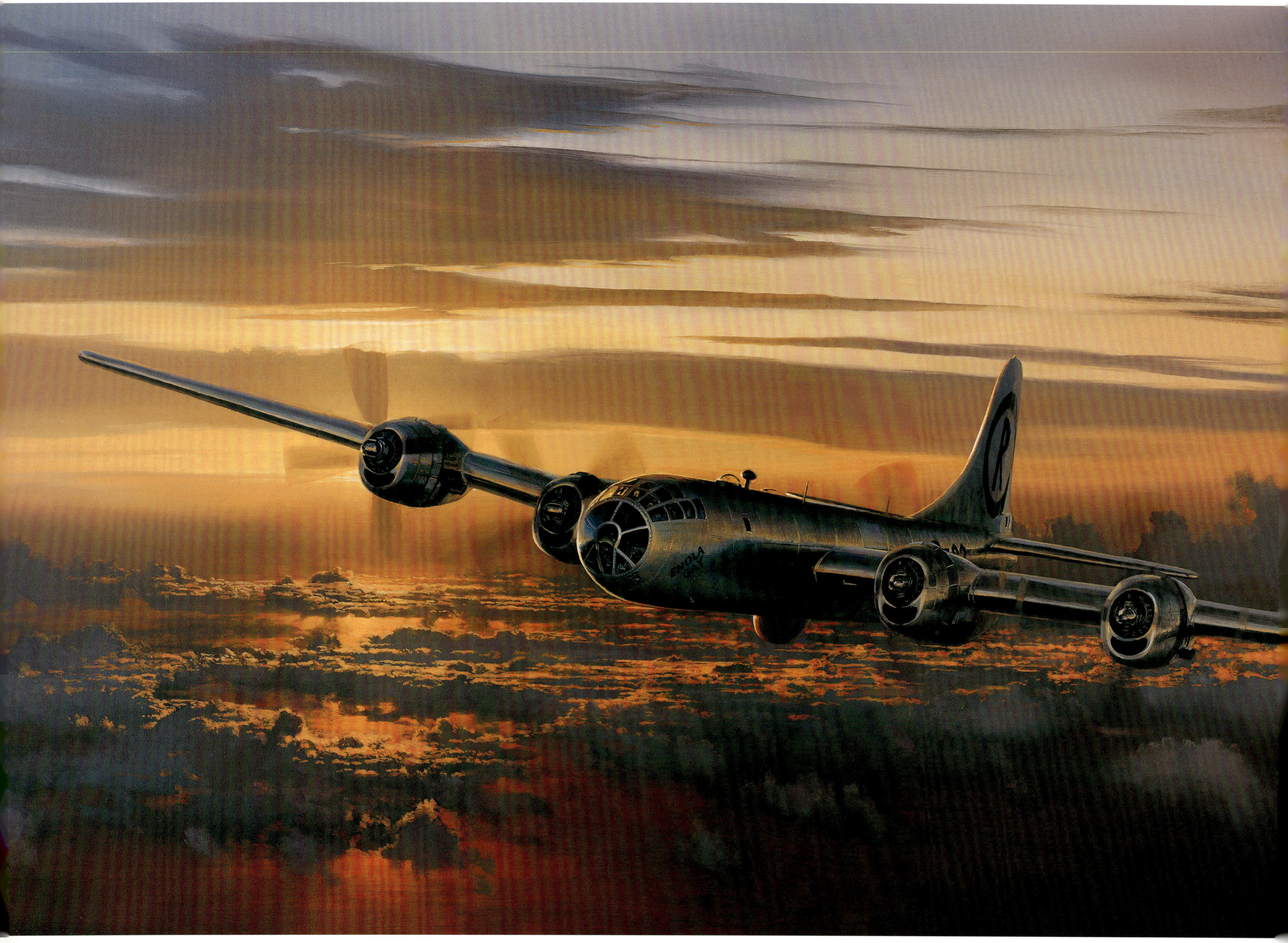
ENOLA
GAY

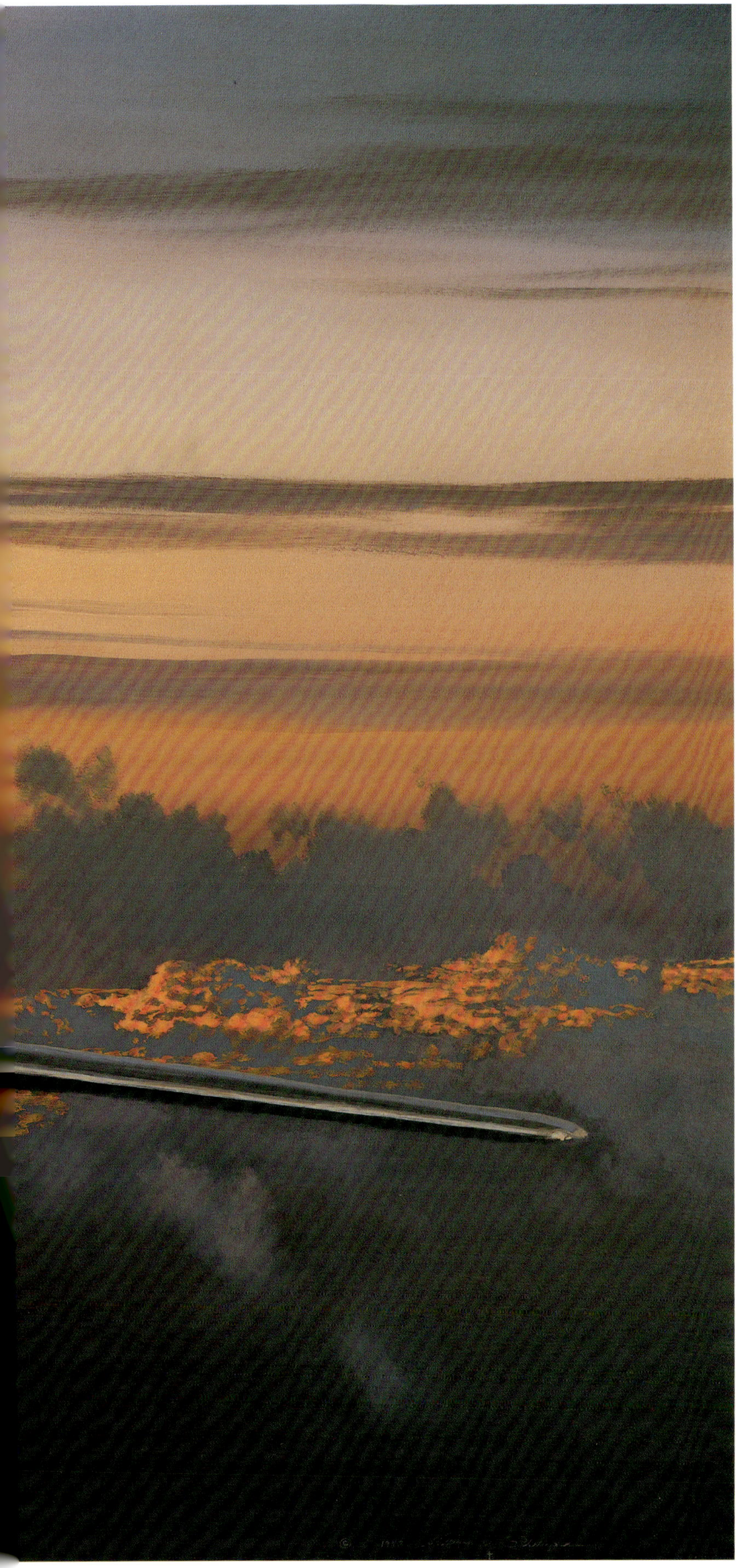

## Dawn, the World Forever Changed

In 1941, Japan's tactical error was initiating a war it could not win. Seeking land and natural resources, it conquered much of China before its December 7, 1941 bombings of a host of Pacific Islands—Midway, Wake, Guam, the Gilberts, the Philippines and Pearl Harbor. By spring, 1945, Japan was defeated, but unconditional surrender represented an unthinkable step to the Samurai heritage. Three choices fell to the Allies: planning and preparation to invade Japan, an aerial firebombing to destroy Japan's ability and will to fight, and a third choice known to very few—the atomic bomb.

Surrender was called for and refused more than once. When U.S. President Truman recognized the loss of Allied lives that would result from a ground offensive of Japan, he made his excruciating decision. The B-29 bomber *Enola Gay* lifted out of Tinian and headed for the home islands on August 6, 1945 with crewmembers that had flown together before. Pilot, Colonel Paul Tibbets; Bombardier, Major Thomas Ferebee; and Navigator, Captain Theodore "Dutch" Van Kirk served together in 1942 as a B-17 crew in the 97th Bomb Group. When Tibbets was handpicked to lead the 509th Composite Group of B-29s, he brought together his core team and they trained for the special mission, practicing high-altitude bombing with a breakaway maneuver to follow release of the payload. In July 1945, the world's first atomic bomb was exploded in the New Mexico desert. In August, the second and third brought an end to the war.

The Japanese surrender came on September 2, and the U.S. Army occupation forces began to democratize and rebuild Japan. Land reform, public education, the rights for women to vote and to run for office, and a new constitution were put into effect in Japan with the assistance of the United States.

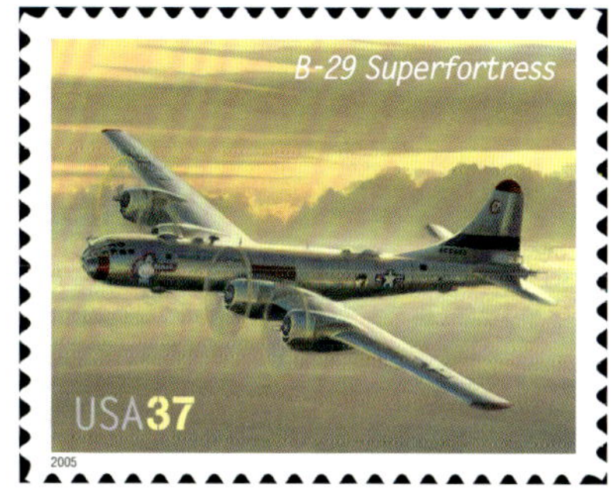


## B-29 Superfortress

U.S. Postal Service, Advances in American Aviation series, 2005

With a total of 3,900 built, the Boeing B-29 Superfortress first flew in 1942, and served in WWII and in Korea. Complex, expensive, and ultimately effective, B-29s operated arduous routes out of India to Japan's islands in 1944. She flew from China, Saipan, Guam, and Tinian and ultimately carried the atomic bomb in a desperate decision to end the war.

## Heartland

In the painting at right, artist Bill Phillips reflects on the huge contribution made by those left behind when U.S. servicemen and women are called into conflict, as he reminds us of the homeland and the values they are fighting to preserve. His image honors the families, defense workers, those whose jobs directly impact our nation's defense, all whose efforts are important and essential and, as depicted here, the farmers who exert tireless efforts to feed our nation's people and to export the fruits of those labors to the less fortunate. In this scene, which could be found in rural areas throughout our nation, we are reminded of the quintessential America, its rich agricultural heritage, and strong work ethic. With the arching rainbow of hope and tiny Piper Cub waiting for flight in the distant farmyard, two North American P-51 Mustangs symbolize those who enter into harm's way to preserve the liberties we sometimes take for granted.

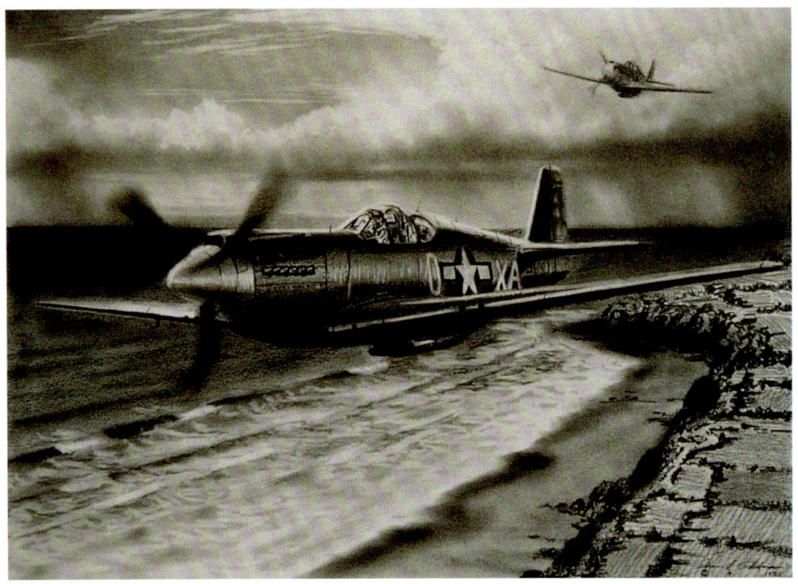

A P-51 (F-6A) makes a high-speed pass down the beach photographing German defenses prior to the Normandy Invasion.

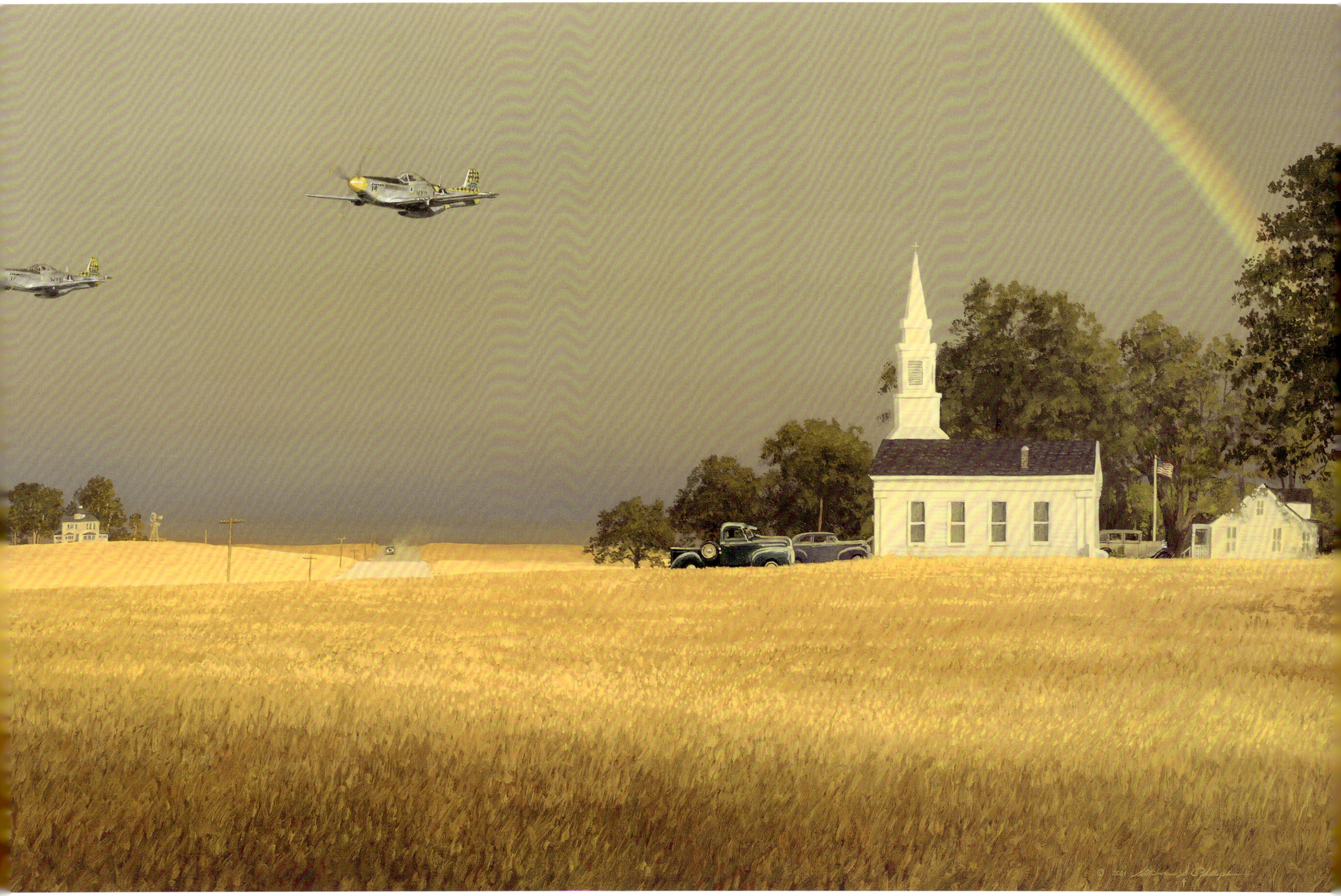

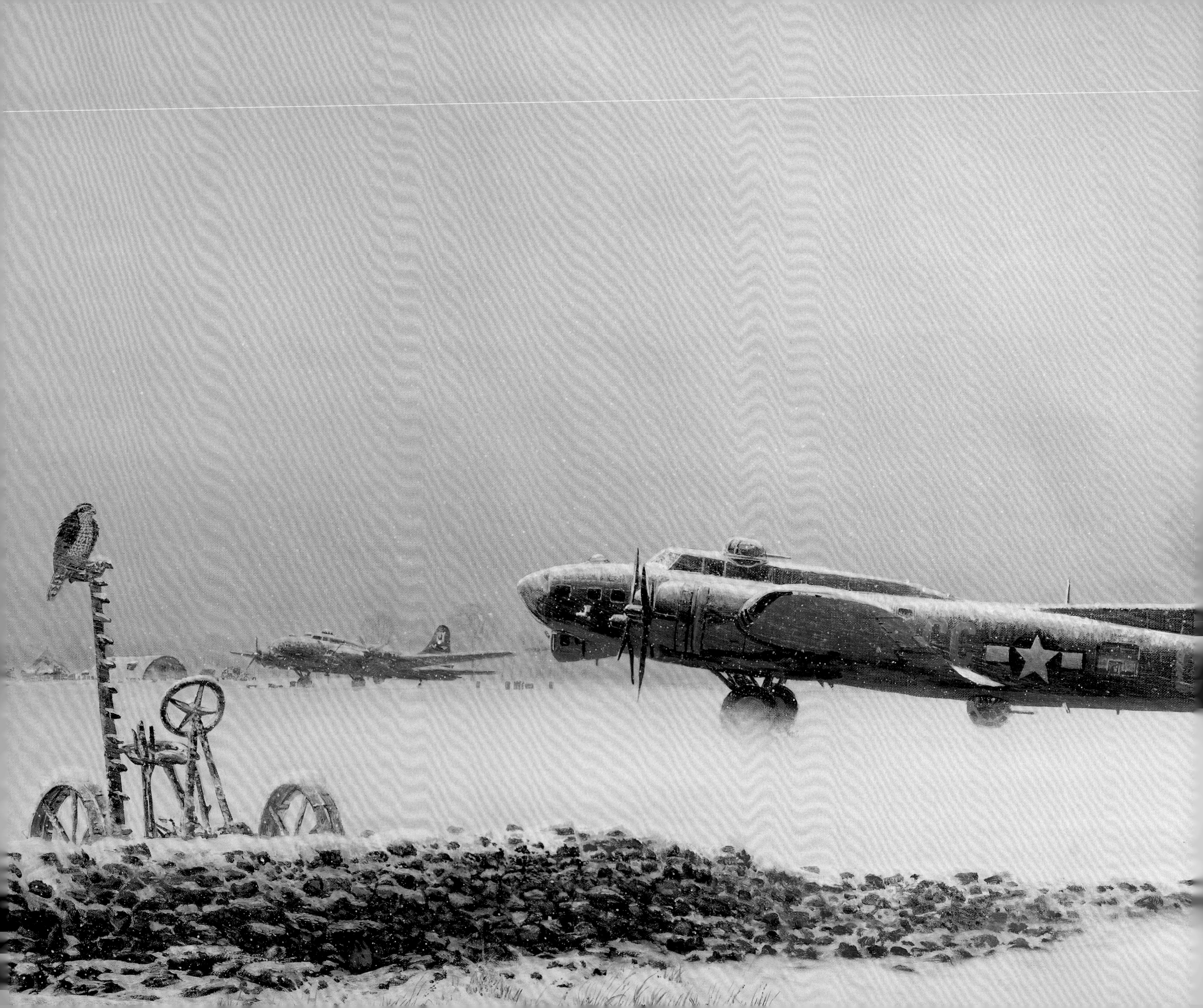

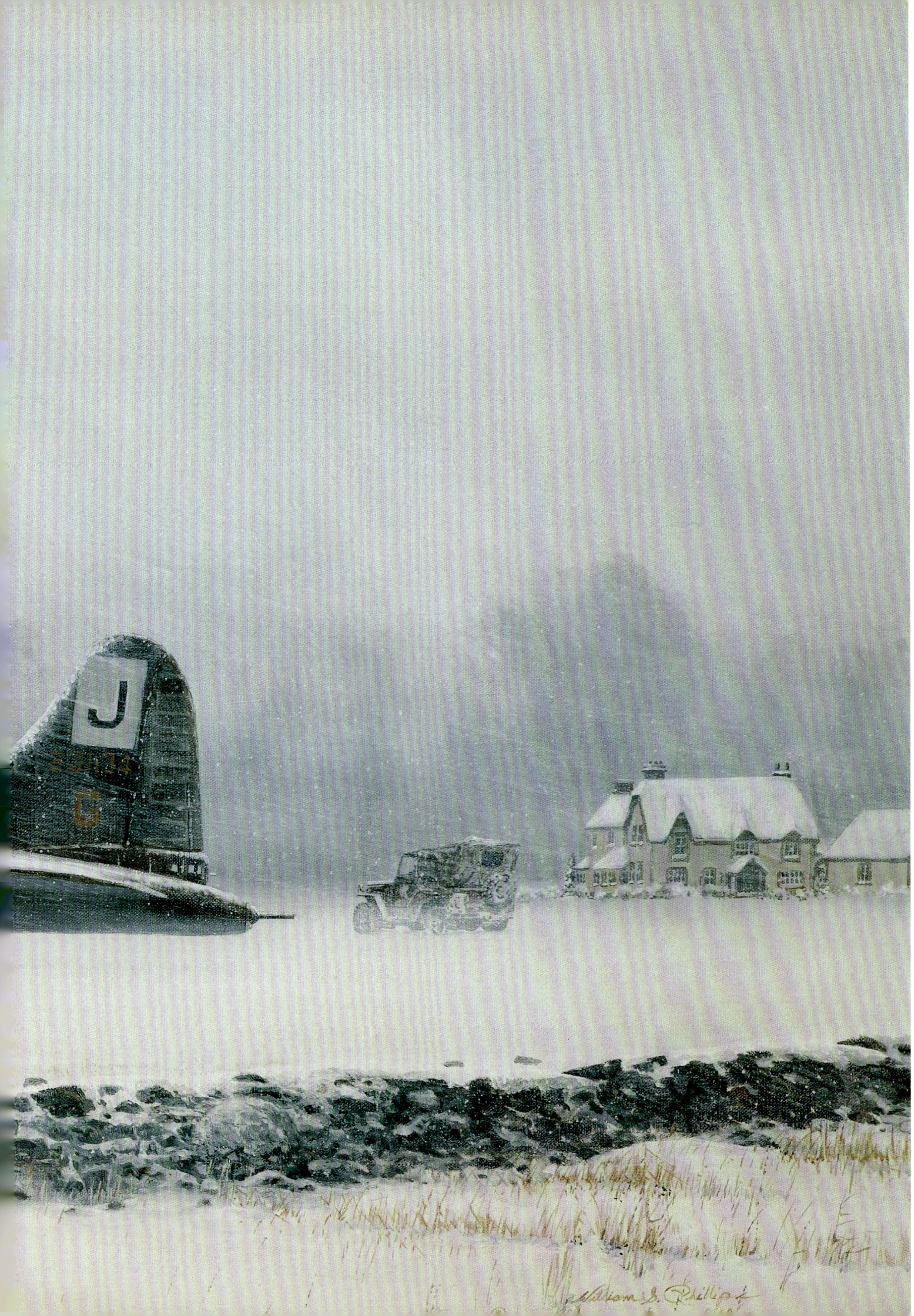

## No Flying Today

In 1934 the Army issued specifications for a multi-engine bomber. Working from those specs, Boeing engineers designed a four-engine craft to replace the Martin B-10—a craft that would prove heavier, faster, capable of higher altitudes and longer range. While the United States recovered from the Great Depression, most of her citizens were convinced that the oceans would separate North America from advances of Nazi Germany, and they sought isolation from the European war. But heat from the war fires stoked by Adolf Hitler was felt across the ocean.

In 1937, the 2nd Bombardment Group was equipped with Boeing's B-17, an outgrowth of its Model 299 (XB-17) that first flew in 1935. The real threat of Nazi submarines forced further developments. As is common in war, actions required reactions. Each action by one model of B-17 demonstrated the need for subsequent upgrades and to the first variant of the B-17 were added improvements in armor, armament, engines, and performance. The first B-17D flew in 1941 and, as war broke out across the Pacific Theater, most were assigned to Hawaii and the Philippines.

Carrying a crew that consisted of pilot, copilot, bombardier, navigator, flight engineer, radio operator, tail gunner, belly gunner, and two waist gunners, B-17Es out of England began attacking Germany's ability to wage war in the summer of 1942. With the introduction of the F-model, Boeing, Lockheed/Vega and Douglas combined their efforts and B-17 Fs first flew over Germany in 1943. The B-17G introduced the chin turret in the nose, designed to repel head-on fighter attacks.

In all aspects of aviation, machinery can be as highly specialized as possible, but the airplane has to operate in atmospheric conditions. At times, weather patterns invite the most stalwart to fully-utilize instrument capabilities. At other times, fog, low-lying clouds, rain and snow kept even the hawks from flying. Here, hunkering close to the cold ground, the B-17G Flying Fortress belies her name as, in such inclement, gray, impenetrable weather and, following the good example of the hawk, there is no flying for either of these birds.

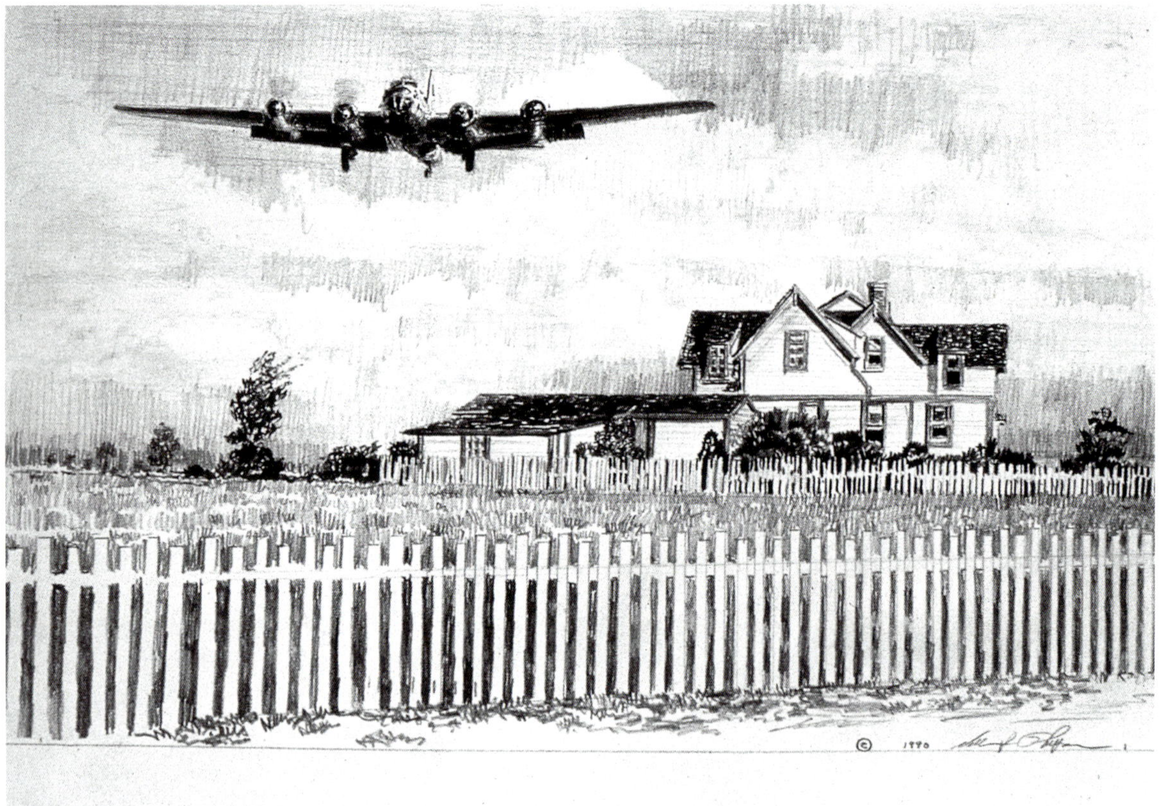

## One Down

As one of the most famous planes in World War II, Boeing B-17 Flying Fortresses first flew in the Pacific Theater and then served the Allies in war-torn skies of Europe. In response to the bombing of five central Pacific Islands on the "infamous" date of December 7, 1941, an outrage coursed throughout the citizenry of the United States and caused a huge growth in military preparedness. Aircraft, ship, and weaponry manufacture went into high gear. More than 12,000 B-17s were built, boasting top speeds of approximately 300 miles per hour, bomb loads of 17,600 pounds and twelve .50 caliber machine guns.

Those newly out of the factory, like this B-17G, followed a ferry route that carried them from Labrador, Greenland and Iceland to England to join the Eighth Air Force, the "Mighty Eighth." The first group of eighteen B-17s departed on June 23, 1942, and reached England on the first of July. By the end of August, 119 had successfully crossed the Atlantic Ocean. Consider the significance: at that time, only fifteen years had passed since Charles Lindbergh had made the first solo crossing of the Atlantic.

The B-17G, with pairs of guns mounted in each of the chin, dorsal, ventral and tail turrets, four in the cheek and waist positions and, on top of the fuselage, a rear-firing gun, bristled with the firepower of thirteen guns. In a combined effort of Boeing, Lockheed (at its Vega plant in Burbank, California) and Douglas, more G-models were built than any of the other B-17 variants, a total of 8,680.

Shown here, equipped with the chin turret that makes the G-model highly recognizable, this newly arrived Flying Fortress has just landed after its first mission. The young crew can steel itself for, perhaps, forty-nine more. Straight from the factory, the aircraft is not yet adorned with its identifying nose art or squadron markings. It will retain the unpainted aluminum exterior, although, it undoubtedly will begin to show the wear and tear of combat. We can't help but hope that it will survive the war.

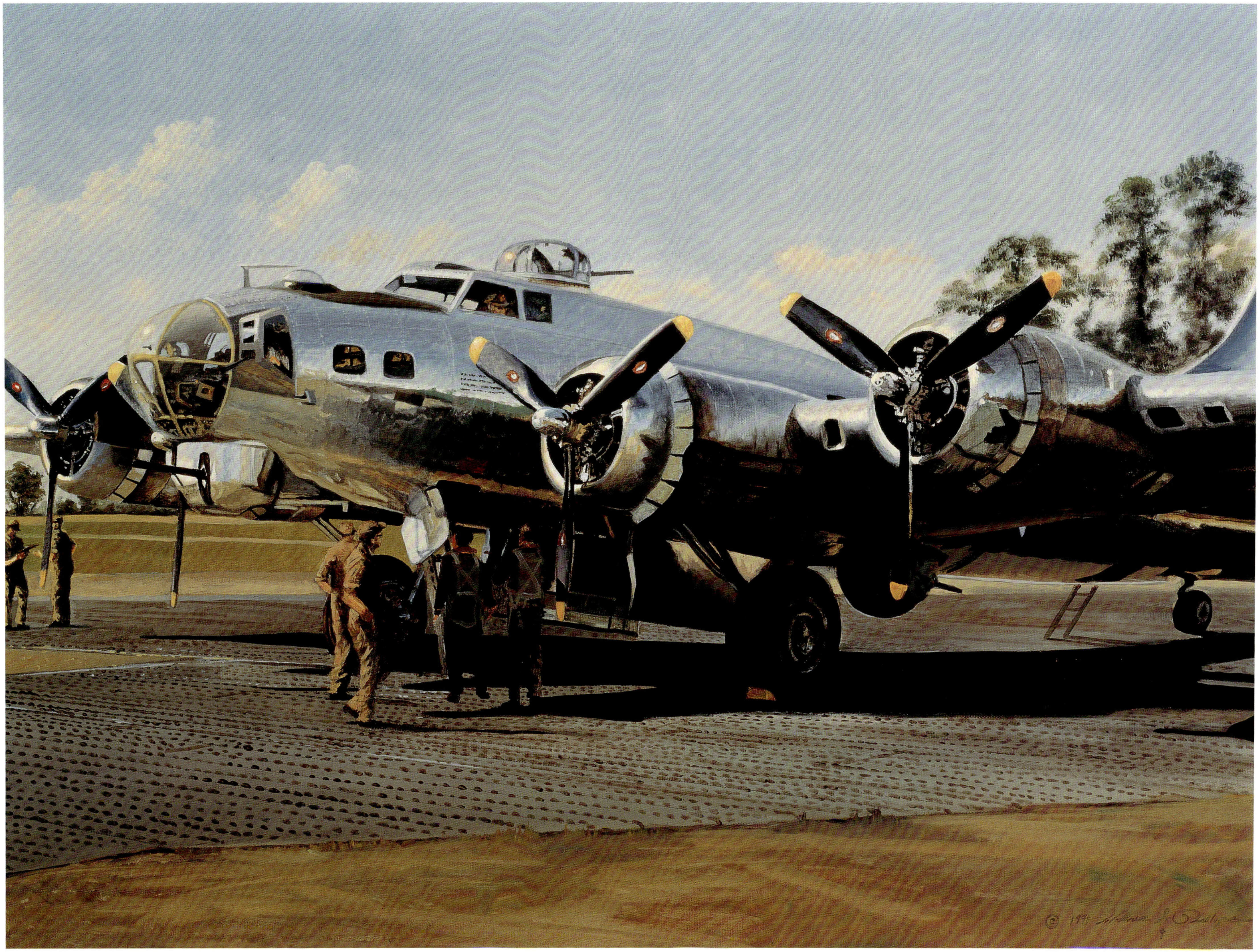

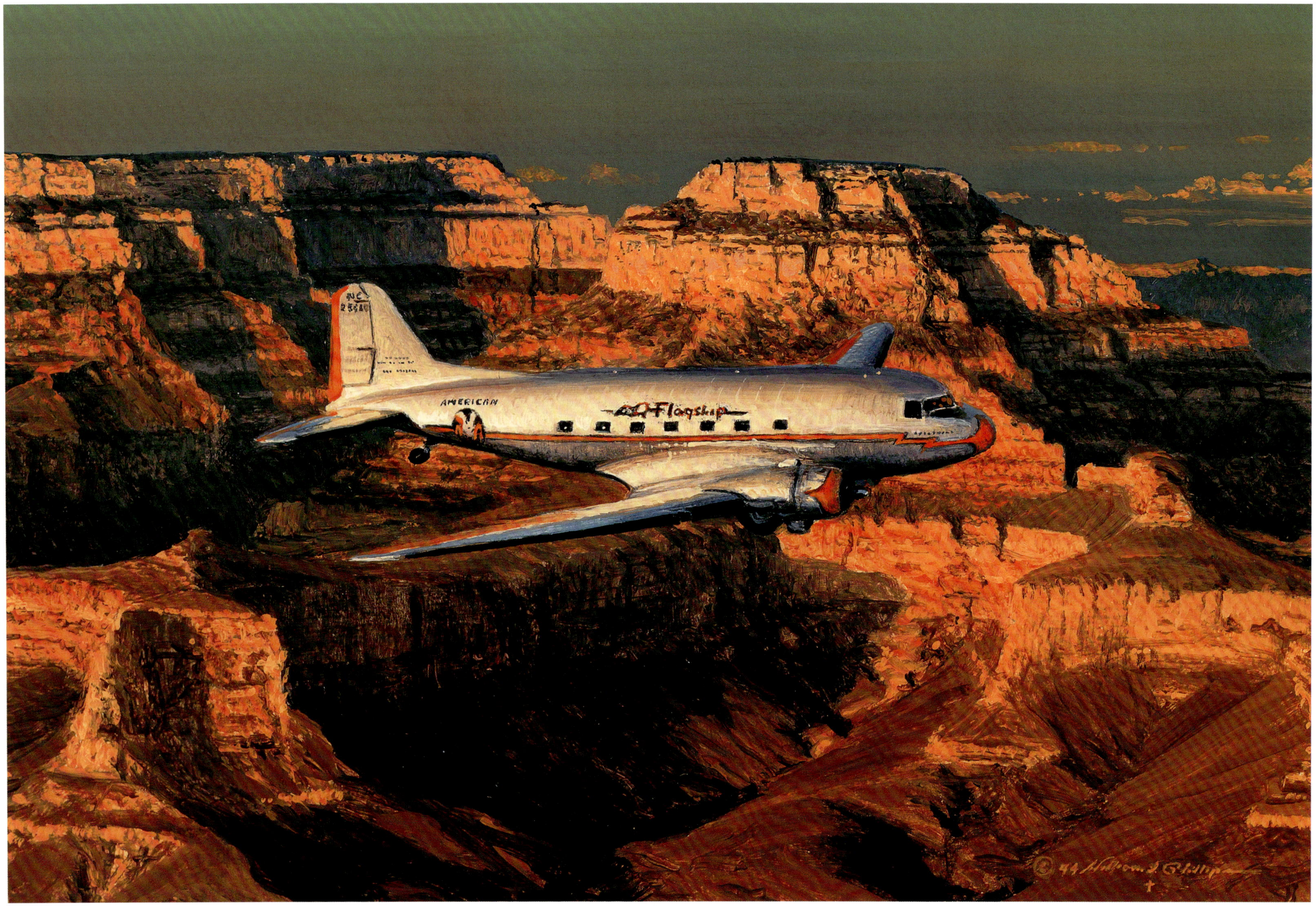
AMERICAN
Flagship

## Grand Canyon Flagship

Standing out in contrast to the heavily-shadowed Grand Canyon beneath its wings, the silver American Airlines DC-3 Grand Canyon Flagship offers her fortunate passengers and crewmembers a view of the vast chasm carved by the Colorado River and eroded by weather and time. The "Grand Old Lady of the Skies" was born in 1935 and, although most aircraft of her era have been relegated to museums or discarded and forgotten, not the DC-3. Many still fly in her original form and a company in Wisconsin, Basler, has been retrofitting DC-3s with turbocharged engines. She may go on forever.

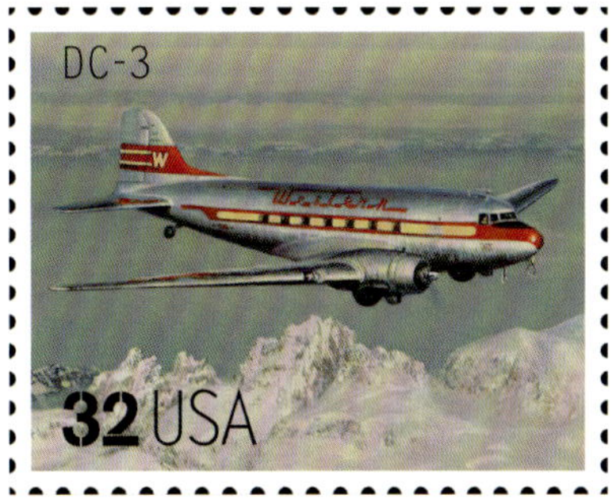


## Douglas DC-3

U.S. Postal Service, Classic American Aircraft series, 1997

Winging over a craggy, snow-powdered range of the Rocky Mountains, this Douglas DC-3 is owned by Western Airlines. The most successful piston-engine transport in history, over 10,000 DC-3s (affectionately known as Gooney Birds) served in a variety of roles during WWII. Remarkably, DC-3s still carry freight and passengers to destinations around the world.

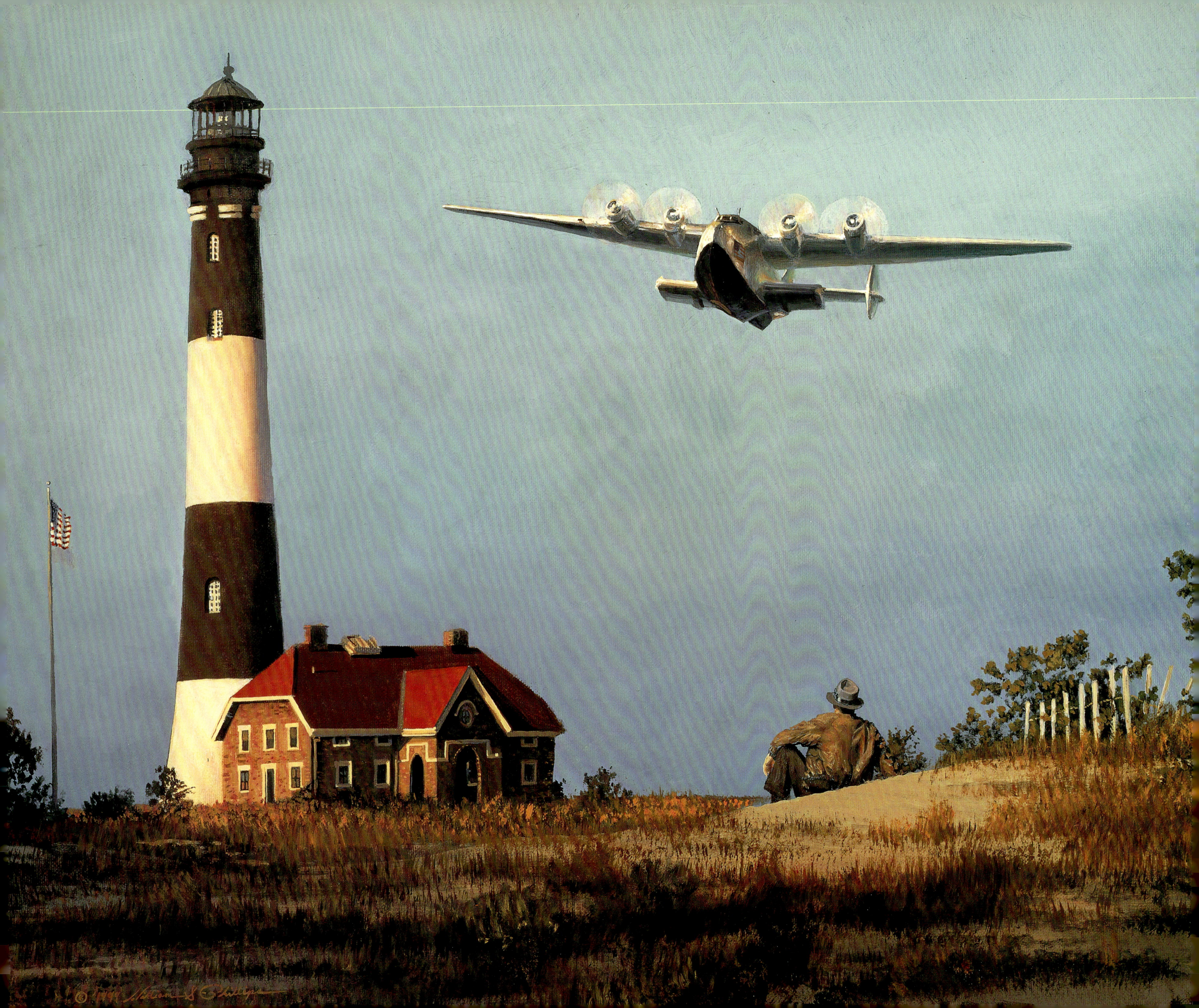

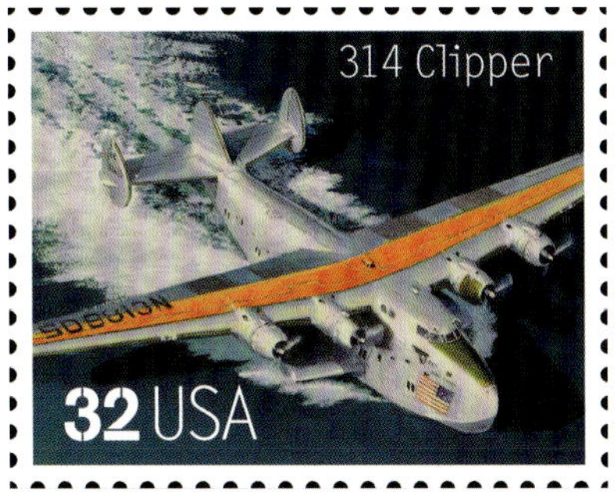


## 314 Clipper

U.S. Postal Service, Classic American Aircraft series, 1997

Boeing's 314 Clipper traced to the 1937 XB-15, the largest, heaviest U.S. airplane. Experimental, its wings and engine nacelles were used on the Clipper of 1938. Its connection with the mysteries of exotic travel spawned movies, books and myth. The mystery deepened when, on July 29, 1938, a Hawaii Clipper was reputed to have been hijacked by Japanese Naval Aviators. The plane, its passengers, and crew were never seen again.

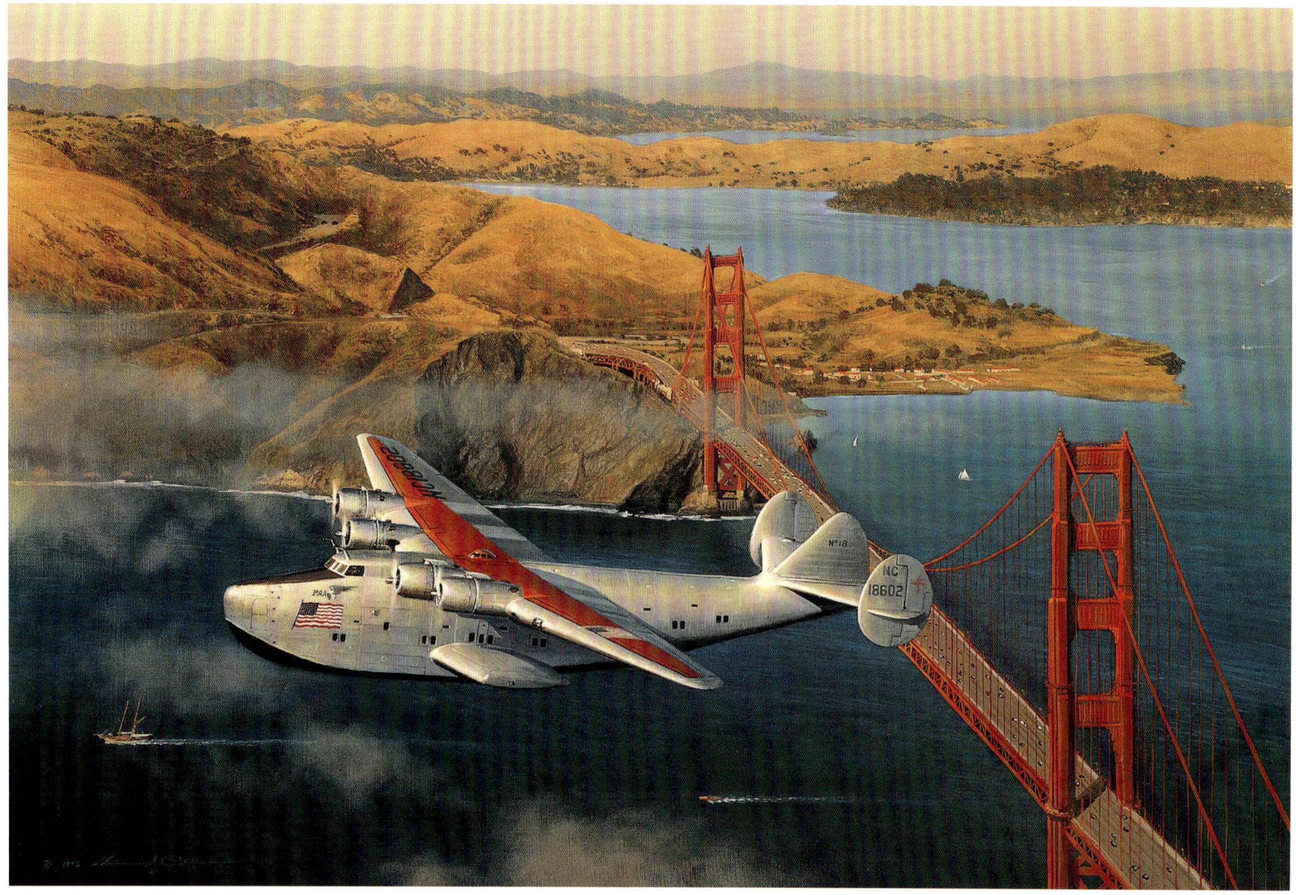

## Afternoon to Remember

This Art History Award winner in the Art for the Parks competition features Fire Island Lighthouse on Long Island, New York's eastern shore. A 314 Clipper comes home on the beam from someplace across the water, either Great South Bay or the Atlantic.

## Clipper at the Gate

In spite of the Great Depression, and before World War II brought foreign passenger travel to a halt, an adventurous spirit of air travel spread throughout the nation in the early 1930s. In response, Pan American Airways wanted a four-engine flying boat that could transport passengers safely across oceans to lands previously only accessible by ship. Glenn Martin Company and Sikorsky Aircraft completed three China Clipper flying boats in 1933 that answered specifications and proved to be the first intercontinental airliners. Those successful flying boats—named in honor of the eighteenth and nineteenth century sailing ships whose billowing sails evoked speed and adventure—were followed by requests for larger designs.

In 1936, Pan Am selected the Boeing 314, and the 82,000-pound behemoth, the jumbo of its day and one of twelve built, first flew on June 7, 1938. The amphibian could carry ten crewmembers and seventy-four passengers at 199 mph with its four Wright Twin Cyclone engines. To facilitate rescue if downed at sea, its wings were striped with the highly visible red-orange paint. Pan Am's crewmembers wore uniforms bearing naval styles and rank insignia. Even the hours aboard the craft were tolled with ship's bells. Outfitted for luxury, passenger seats converted into bunks for overnight travel, dressing rooms were provided, and a dining salon featured gourmet food prepared by excellent chefs. U.S. President Franklin Roosevelt and Britain's Prime Minister Winston Churchill flew aboard B-314s during WWII.

Shown here is the California Clipper NC18602, saluting the newly opened Golden Gate Bridge while departing San Francisco. *Clipper at the Gate* preserves an important point in aviation history and champions Pan Am's pioneer spirit.

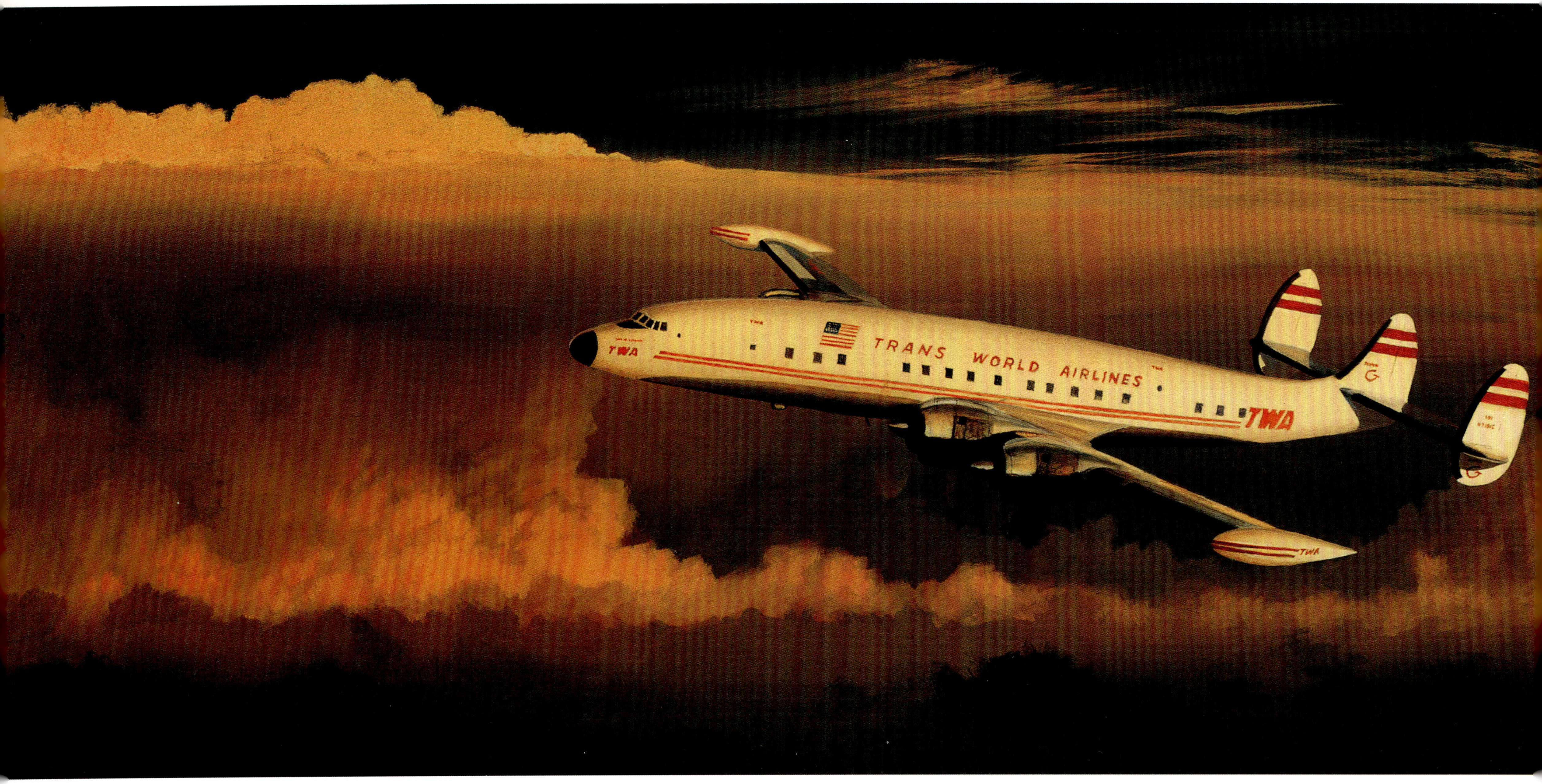


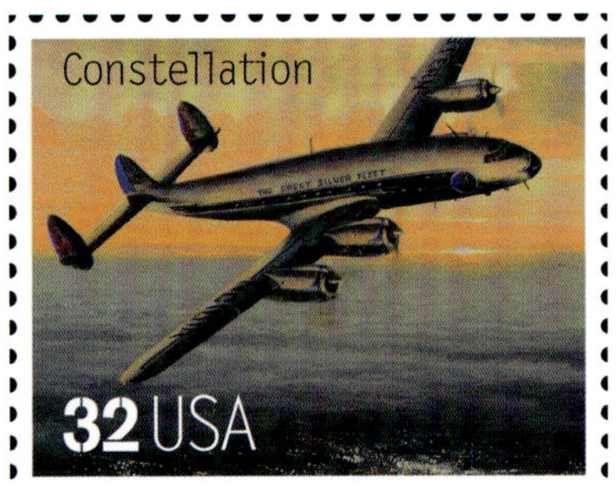


## Constellation

U.S. Postal Service, Classic American Aircraft series, 1997

Bearing Eastern Airlines' livery, the beautiful "Connie" was the first of pressurized airliners and reflective of the talents of designer Clarence "Kelly" Johnson and his team. The 856 Constellations, in four principal models, served President Eisenhower as the chief executive's airplane, the U.S. Air Force and U. S. Navy in Airborne Early Warning and Fighter Control, and brought air travel to millions.

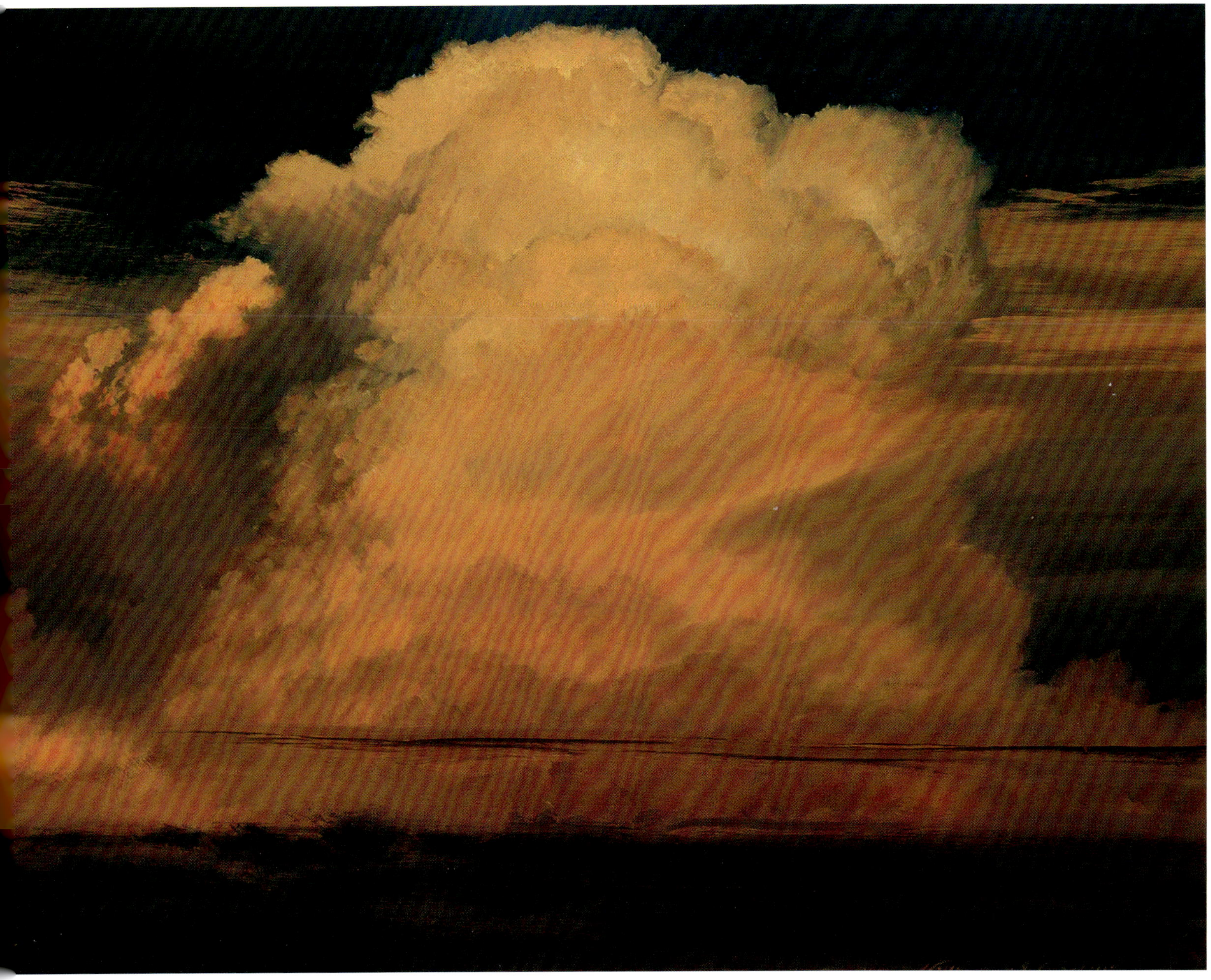

## Skirting the Front

In handsome red-and-white Trans World Airlines livery, a Super G Constellation diverts from its course to avoid the developing thunderstorms. At its ceiling of 22,800 feet, weather was a serious consideration and the wise airline captain avoided turbulent storms. In the postwar period, airliner competition became keen—speed, payload, and range were the benchmarks—and airlines in the United States and overseas sought the best "bangs for the buck." Douglas' entries, the DC-6B and the DC-7, and Boeing's Model 377 Stratocruiser vied with the Lockheed L-1049 model Super Connies for airline business.

As new model Constellations were introduced and sold, their piston-engine competition increased. The U.S. military helped spur that competition as it purchased, then decommissioned, Connies in its transport arm, the Military Air Transport Service (MATS). Even President Eisenhower contributed. As NATO Chief of Staff, he moved up to a Connie, designated VC-121A, and named *Columbine*. As U.S. President, Eisenhower was flown in a VC-121B, *Columbine II* and, later, a VC-121C named *Columbine III*.

Super Constellations were modified for use as airborne early warning surveillance and control aircraft by both the Navy and the Air Force, with a large search radar antenna mounted beneath the wings and a height finder antenna mounted atop the fuselage. In Vietnam, that EC-121 provided command and control and fighter direction to the target. Lockheed, like Douglas and Boeing, responded to increasing expectations of military need and airline demands.

The second front this beautiful four-engine piston airplane was skirting was the development of the jet engine-powered airliner, which made its debut as the deHavilland Comet in July 1949. Comfortable and reliable as the Constellations had become, and despite their speed and distance capability, the end was in sight

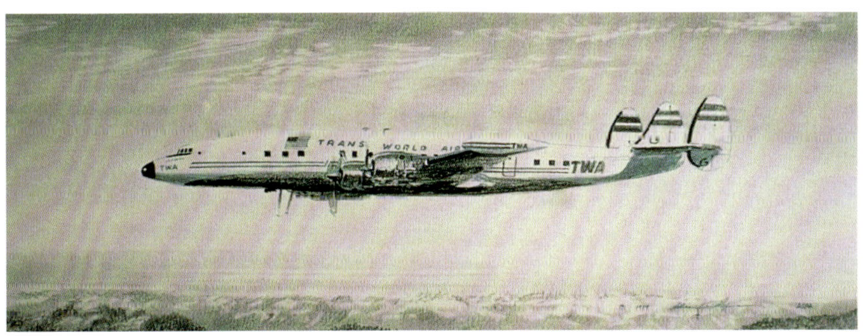

## Time to Head Home

The USS *Boxer* (CV 21) an Essex class aircraft carrier commissioned in 1945, made a significant contribution in the Korean conflict. In June 1950, it rushed U.S. Navy and Air Force personnel and aircraft to the war zone in a record Pacific crossing and, the same fall, provided air support for United Nations' forces. Between 1950 and 1953, *Boxer* served off Korea, surviving damage and casualties in 1952 that resulted from a hangar deck fire that took her out of action for two weeks.

Missions for the carrier's aircraft, like this F4U-4 of VF-791 "The Fighting Rebels," a Navy reserve squadron from Memphis, Tennessee, included close air support for ground troops and firing upon transportation and infrastructure targets. Here, the VF-791 Corsair expends the last of its rockets on the North Korean target and its pilot thinks of turning for the waiting deck of the *Boxer*—home away from home.

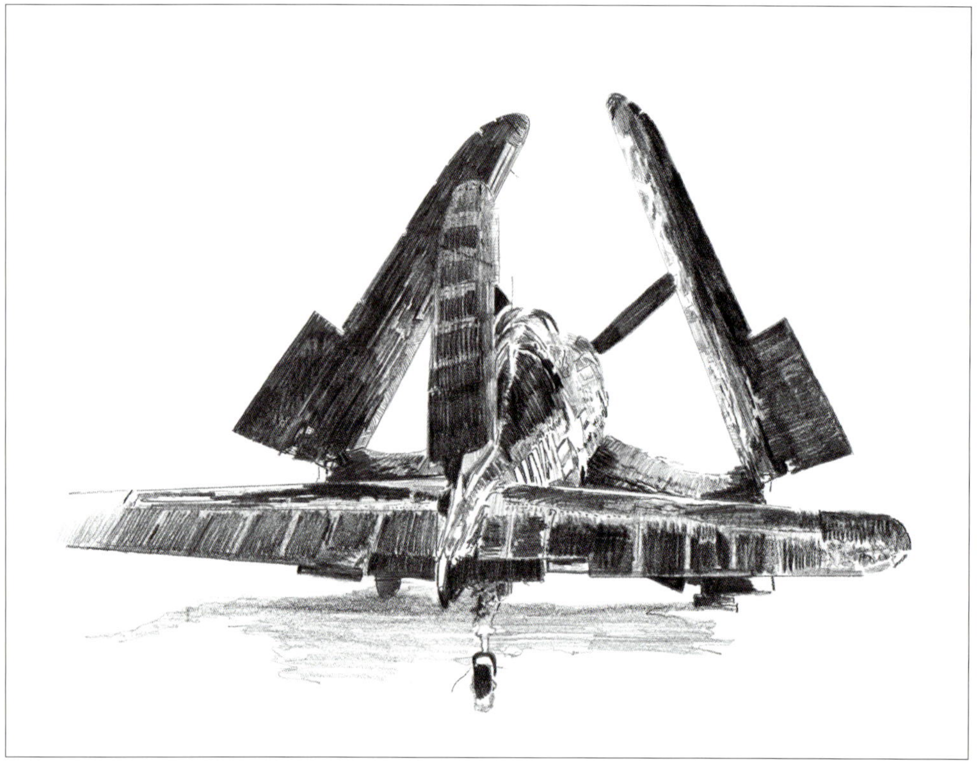

The artist sketched this F4U-4 as it sat in the grass at Harlingen, Texas. The F4U-4 had a wingfold capability for ease of placement aboard aircraft carriers.

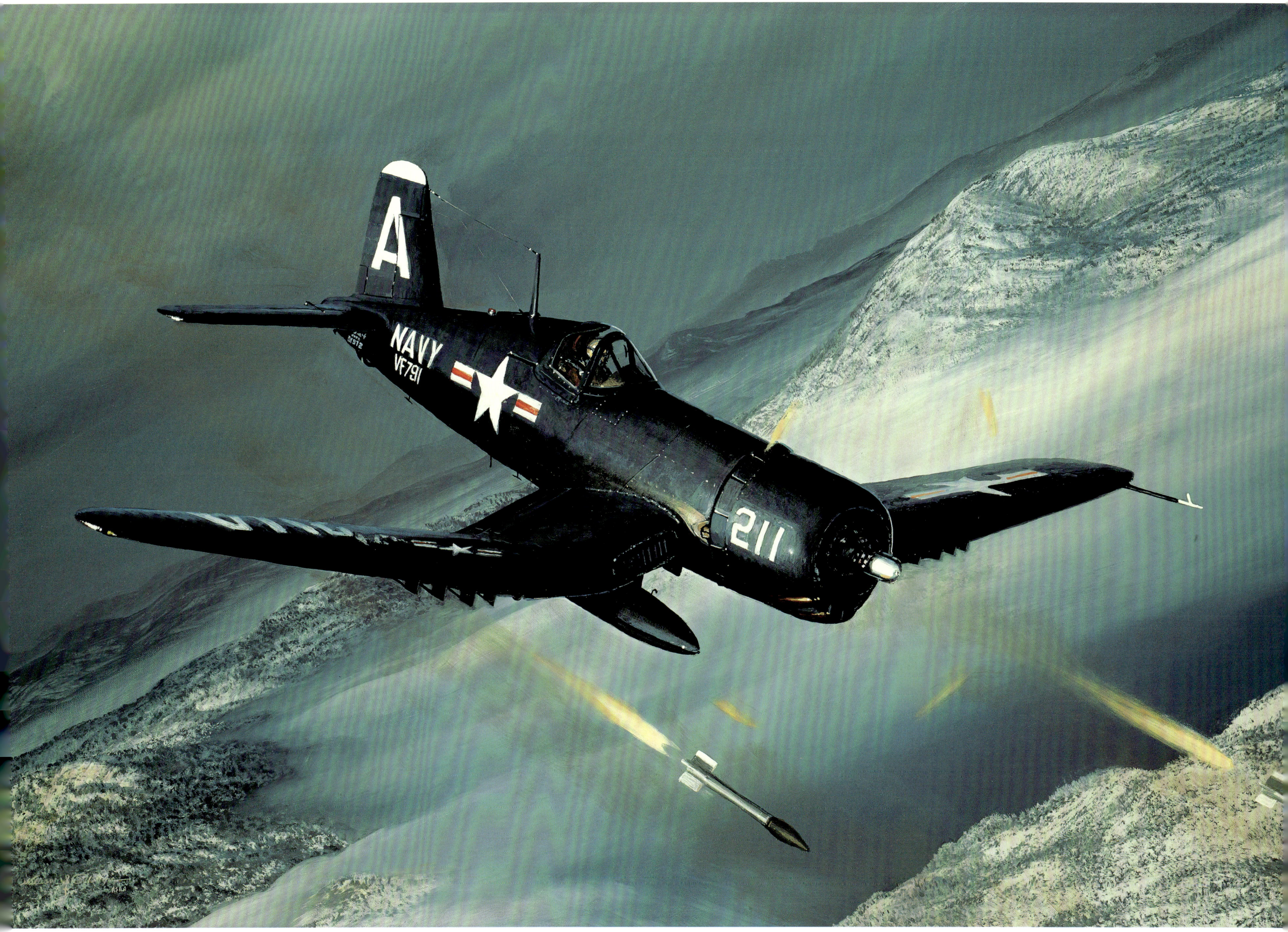
A
NAVY
VF791
211

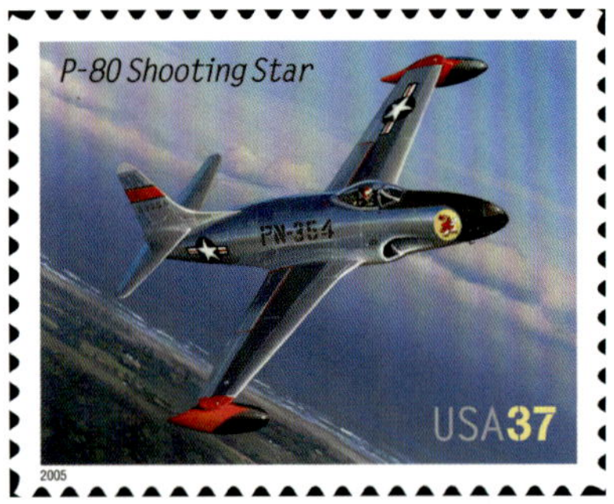


## P-80

U.S. Postal Service, Advances in American Aviation series, 2005

First flown in mid-1944, Lockheed's P-80 Shooting Star lived up to its name, reaching speeds that were 100 to 150 mph faster than the swiftest piston-engine fighters. In Korea, a P-80 scored the first jet combat victory, downing a MiG-15.

## Bell X-1

Before an unconditional surrender brought World War II to a close in 1945, a contract was awarded to Bell Aircraft Corporation, Buffalo, New York, by the U.S. Army Air Technical Service Command to develop three research aircraft capable of transonic and supersonic flight. Designated the XS-1 for Experimental Sonic 1, Bell's resulting aircraft made aviation history.

On January 19, 1946, the first XS-1was air-launched—released to glide without power—from the bomb bay of a Boeing B-29 Superfortress. The first powered flight, with Bell test pilot Chalmers Goodlin, followed almost a year of engineering efforts and took place on December 9, 1946. Throughout the testing, as engineers reacted to the dynamics of each flight, structural and aerodynamic alterations improved each subsequent test vehicle. Extremely thin, strong wing sections were adapted to the Bell X-1, as well as a vertically adjustable horizontal stabilizer for enhanced control as the craft reached transonic speeds (prior to achieving the speed of sound). Technological advances transferred to subsequent transonic and supersonic military aircraft. Shaped like a .50 caliber bullet, including its flared windscreen, the X-1 was equipped with 500 pounds of flight test instruments and turbo-driven fuel pumps, among other advances.

On October 14, 1947, with then-Captain Charles E. "Chuck" Yeager at the controls, a Bell X-1 was released from its mother ship at 23,000 feet of altitude. He rocketed 20,000 feet and, at an altitude of 43,000 feet, achieved man's first supersonic speed of 700 miles per hour—Mach 1.06—in level flight. Eighteen pilots were involved with its testing. Yeager again made aviation history during another of the X-1's 78 flights, when he attained a speed of 957 miles per hour—Mach 1.45—at an altitude of 71,900 feet. This represented manned flight's highest velocity and altitude at the time.

Artist Bill Phillips depicts the awesome streaking bullet with its characteristic "mach diamond" exhaust reflecting the four-chamber configuration of the rocket engine—a Reaction Motors' XLR-11-RM-3 rated at 26,500 newtons or 6,000 pounds static thrust. It is "chased" by Lockheed's P-80 Shooting Star, the U.S. Air Force's first combat jet aircraft.

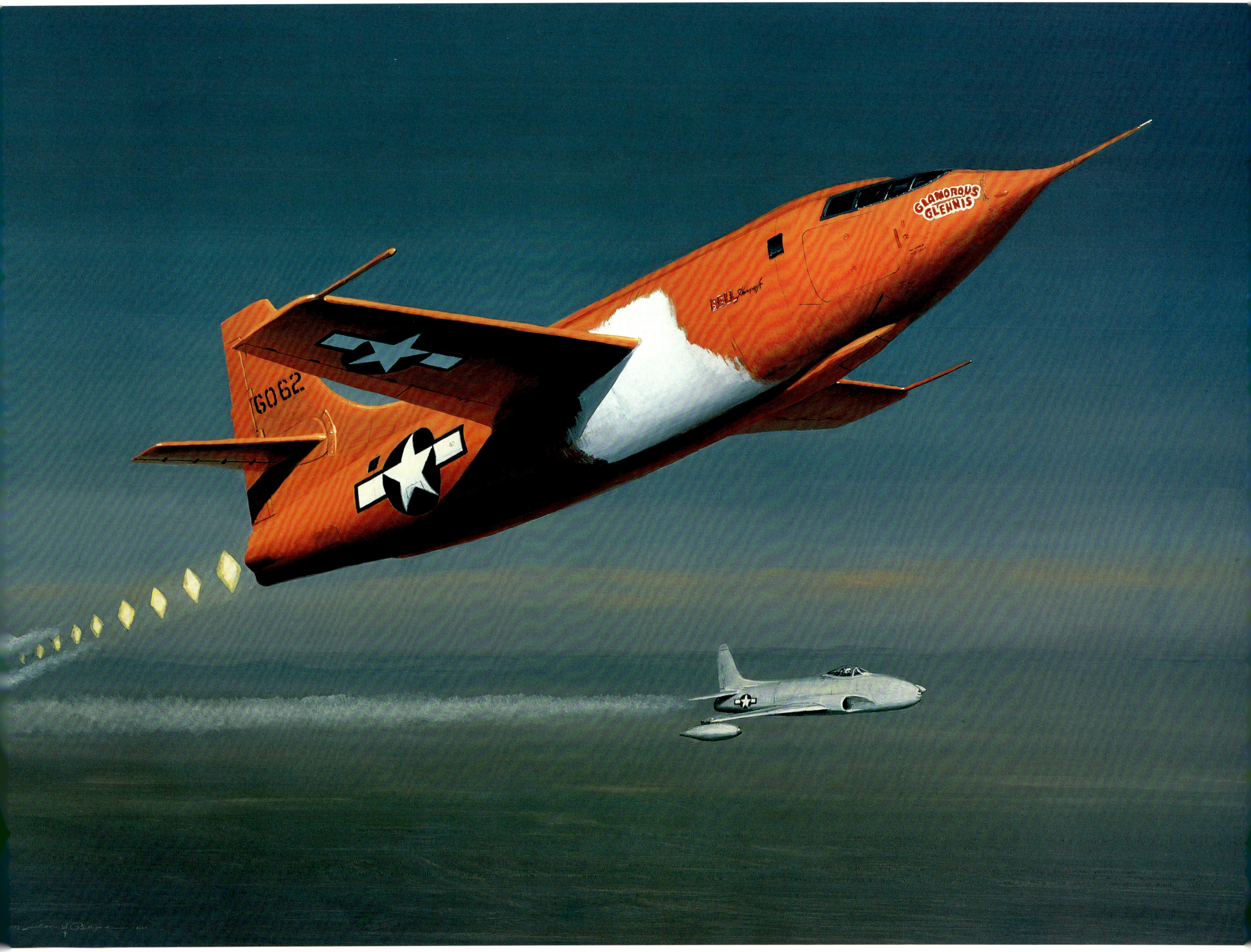
GLAMOROUS GLENNIS
BELL
6062

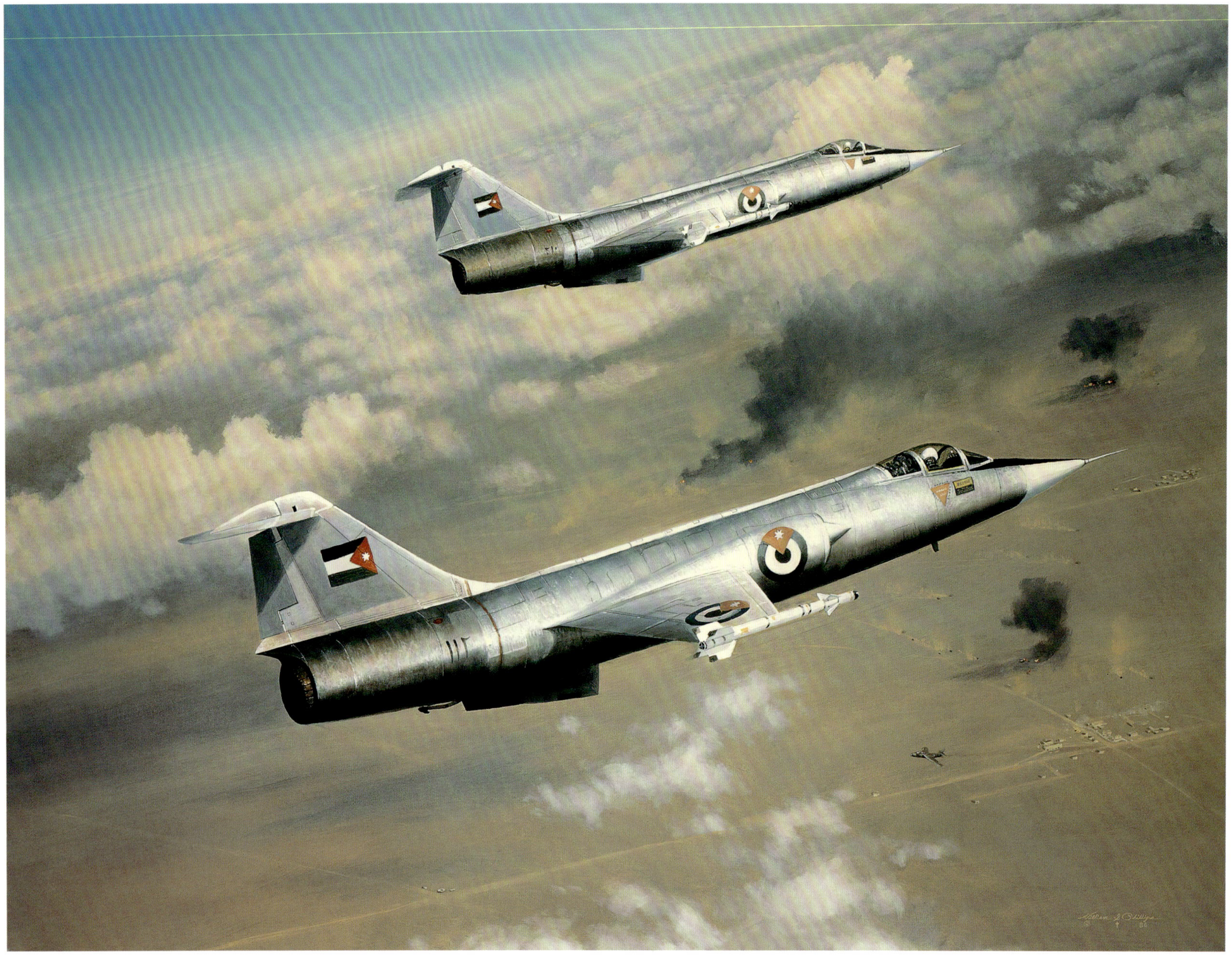

## Jordanian F-104 Combat Air Patrol

Jordan acquired the Lockheed F-104 Starfighter from the U.S. Air Force in 1969 after Jordan's serviceable combat planes were destroyed during the June 1967 War. The F-104s, armed with AIM-9 air-to-air guided missiles, included four, two-seat, B-model trainers and provided the first Jordanian fighter-interceptor squadron. The Starfighters were assigned to Number 9 Squadron and to Number 25 Squadron and Jordan also installed surveillance radars, protective shelters for the aircraft, and issued emergency dispersal plans. Here, providing combat air patrol, CAP, over a desert battle in Pakistan during the 1971 India-Pakistan War, Jordanian F-104s were a threat to the Indian Air Force.

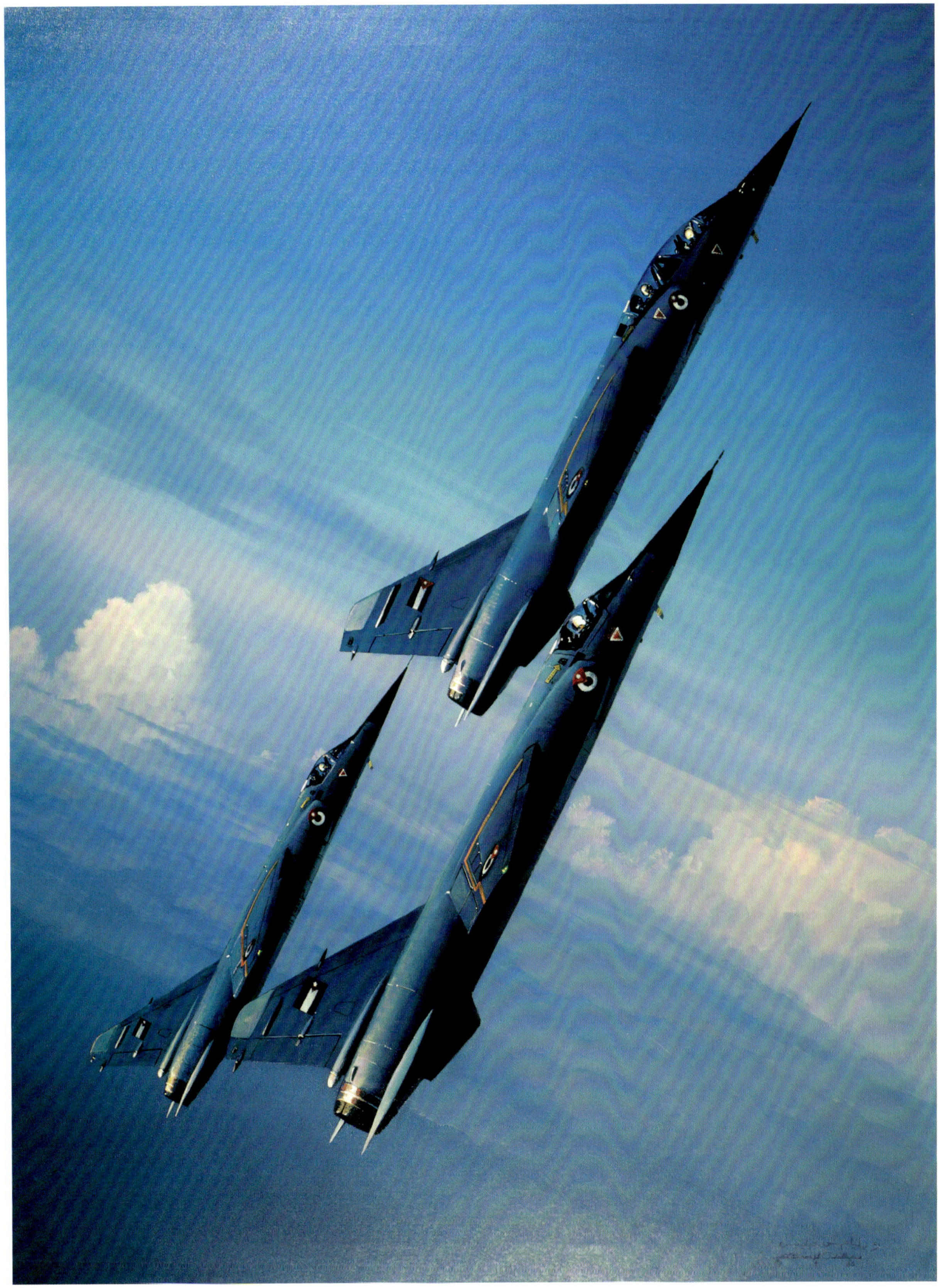

## Taking the High Ground

In a vertical climb over the Jordanian desert, these Dassault-Breguet Mirage F1s demonstrate the capability that has made the French interceptor so popular at home and abroad. This trio, from Number 25 Squadron of the Royal Jordanian Air Force, includes one F1BJ (the two-seat trainer in which the artist flew missions) and two F1CJs, the single-seat fighter. Jordan acquired 34 Mirage F1s in 1979 when the United States declined to sell F-16s to the Kingdom. Powered by the SNECMA Atar 9K50 jet engine with afterburner, the Mirage F1 has a top speed of Mach 2.2 and a ceiling of 52,000 feet. With in-flight refueling and complete avionics and armament suites, the Mirage earned its outstanding reputation as a fighter.

# Guardians of an Unforgiving Shore

From Heceta Head Lighthouse, sunsets start slowly and expand into glorious reds and golds, unless the Pacific Ocean is provoked into a frothing monster by storms and thunders against the rocks up toward the light. Those in danger are protected by the lighthouse from the land and by the helicopter over the sea.

The HH-3F Pelican joined the U.S. Coast Guard's admirable fleet of aircraft in 1967 and promised to extend offshore helicopter coverage to three hundred miles. With its sophisticated avionics package, the twin engine amphibian could carry up to twenty passengers plus its normal crew of four—pilot, co-pilot, navigator, and flight mechanic. A highly successful search and rescue craft, the Pelican was also used in marine environmental protection, reconnaissance support, defense readiness, drug interdiction, and enforcement of laws and treaties—a big job handled well by the small, conscientious crew.

A HH-3F Pelican achieved fame in the 1980 rescue operation that followed the disastrous fire in a cruise ship in the Gulf of Alaska—hundreds of evacuees bobbed in lifeboats two hundred miles out and owed their lives to the crewmembers who plucked them to safety. During its career with the Coast Guard, a fleet of forty Pelicans was responsible for saving 23,169 lives and assisting an amazing 65,377 others. No wonder the artist chose this remarkable craft to represent those who guard an otherwise treacherous and unforgiving shore.

Artist Bill Phillips' choice of "land-based protector" is on Oregon's rocky coast. The lighthouse area today is called Heceta Head Lighthouse State Scenic Viewpoint. Named for Don Bruno Heceta, who sailed for the Royal Spanish Navy in 1775, the light was built between 1892 and 1894. Heceta intended to travel from Mexico to the Arctic Circle, claiming land for Spain, but he turned near the headland that now bears his name because his sailors were stricken with scurvy.

Towering fifty-six feet in height, Heceta Head Lighthouse is the most powerful light on the Oregon coast. Its beams are thrown by a Fresnel lens with eight panels, 640 prisms, each two inches thick and can be seen twenty-one miles out to sea. The light ceased shining once, in June 2000, when the Coast Guard disassembled the lens for repair. Restoration consumed almost a year before the light could be reactivated in March, 2001. (The Coast Guard agreed to repair the lens before relinquishing maintenance of the lighthouse to the Oregon state parks department.) That was only the second time in the light's 106-year history that the lens stopped. It failed once, February 12, 1961, when a rockslide snapped electrical wires and the generator failed. The keeper and two Coast Guard assistants used a manually-operated lamp, turning the lens by hand as they circled the interior of the lamp room by foot, keeping their vigil from midnight until 7:30 the next morning. Such is the commitment to the safety of those who are at risk on a tempestuous sea.

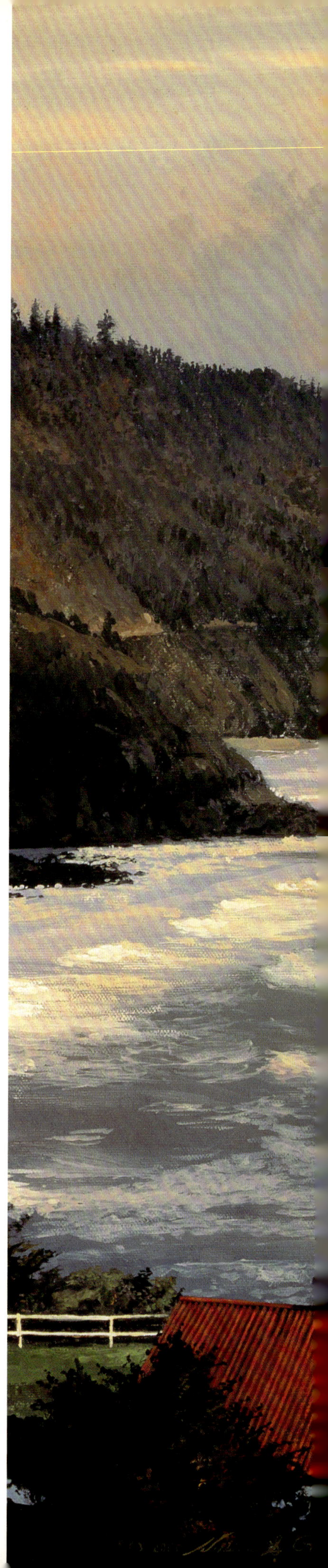

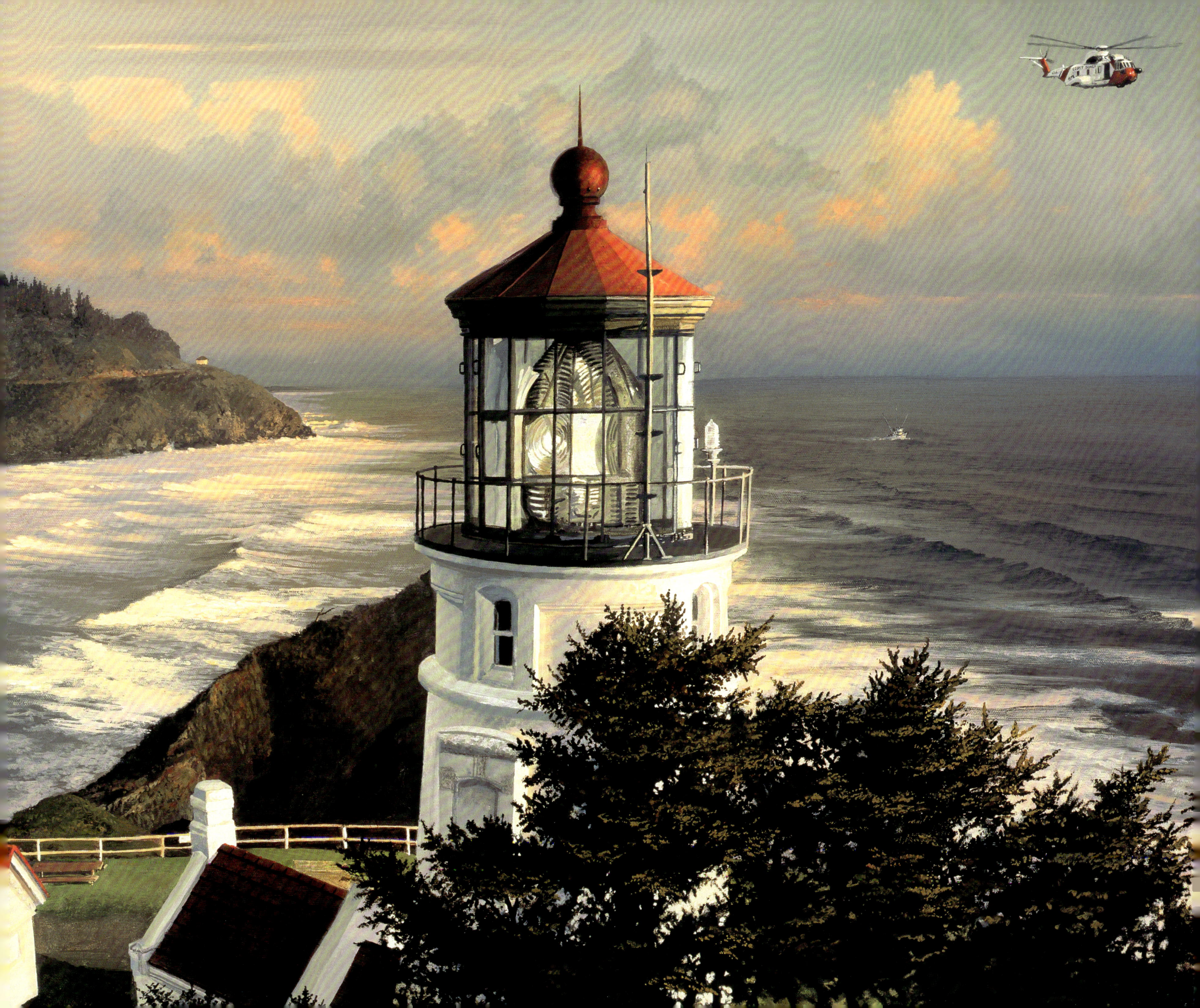

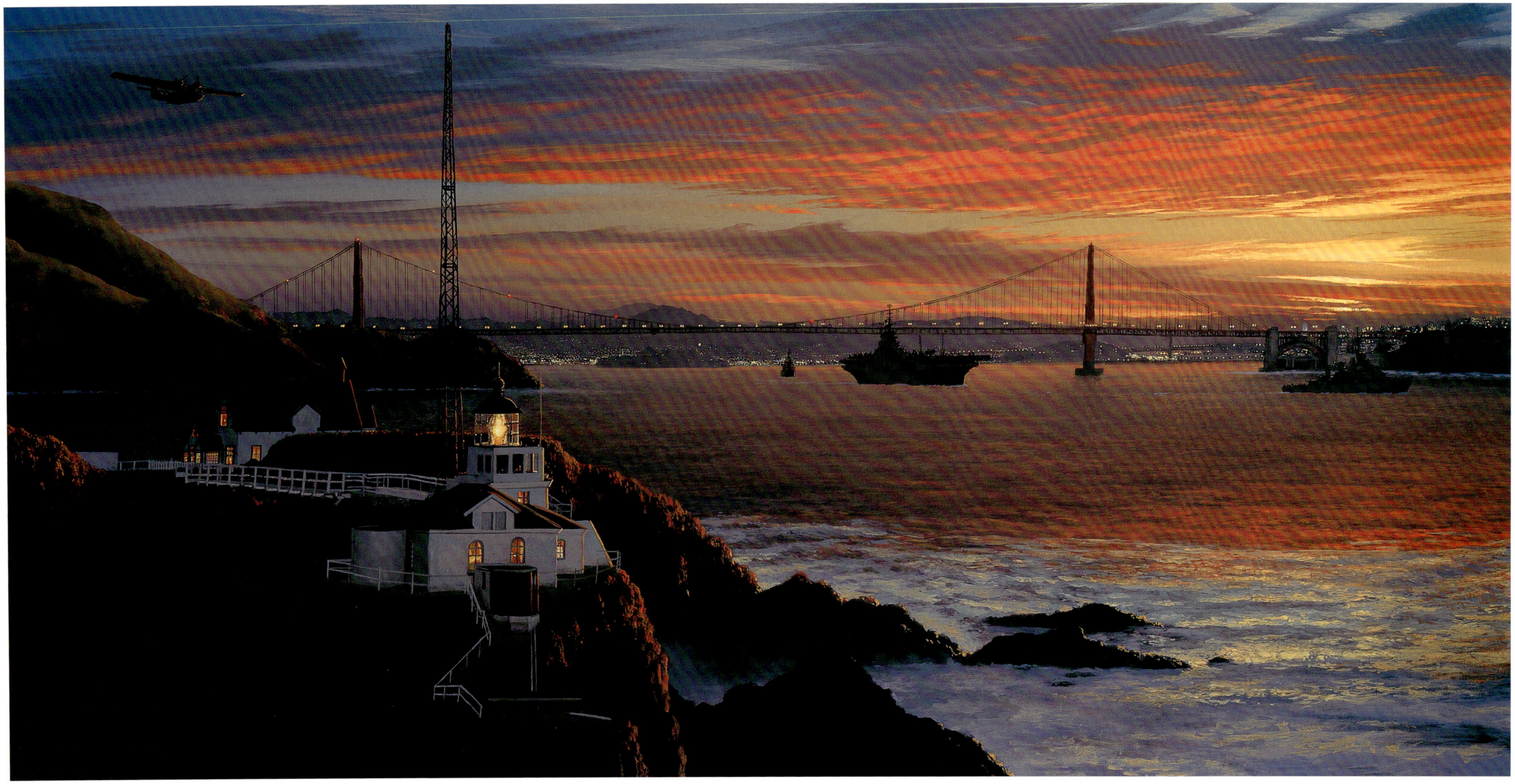

## Point Bonita: First Light

Jutting out into the Pacific Ocean west of the famed Golden Gate Bridge, the Point Bonita Lighthouse is shown above as it appeared in the early 1940s. The last manned lighthouse in California, it guarded the northern entrance to San Francisco Bay. Although its construction began in the early 1850s, it didn't save the steamship *Tennessee,* which ran aground in 1853, or the clipper ship *San Francisco,* which struck the rocks and sank in 1854. In first light, against the splendid glow of the eastern sky, the silhouetted Golden Gate Bridge looms over an escorted aircraft carrier departing for war in the Pacific.

## Point Bonita: Last Light

The guardian sentinel Point Bonita Lighthouse, its gleaming buildings, and suspension bridge, are all brilliantly bathed in twilight. This stunning maritime landmark is only accessible by a narrow half-mile trail that follows a tunnel and a bridge (limited to only five persons at a time and built after a particularly violent storm destroyed the path in the 1940s). In contrast to *Point Bonita: First Light*, a modern aircraft carrier enters the bay, there are modernizations to the lighthouse, and overhead, a Coast Guard aircraft doubles the protection.

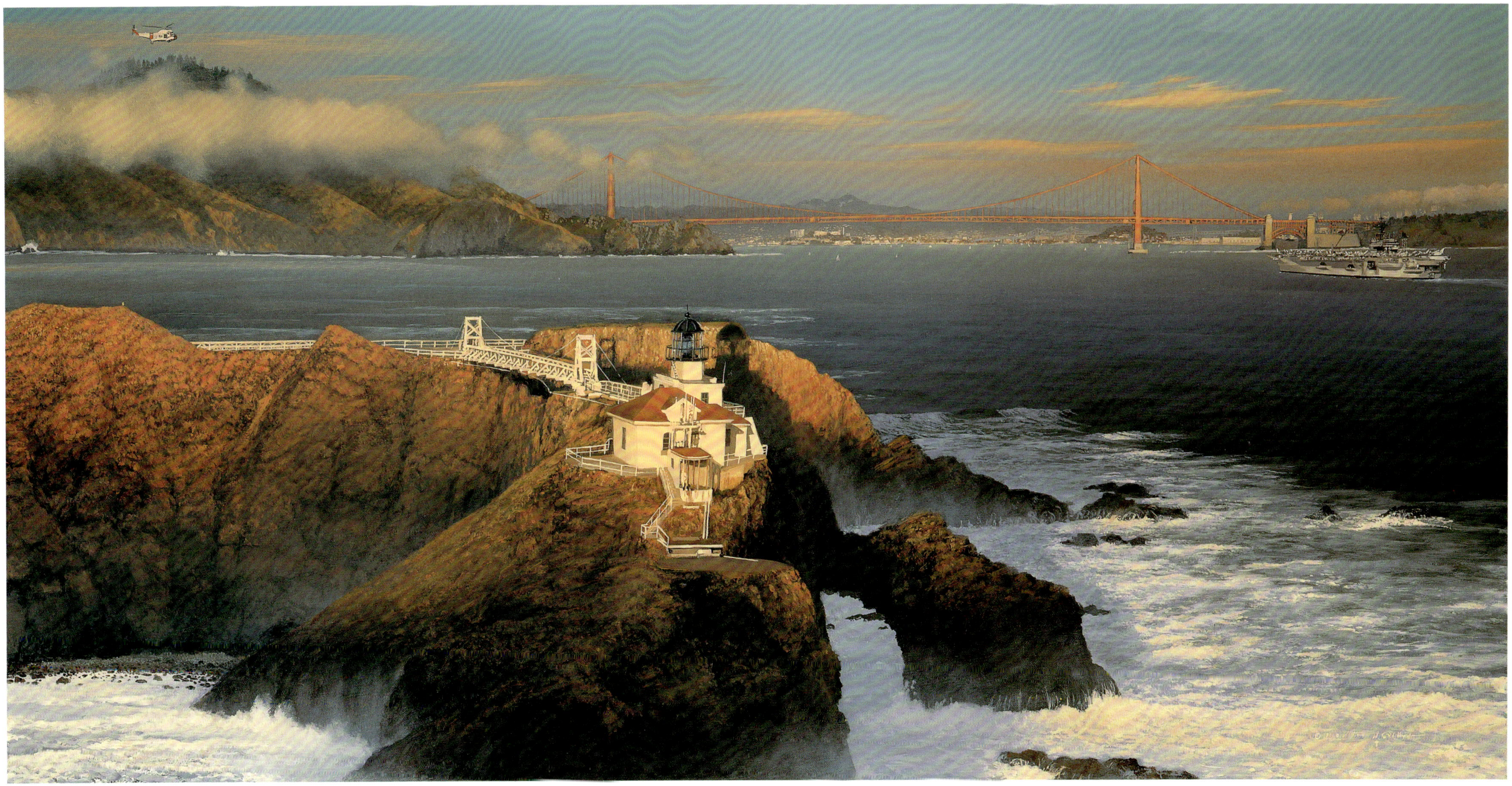

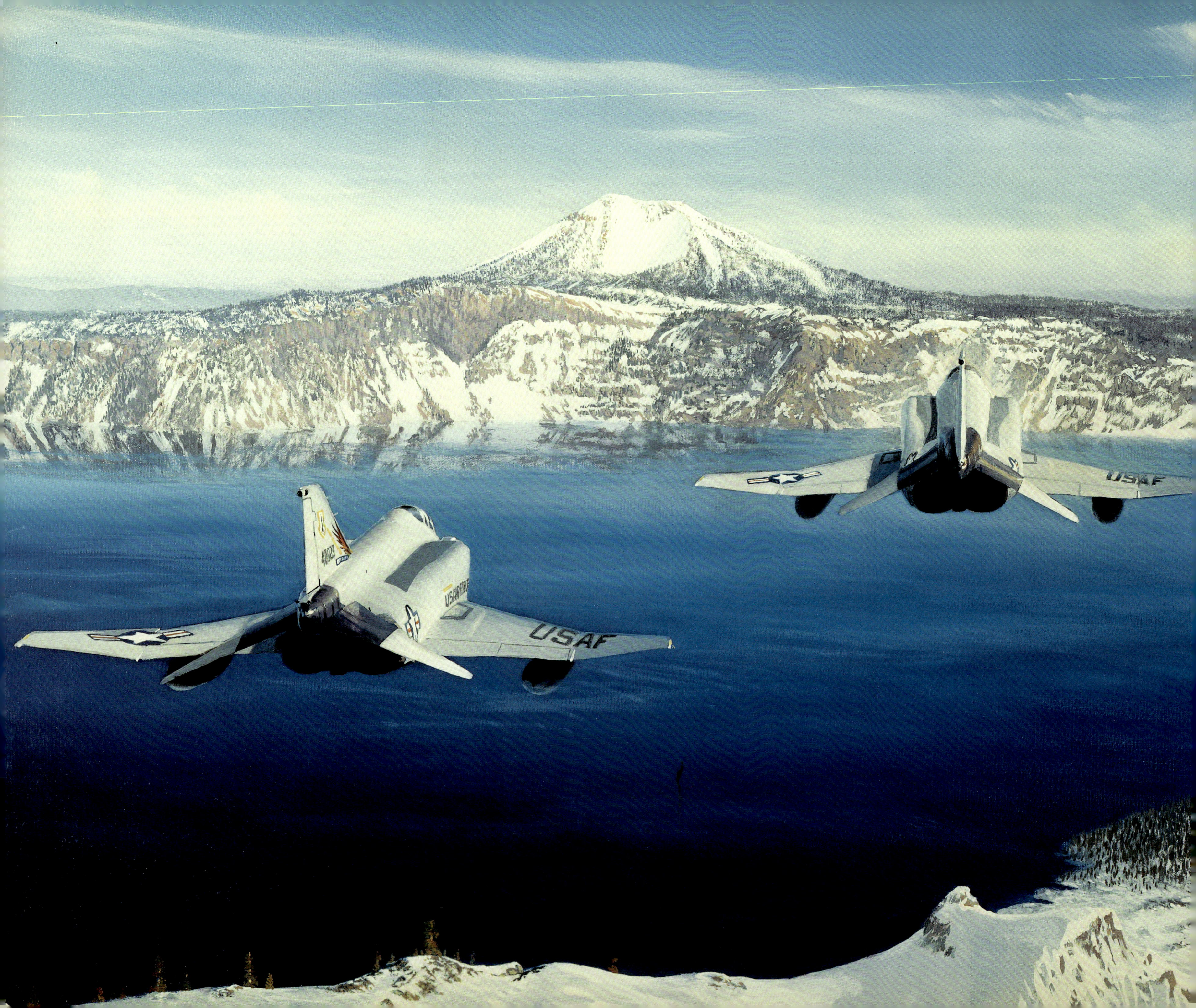
USAF
USAF

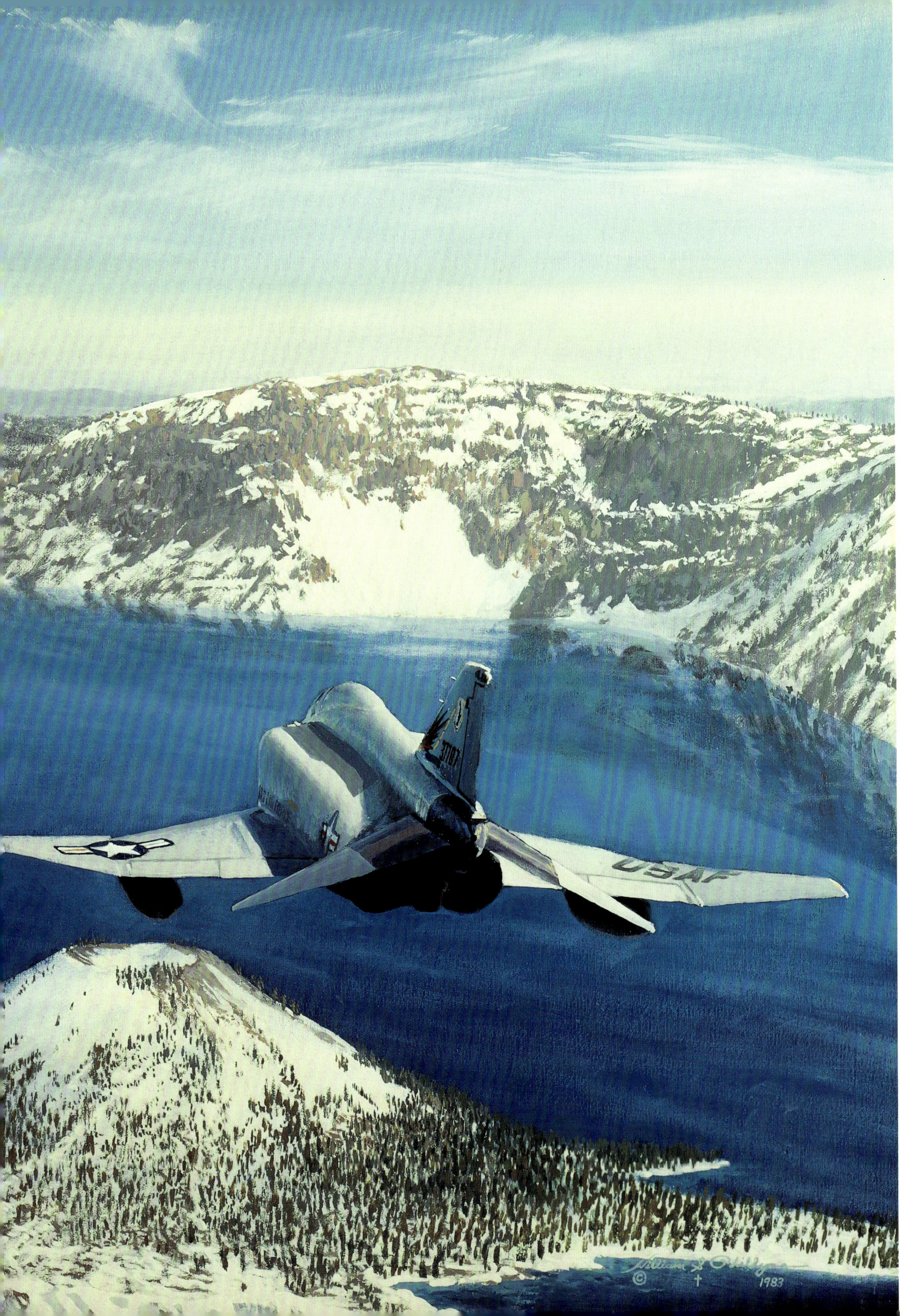


## Phantoms and the Wizard

High in the Oregon Cascades, its rim at 7,100 feet, the spectacular Crater Lake came into being after a cataclysmic eruption of a mountain, later christened Mount Mazama. The native peoples of Oregon witnessed the spectacle and the implosion must have presented a terrifying change to the area where they picked huckleberries and hunted game.

Gold prospectors were the first to discover this sapphire jewel—the deepest lake in the United States at a maximum depth of 1,932 feet—and members of a geological survey group explored the waters. But it was William Gladstone Steel who created Crater Lake National Park. Steel, at age sixteen, read about the lake in newspapers that wrapped his lunch. Intrigued, he ventured to the Cascade Mountains in 1885 and devoted a half-century of his life to the area's preservation. It was Steel who named the small volcanic island in the crater, Wizard Island, which, in turn, inspired the title of this painting.

The Phantoms, McDonnell (later McDonnell Douglas) F-4s, were introduced in 1958 and, although now retired from the U.S. military, still fly in defense of eight nations. Here they cruise in fingertip formation from the west toward snow-covered Mount Scott that reaches almost 9,000 feet into the sky. At the time the artist flew over Crater Lake, the Pacific Northwest was guarded from nearby Kingsley Field, Klamath Falls, Oregon's joint-use airfield co-located with the municipal airport and civilian operations.

Then, the F-4C Phantoms were assigned to the 114th Tactical Fighter Training Squadron, TFTS, a training outfit for all F-4 pilots in various units of the Air National Guard. As the first multi service aircraft flying concurrently with the U.S. Navy, Air Force, and Marines, it served every traditional military mission: air superiority, close air support, interception, air defense suppression, long range strike, fleet defense, attack, and reconnaissance.

Phillips took artistic license by dropping the jets down close, and dipping them dramatically over the natural wonders. Yesterday's F-4s and today's F-15 Eagles, which now fly out of Klamath Falls, would maintain greater terrain separation.

# Home Is the Hunter

On station in the South China Sea, the USS *Constellation* recovers its aircraft in a setting that belies the dangers inherent in carrier operations. Its story begins with the Kitty Hawk class carrier, USS *Constellation* (CVA/CV 64), built in the New York Naval Shipyard and christened on October 8, 1960. Damaged before completion by a fire on the hangar deck that killed fifty civilian workers, *Constellation* more than overcame the challenge and served proudly and well.

The carrier was assigned to San Diego throughout its service; yet it was much more at home on the high seas. Between May 1964 and 1972, *Constellation* deployed to the western Pacific six times. It was in the sixth deployment, departing October 1971, that *Constellation*, USS *Oriskany*, and USS *Enterprise* alternated on Yankee Station. Three months later, in a F-4 Phantom II off *Constellation*, Lieutenants Randy Cunningham and Willie Driscoll claimed a MiG-21, the first enemy aircraft downed since March 1970. In May, the two engaged in a vicious dogfight in which they were vastly outnumbered, and Cunningham and Driscoll counted three enemy fighters. Their foes were the first downed by air-to-air missiles, elevating the duo to the first U.S. aces of the Vietnam conflict—five enemy aircraft in four months.

The artist depicts the Phantom, with its characteristic smoke trails signaling minute power adjustments, descending to make the trap. Warm evening light welcomes the aircrew home and the hunt is over, for a while.



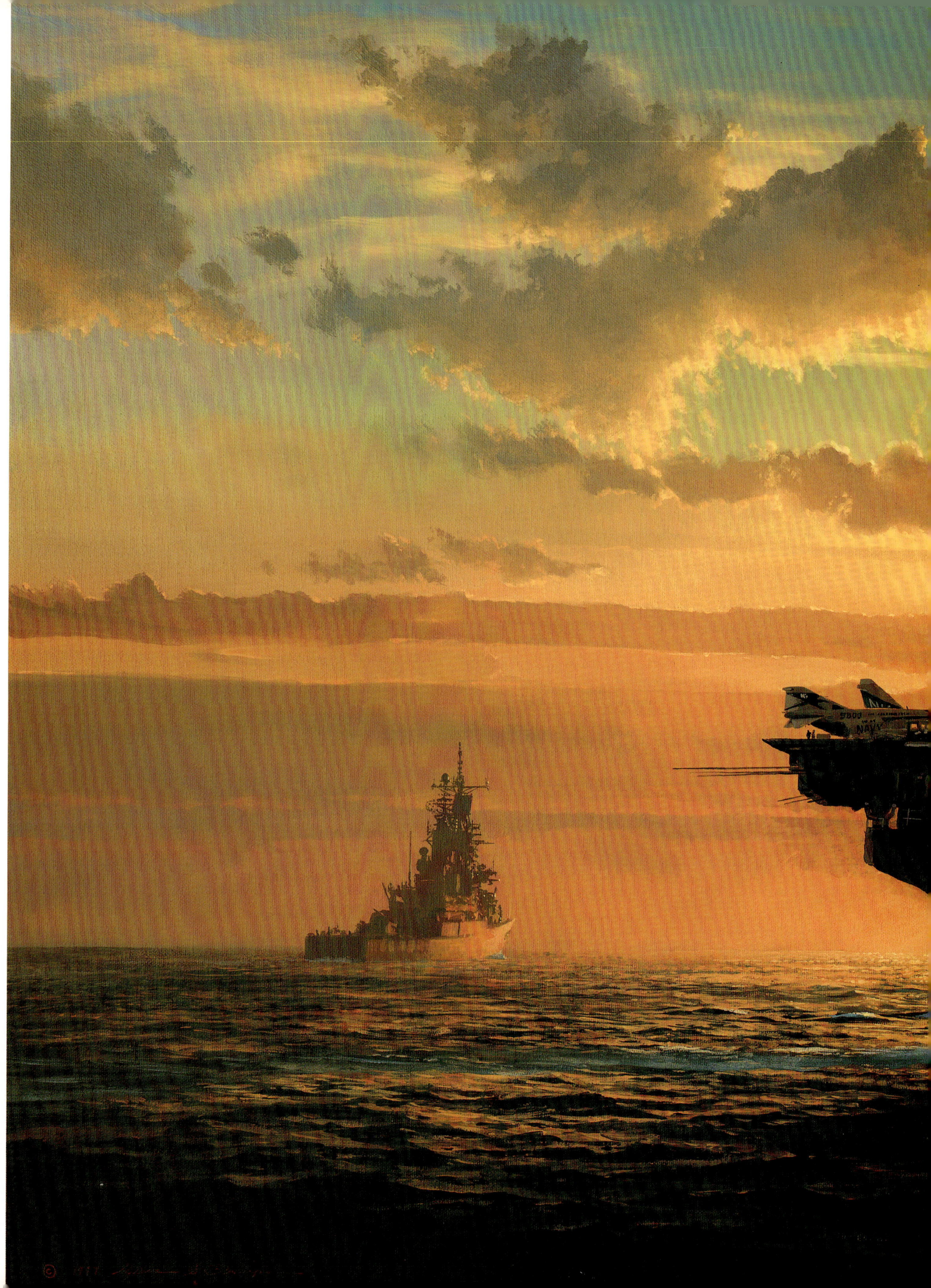

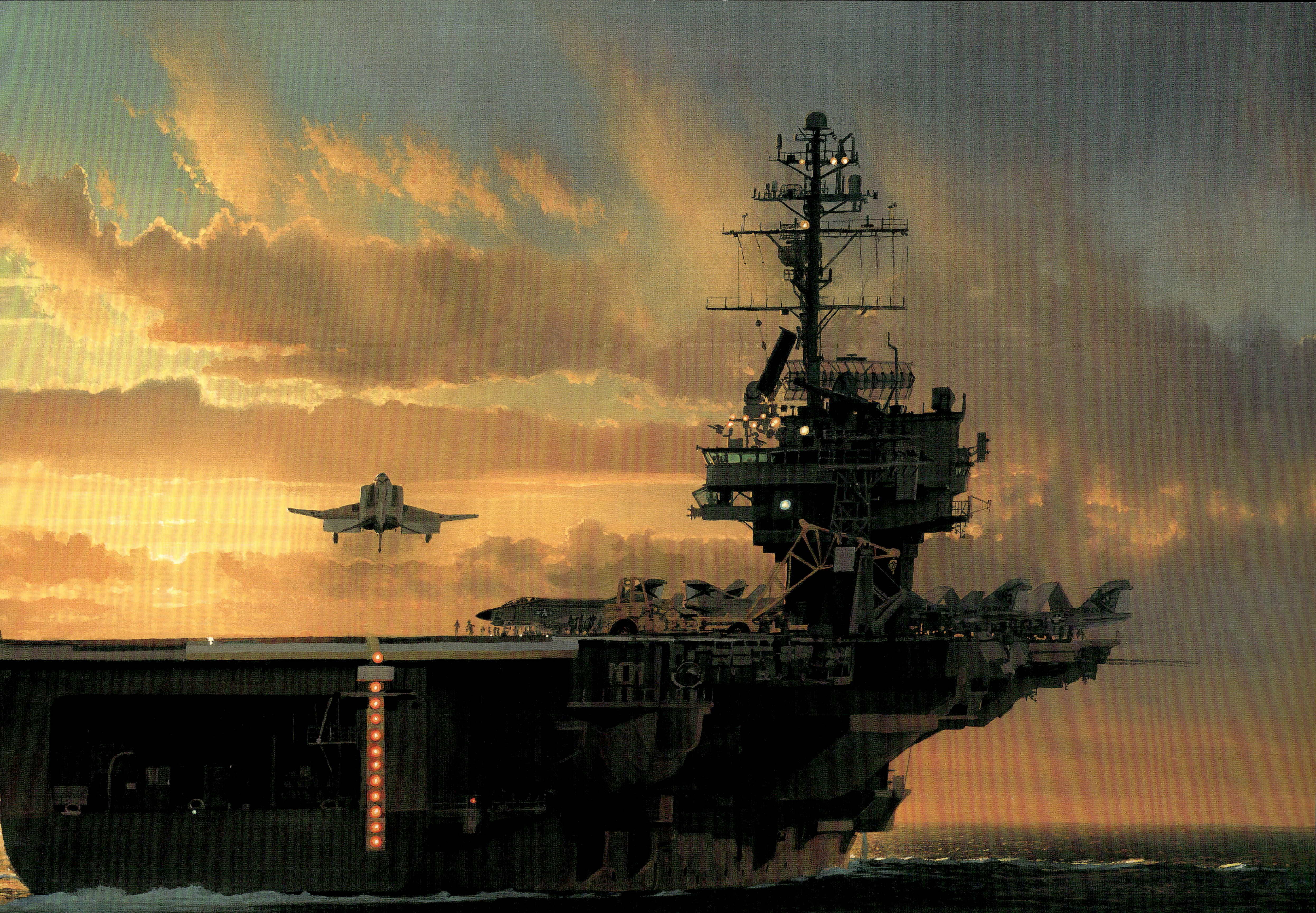

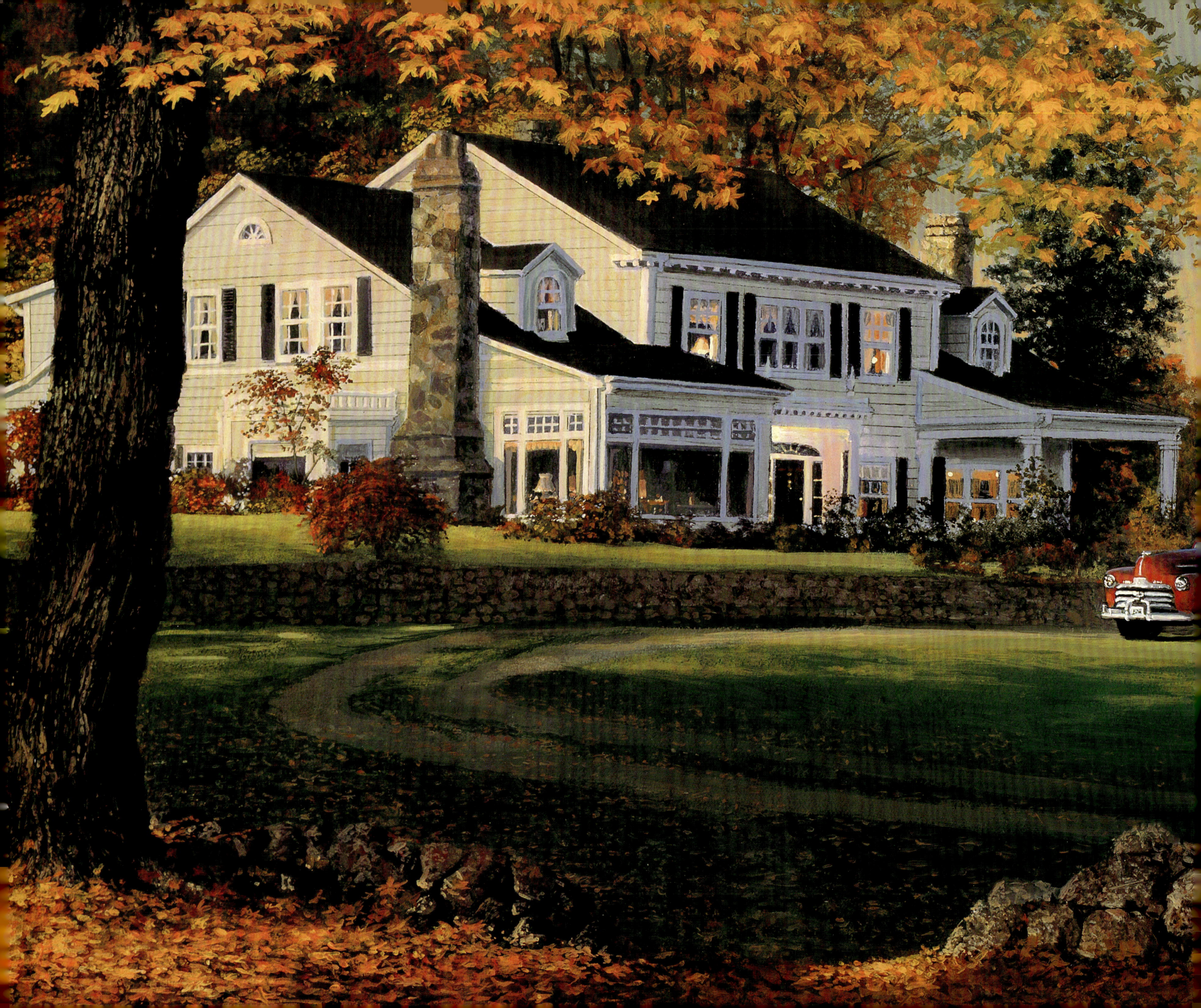

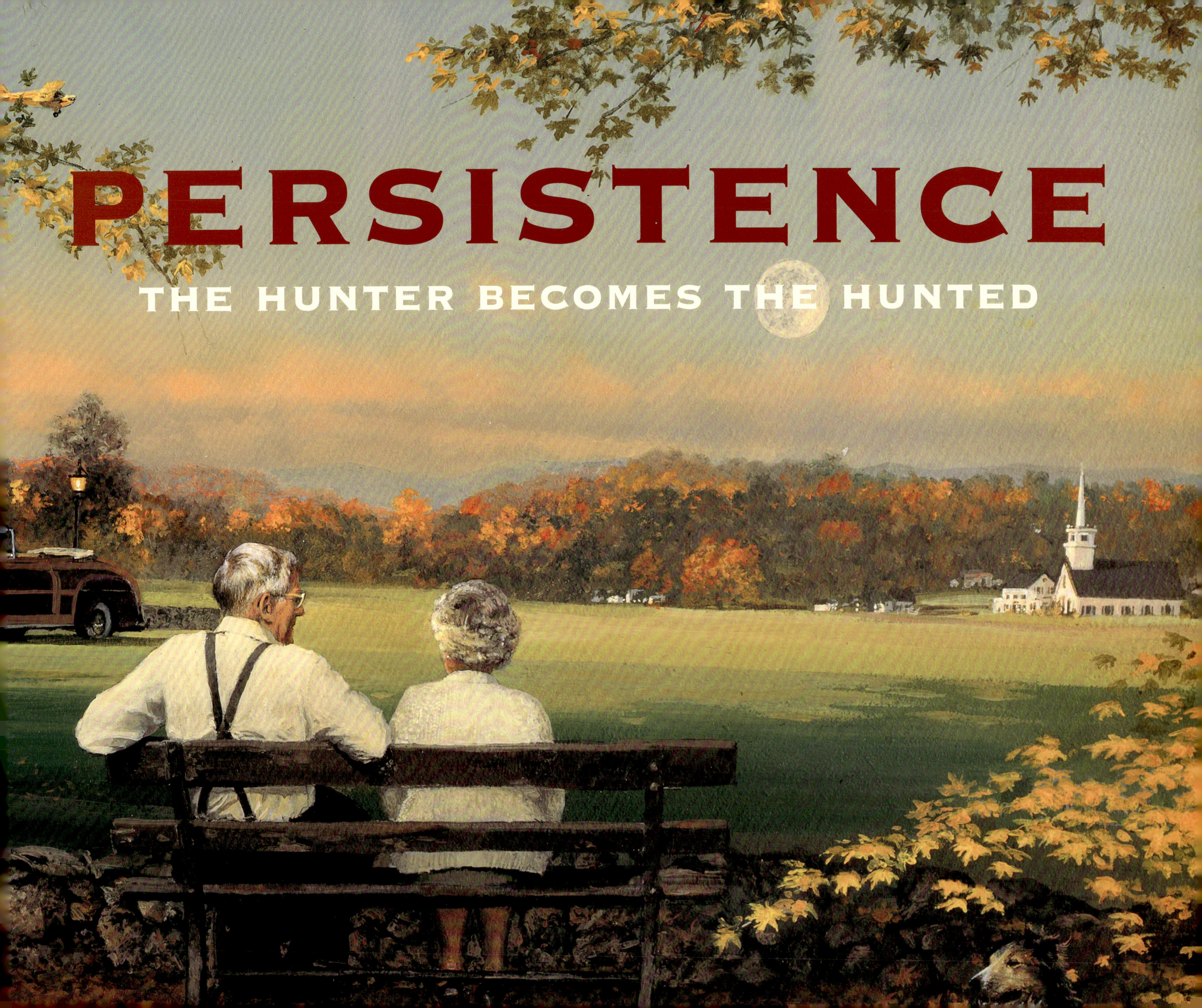
PERSISTENCE
THE HUNTER BECOMES THE HUNTED

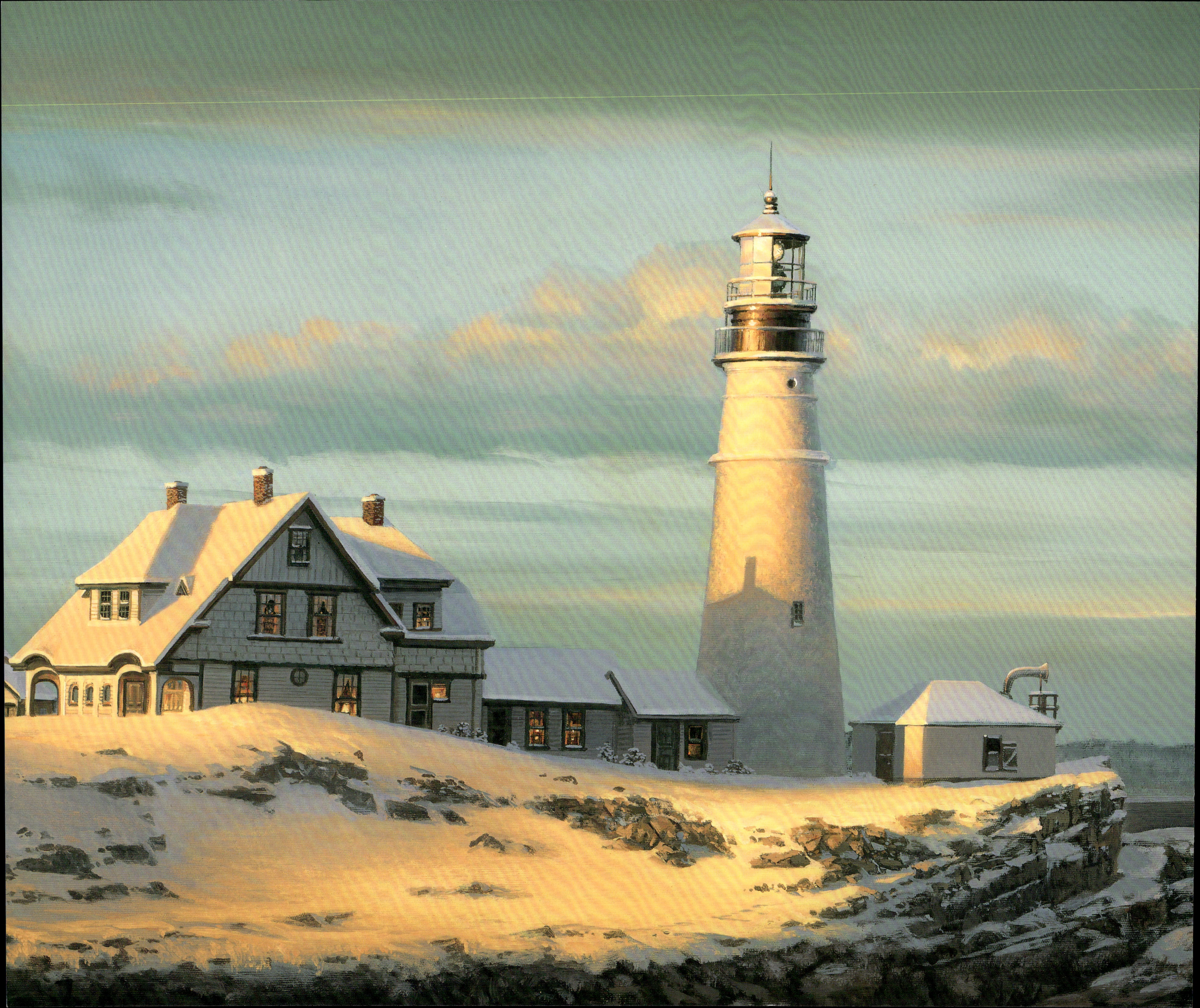

*I know the price of success: dedication, hard work, and an unremitting devotion to the things you want to see happen.*

FRANK LLOYD WRIGHT

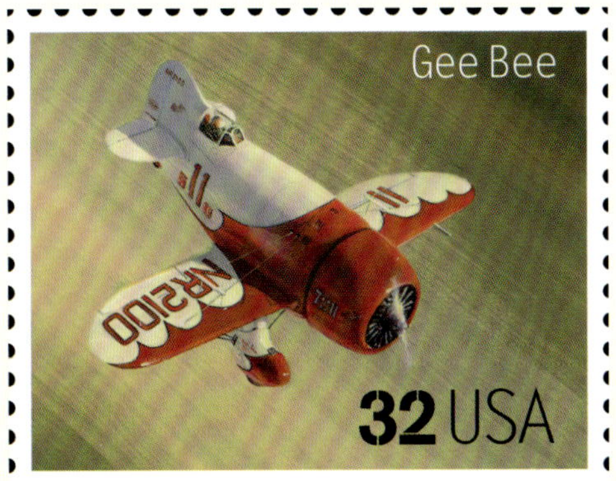


**GEE BEE**

U.S. Postal Service, Classic American Aircraft series, 1997

The five Granville brothers of Springfield, Massachusetts, gave their initials to the Gee Bee aircraft designs—twenty-two in total—that came from their drawing boards. This barrel-shaped Model R1 was built to win the 1932 Thompson Trophy pylon race and, at the hands of master pilot Jimmy Doolittle, win it did!

Bill remembers the thrill of his first ride in a military jet trainer. "Fields of wheat stretched northward toward Washington's Palouse Hills and August's hot breath kicked tendrils of dust along the edge of the tarmac," he recalled, "but all I could concentrate on as Captain John Williams and I walked across the ramp at Walla Walla Airport was the glistening T-33 assigned to the 318th Fighter Interceptor Squadron. I had dreamed of flying in fighter aircraft since my boyhood in the San Fernando Valley. And now, after four years in the Air Force and thanks to Captain Williams, my dreams were about to come true. As I climbed into the shiny T-33, it occurred to me that this would be my only chance to fly with the military in a high performance aircraft. Now, many years later, I smile as I remember my aerobatic ride in that T-bird. That ride was just the beginning. Persistence would be key to many more."

In 1975, the United States Air Force Academy commissioned Bill to paint the Air Force's new F-15 Eagle, sending him to Luke Air Force Base in Arizona to document the Triple Nickel, the 555th Tactical Fighter Training Squadron. When he presented a finished painting to Colonel Ted Laudice, the 555th Commander, he was introduced to a General Officer who was on an inspection tour for the Pentagon. Surprisingly, the General was seated next to Bill on the departure flight out of Phoenix the next morning. He mentioned the Air Force Art Program and promised to bring Bill's art to the attention of those in charge. Within a week, Alice Price, the highly respected program manager, called to request examples of his work and, in 1976, she offered Bill his first opportunity to travel overseas and fly in some of the Air Force's finest fighters. "Through the Air Force Art Program," Bill said, "I was given the unique opportunity to rejoin the military aviation community I so admired."

Flight was research. In the exhilarating, gut-wrenching, high-G environment, Bill used all of his senses: hearing the engine's throaty growl, seeing the ground blur, smelling the odor of jet fuel, tasting 100% oxygen inside a tightly fitting mask, and touching the controls with gloved hands. He sought opportunities, offering artwork in exchange for chances to fly. He exulted in flights in a wide variety of high performance aircraft around the globe and translated his experiences to his art.

Bill became the first civilian to fly a Red Flag mission at Nellis Air Force Base, Nevada. He was invited to fly with the Air Force Air Demonstration Squadron, the Thunderbirds, and was one of the first people to fly into the devastated zone around Mt. St. Helens to document rescue operations after the 1980 volcanic eruption. He flew with the Navy Flight Demonstration Team, the Blue Angels, and with pilots of the Navy's "Top Gun" school at Miramar Naval Air Station, California. "The people with whom I have been privileged to work and the aircraft I have flown are unforgettable. I will be forever indebted to the U.S. Air Force and the U.S. Navy for years of 'hands on' historical research."

Networking was as vital as persistence. Bill cherishes his friendships with Alice Price and with Virginia Bader, two dynamic and talented women. Virginia, an avid representative of aviation art, shared Bill's opinion that aviation paintings could be marketed as fine art. She traveled throughout the United States representing several artists, including Bill. "I accompanied Virginia to several air shows," the artist recalled, "and endured the heat and humidity of Harlingen, Texas, had display boards blown over during thunderstorms and was overwhelmed by the responses of people who share the love of flight."

Virginia was instrumental in two of Bill's pivotal opportunities: a meeting with David Usher, the president and publisher of The Greenwich Workshop, and an historic interview with General Jimmy Doolittle. In 1982, Virginia told Bill that fine art publisher David Usher was interested in discussing limited edition aviation art prints with him. They would meet at the Air Force Art Program Presentation Banquet. Bill's career as an aviation artist was well underway and, despite a backlog of commissions for wildlife and aviation paintings, he felt compelled to create a large painting of an F-16 flying above the Grand Canyon as his donation to the Air Force Art Program. *Canyon Falcon* proved to be the collection's largest painting that year and was hung in a prominent, well-illuminated spot. David Usher appreciated the painting and invited Bill, Keith Ferris, Frank Wootton, and Wilson Hurley to anchor his new program. Dave was not sure how successful aviation art would be, but he was willing to give it a try. In the ensuing twenty-three years, Greenwich Workshop has produced a very successful aviation art program.

On the strength of his publishing success with The Greenwich Workshop, Bill left his job as a fire fighter to focus his prodigious energies on producing fine art. In 1985, Kristi left her job and began to manage his business affairs. This remarkable team continues to this day. Kristi's assistance, combined with Bill's accumulated experience and maturity as an artist, has created an amazing collection of work. Bill went on to become one of America's most successful aviation artists. Known for his dramatic renderings of the sky, he expresses eloquently the beauty and glory of winging through the grandeur of the air.

Shortly after the start of his association with Greenwich, Bill gave a presentation at the Smithsonian's National Air and Space Museum that was followed by a dinner with Virginia Bader and The Greenwich Workshop's Charlie Rhudy. Virginia asked the artist if he could paint any aviation subject,

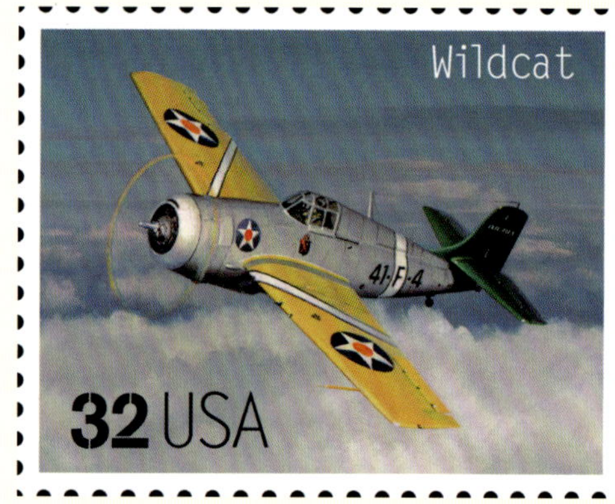


**WILDCAT**

U.S. Postal Service, Classic American Aircraft series, 1997

In a paint scheme of pre-war U.S. Navy markings, the Grumman F4F Wildcat, the first carrier-based monoplane created by the "Iron Works," became the seaborne first line of defense. Its Navy pilots learned to outmaneuver the faster, more agile Japanese Zero in critical battles over the Coral Sea and Guadalcanal, and at Midway.

what would it be? "My answer was swift. My father had played the role of Don Smith in the movie *Thirty Seconds Over Tokyo* and I thought that the Doolittle Raid would be a great challenge. I wanted to paint Jimmy Doolittle's airplane as it approached Japan and I would have as many original crewmembers as possible sign the artwork with me. But then I wondered if that might prove too difficult. Virginia's eyes lit up and she exclaimed, 'I know General Doolittle. If I could get the two of you together, would you do that painting?' Charlie Rhudy offered that The Greenwich Workshop would produce a limited edition print of the painting, and the evening took on an almost electric quality. The air filled with excitement and, as I recall, no one finished dinner." At the time, limited edition prints co-signed by historical figures were rare and *The Giant Begins to Stir*, portraying the Tokyo Raiders and signed by all the crew members attending their reunion in Carmel, California, became an instant sellout. It hit the secondary market with such force that it was written up in *Business Week* magazine.

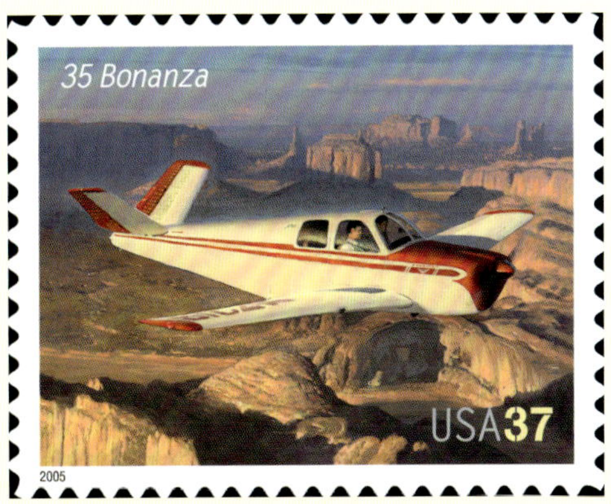


**BEECHCRAFT 35 BONANZA**
U.S. Postal Service, Advances in American Aviation series, 2005

When Walter Beech wanted a four-place airplane to carry passengers in car-like comfort, the result, a Beechcraft V-tail Model 35 Bonanza, soon became a classic. In March 1949, a modified Bonanza flew 5,273 miles from Hawaii to New Jersey in approximately thirty-six hours. Its straight tail version continues to be produced today, and Bonanzas are among the most successful aircraft ever built—more than 17,000 in fifty-seven years of continuous manufacture.

Bill's paintings earned publicity, awards, and invitations that took him beyond his dreams. Alice Price assigned him to travel to Guam and Okinawa with the Air Force Art Program. Later, thanks to Keith Ferris and Northrop, he traveled to Saudi Arabia to document the Saudi Arabian Air Force. Then, Virginia Bader notified him that Prince Faisal of Jordan commissioned one of his oils as a gift for his father, King Hussein, and, expressing delight with Bill's painting, King Hussein invited Bill and Kristi in 1983 to be guests of his royal family and to document the Royal Jordanian Air Force. The travel enriched their lives. Bill flew with Jordanian pilots and viewed an ancient, historic world from their military's modern cockpits. He presented the royal family with several oil paintings.

In 1988, during the Iran-Iraq War, Bill was asked to serve in a hazardous assignment as a Navy Combat Artist in the Persian Gulf. Operation Earnest Will was the United States' response to Kuwait's request for protection for her ships from Iranian attacks while transiting the Gulf. Bill departed for the Persian Gulf and, several helicopter rides later, he landed on the USS *Richmond K. Turner*, a missile cruiser. There, he was privileged to research the Special Operations Forces at work.

"In retrospect," he acknowledged, "as I strapped myself into the ejection set of that T-33 way back in 1967, for what I thought would be my one and only flight in a high performance military jet, I knew nothing of the flying adventures that lay ahead for me. Through a God-given talent, the help and support of countless wonderful people and a dogged persistence on my part, my life holds more wonder than I could dare dream."

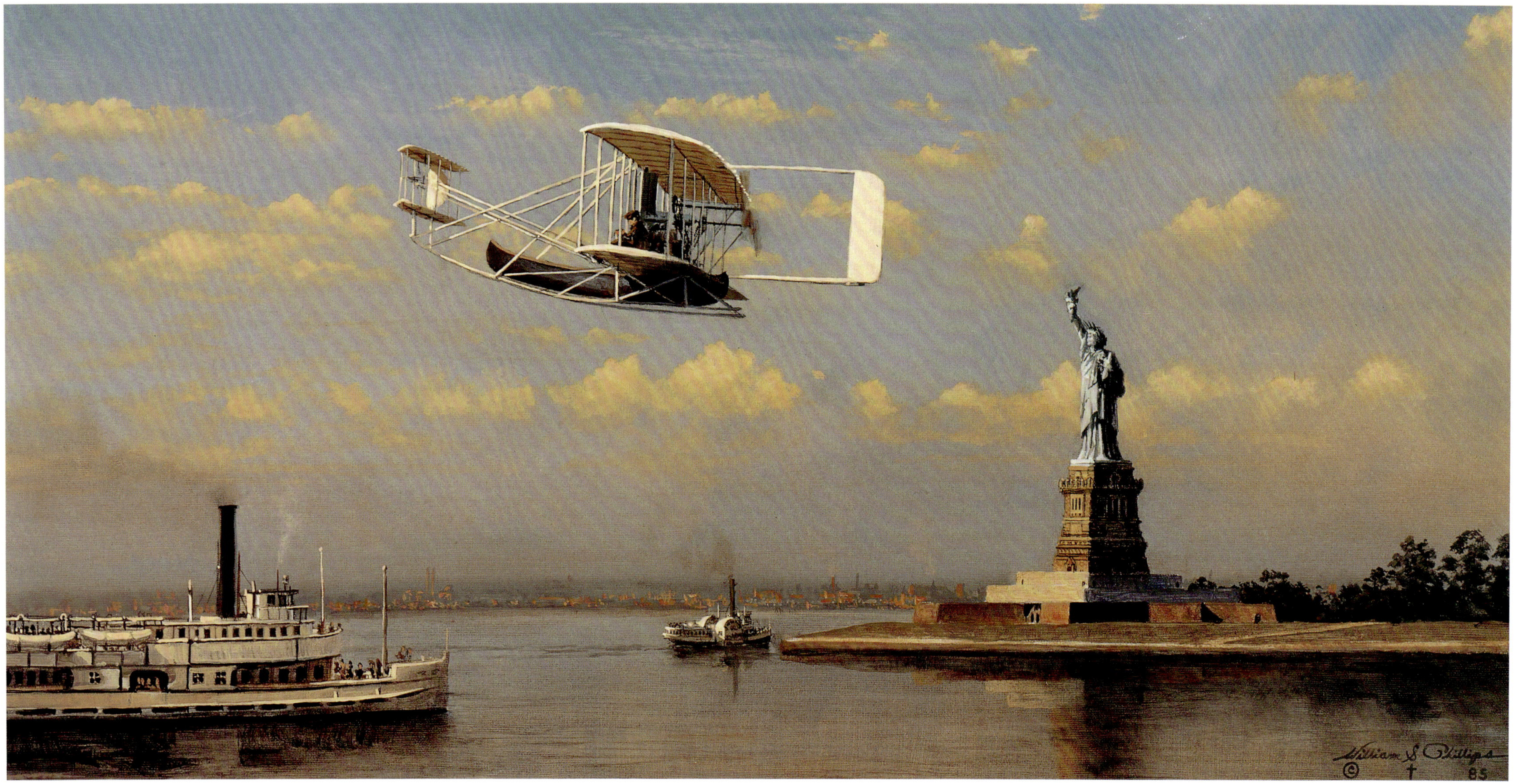

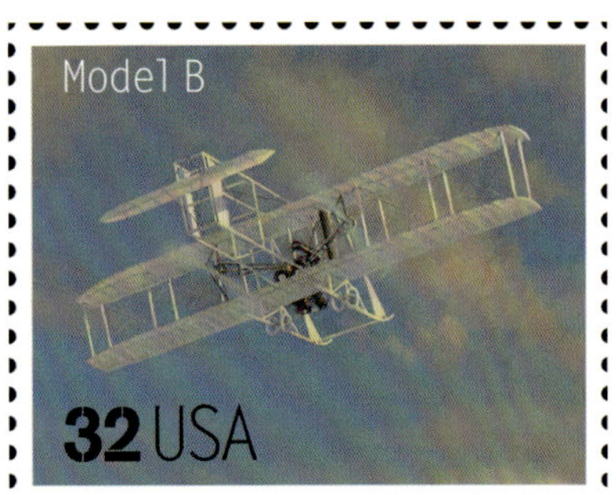


## WRIGHT MODEL B

U.S. Postal Service, Classic American Aircraft series, 1997

The Model B Flyer, Wright's first aircraft to be produced in quantity, became the U.S. Army Signal Corps Airplanes Number 3 and Number 4 in 1911. Used primarily for training pilots at College Park, Maryland, a modified Model B gained fame as the *Vin Fiz* on Cal Rodgers' transcontinental flight.

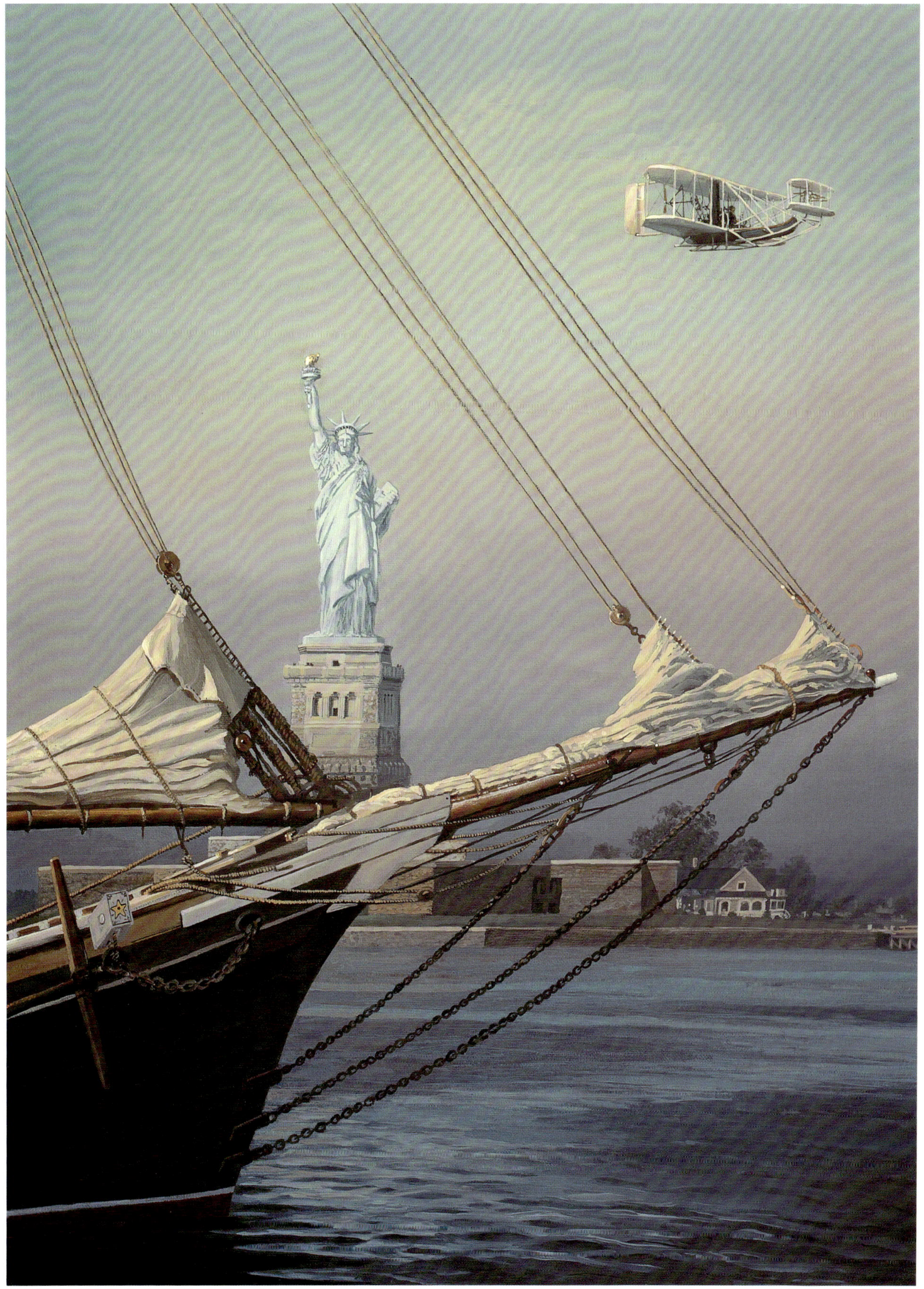

## Wright By Liberty

(right)

## A Wright to Liberty

(opposite)

Wilbur Wright, prone on the lower wing, piloted this 1909 Flyer past the Statue of Liberty during the September Centennial celebration of Robert Fulton's steamboat. Wright took off from Governor's Island in New York Harbor, and wisely provided for his safety in the event of a water landing by mounting an Abercrombie and Fitch canoe beneath the fuselage. The plane later became U.S. Army Signal Corps Airplane No. 1. The canoe, along with the Wright III Flyer, is on prominent display at Carillon Historical Park, part of Dayton Aviation Heritage National Historical Park, in Ohio.

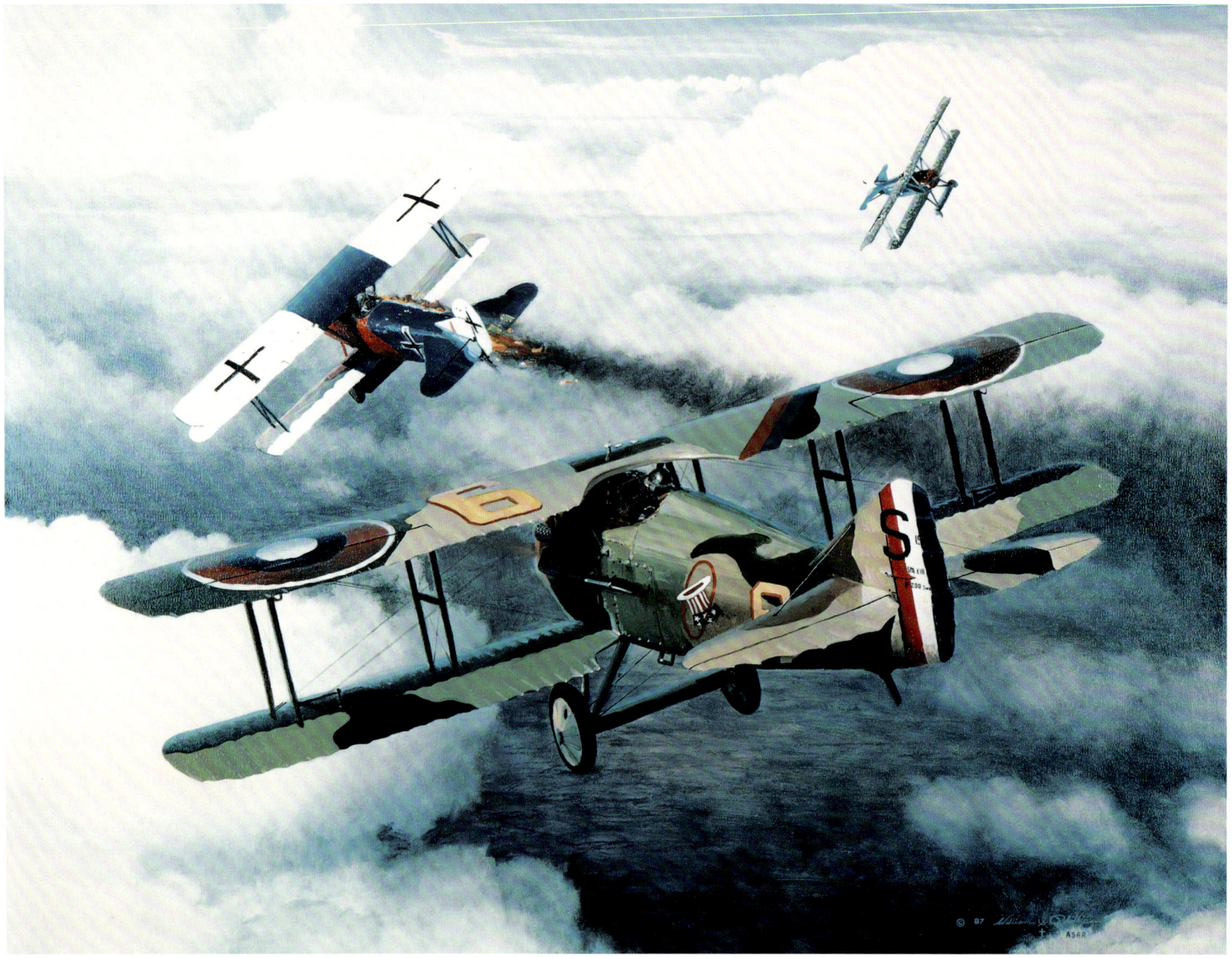

## Two Down to Glory

Chivalrous though it was reputed to be, World War I aerial combat was lethal. In the painting, Major Reed Chambers of the U.S. Air Service, a seven-victory ace, downs a German Fokker D.VII with his 94th Aero Squadron SPAD XIII. Bearing the 94th famed "Hat in the Ring" symbol, the French-designed and built SPAD proved reliable and effective.

A native Kansan, Chambers had served on the Mexican border in 1916 before transferring to the Air Service where he teamed with, among others, Eddie Rickenbacker, America's Ace of Aces, to take down seventy enemy aircraft between April and November 1918. After the war, Chambers and Rickenbacker formed Florida Airways.

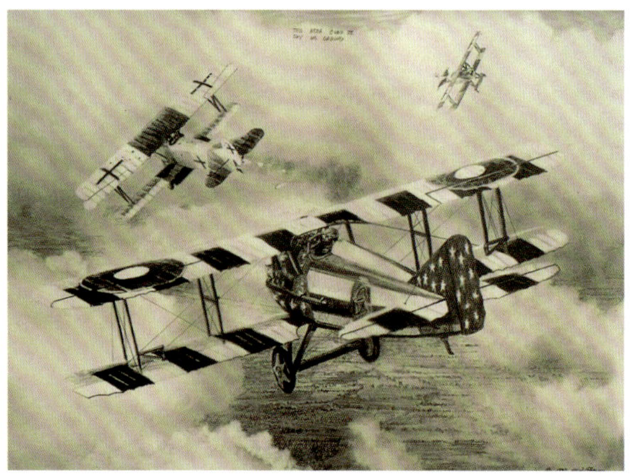

Pencil sketch for *Two Down to Glory*.

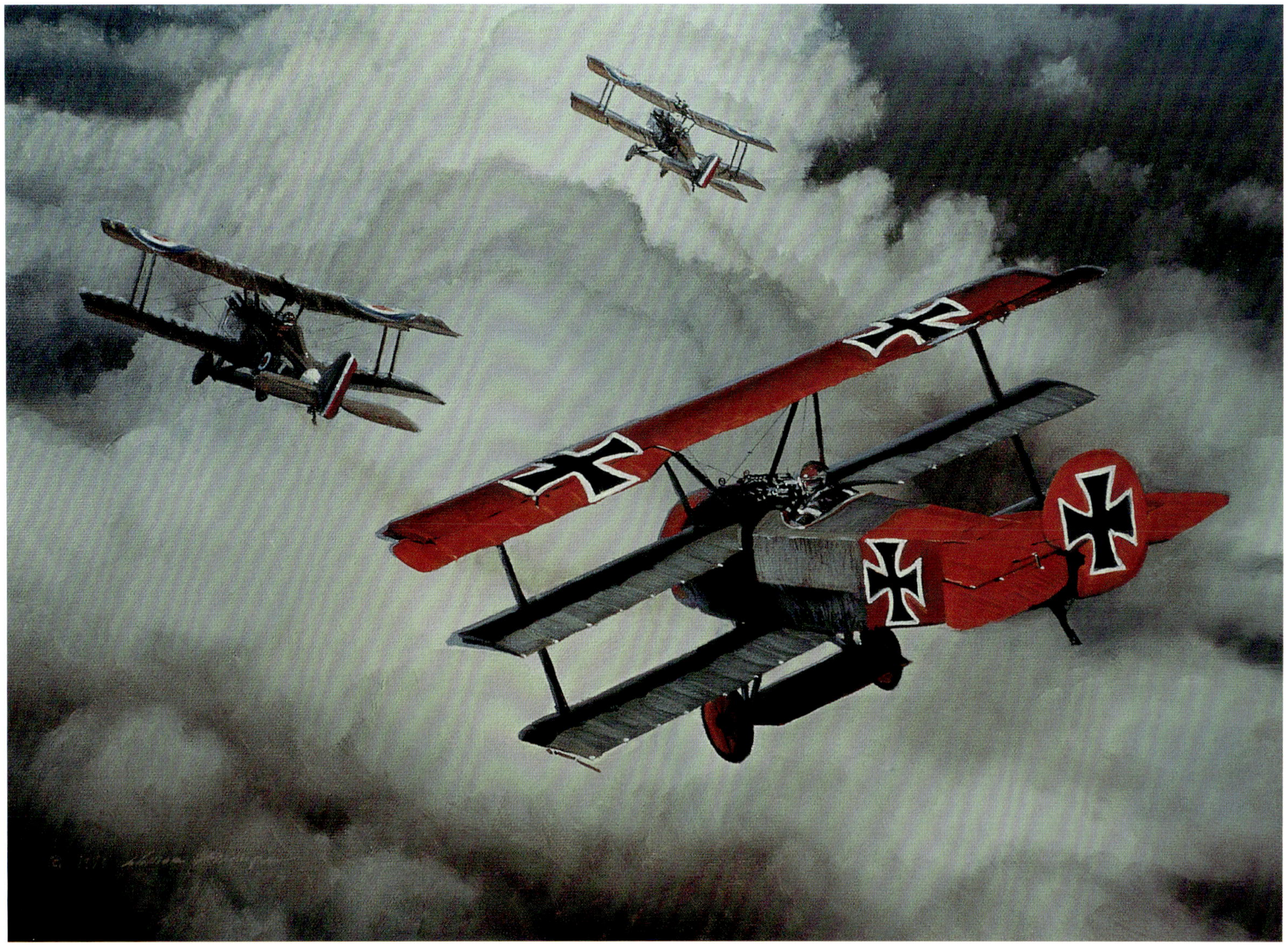

## Closing the Gap

A Fokker Dr.1 of Jadgeschwader 1 closes the gap on two British S.E.5a in the spring of 1918. The Dr.1 triplane was a radical and sometimes dangerous new design, made famous by Manfred von Richthofen. The S.E.5 was flown after the war as part of the 94th Aero Squadron based at Austin, Texas in the 1919–1920 time period.

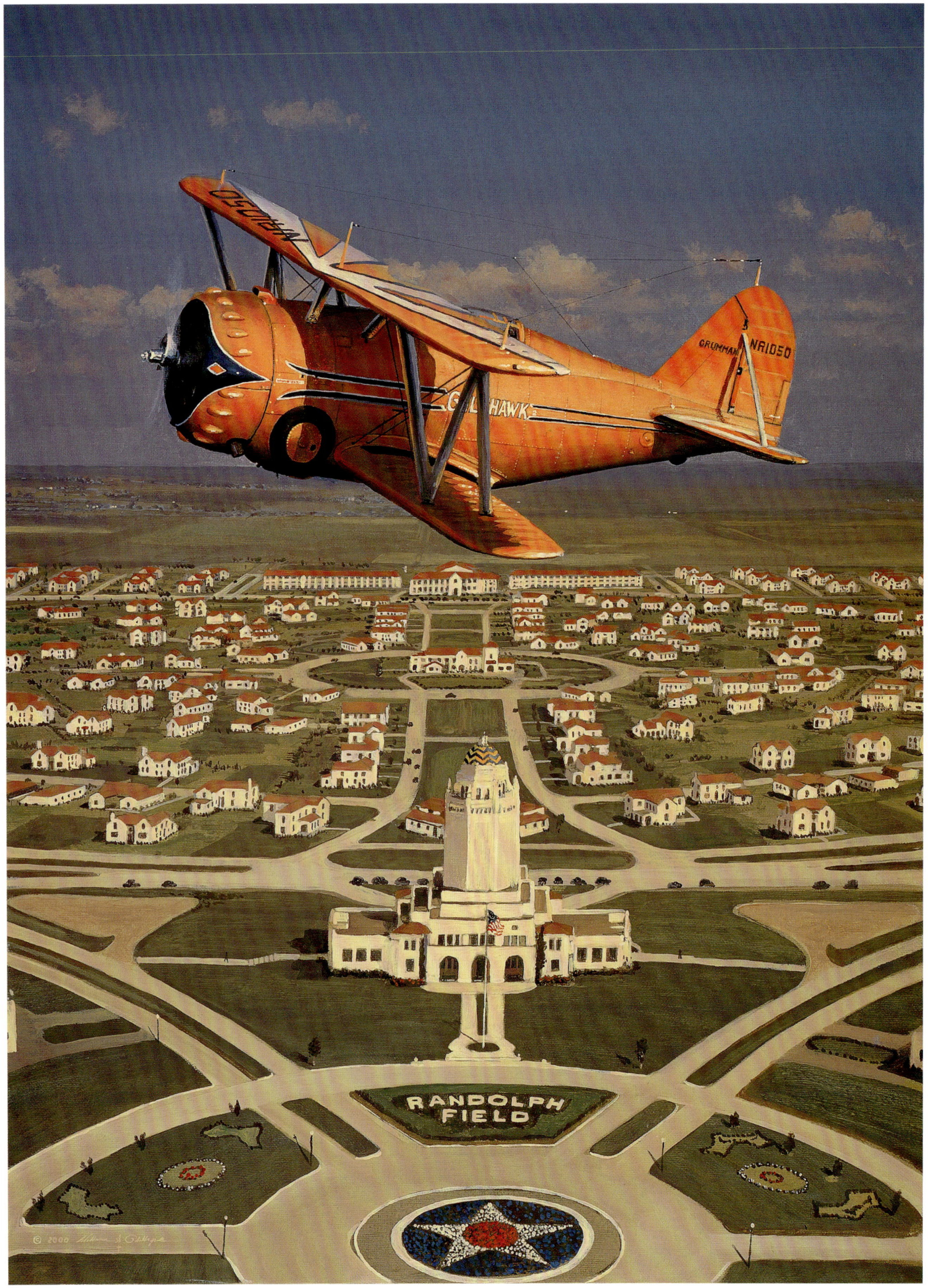


## Randolph Field and the Gulfhawk

In the mid-1920s, the inadequacy of existing flight training facilities called for a new Air Corps Training Center. A site was chosen north of San Antonio, Texas and responsibility for the project fell to General Frank Lahm, head of flight training under the Air Corps Act of 1926. Lt. Harold Clark, a trained architect, presented his conceptual drawings for a perfect "Air City" to General Lahm. Lahm, in turn, was so impressed that he appointed Clark to the project, which became Randolph Field—later known as the West Point of the Air Force—honoring the memory of an Austin pilot, Captain William Randolph, killed in an aircraft accident in 1928.

The construction project was the largest undertaken by the U.S. Army Corps of Engineers since the Panama Canal. Artist Bill Phillips conducted meticulous research in order to paint the field as it looked when dedicated. Randolph Field, originally designed for flight training, continues to provide that service as Randolph Air Force Base. The aircraft and the people have changed, but the commitment to educating first-class military pilots has not.

Major Al Williams, airshow pilot and air speed record holder, was one of Randolph Field's most famous visitors. A former naval aviator, Williams turned to Grumman Aircraft Engineering Corporation on Long Island, New York in 1936 for a new aerobatic airplane. The company responded with a fast, highly maneuverable biplane with retractable landing gear. As a pilot for the Gulf Oil Company, Williams named his airplane, *Gulfhawk*. After *Gulfhawk I* was destroyed in a crash, its replacement *Gulfhawk II*, performed outstandingly throughout the country.

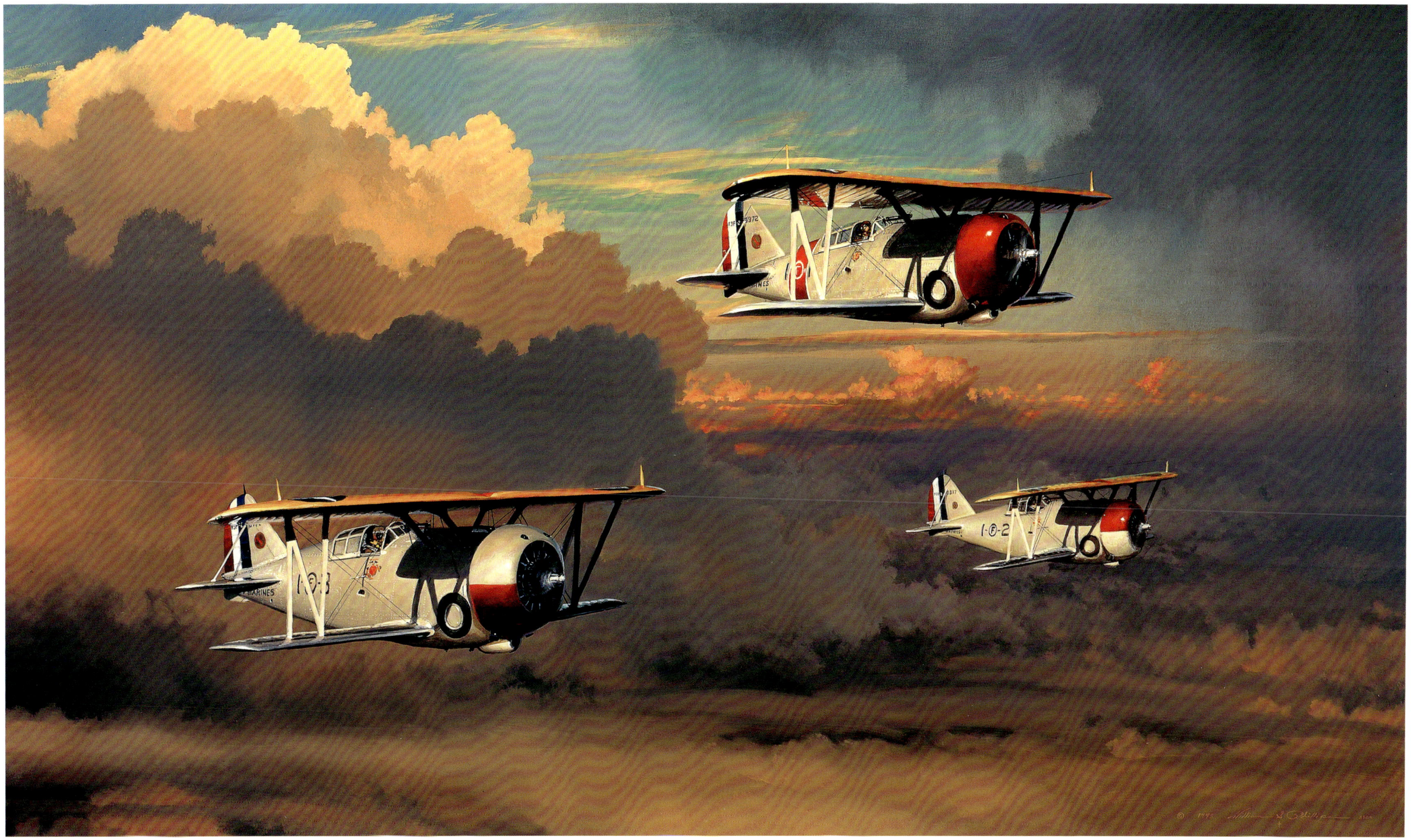

## Thirty's Thunder

In the 1930s, when biplanes ruled the U.S. skies, Grumman Aircraft's F3F-1, a "flying barrel" of an airplane, was ordered into production. Despite the Great Depression, fifty-four robust aircraft came off the assembly lines. The F3F-1 was an aerodynamically refined version of the F2F-1, and powered by a P&W R-1535. By 1936, demand for increased speed and higher service ceilings, resulted in the fitting of a 1,000-hp Wright Cyclone engine that added 25 mph to max speed and 4,000 feet to the F3F-1's ceiling. Eighty-four of these were ordered. Finally, these F3F-3s, ducking among the clouds, were introduced.

This further mod proved to have strength, agility, speed, and maneuverability. Its snap rolls were crisp and three point landings were the rule rather than the exception. Its pilots were proud to fly the ultimate biplane shipboard fighter. With one .50 caliber and one .30 caliber forward-firing machine gun, it was well armed for its time. It joined other Grumman models and, by 1939, all first line U.S. Navy and Marine Corps squadrons were flying Grumman fighters. VMF 2, shown here wearing its pre-war USMC colors, was the last naval aviation fighter unit to operate biplanes. They turned in their F3F-2s in October 1941.

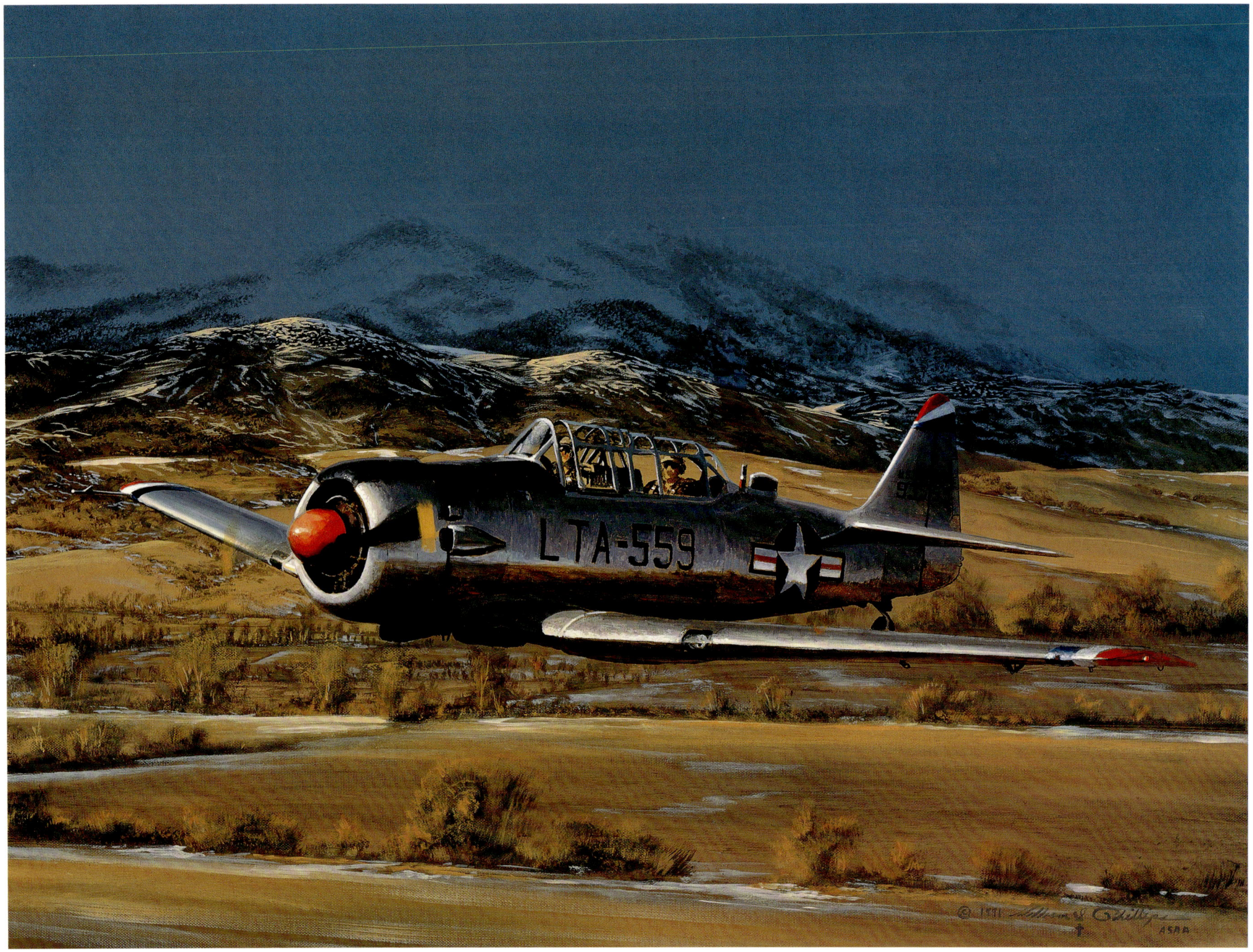


## Trolling for NKs

During the Korean conflict, Forward Air Controllers (FACs) piloted North American AT-6 Texans. Flying low, slow, and vulnerable, FACs provided aerial intelligence to direct U.S. Air Force and Navy pilots and U.S. Army soldiers to selected North Korean (NK) targets by marking the sites. "Mosquitos" of 6147th Tactical Control Group identified targets by firing a bazooka round attached at the front end of a practice rocket. The explosion's smoke was visible for miles. Early T-6s carried six rockets while later models, LT-6Gs, carried twelve. The airborne Mosquito coordinated with a Tactical Air Control Party (TACP) equipped with one or two radio-carrying jeeps manned by experienced FACs and radio operators. By the end of the Korean conflict, Mosquitos had flown 40,000 sorties, and were sometimes targets themselves.

## Hill Country Patrol (opposite)

Piper J-3s were the most numerous small aircraft used in WWII for spotting targets for artillery fire, observing enemy movements, and providing ambulance service. Approximately 5,250 were in service, known to the Army as L4, to the Navy as NE-1 and to the Army Air Corps (AAC) as O-59. Their contributions were considered so damaging that German soldiers in Italy were granted fifteen days leave for shooting one down.

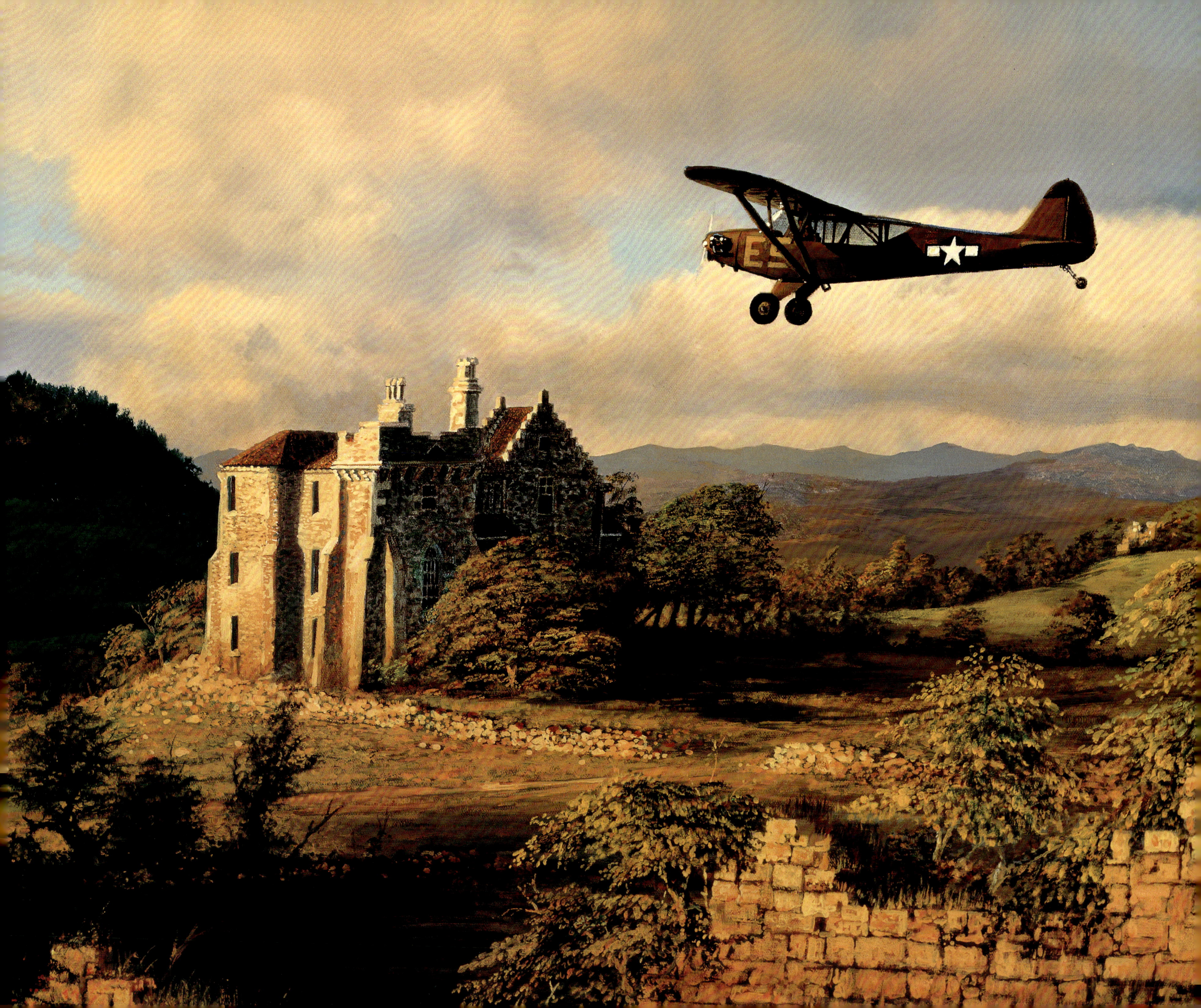

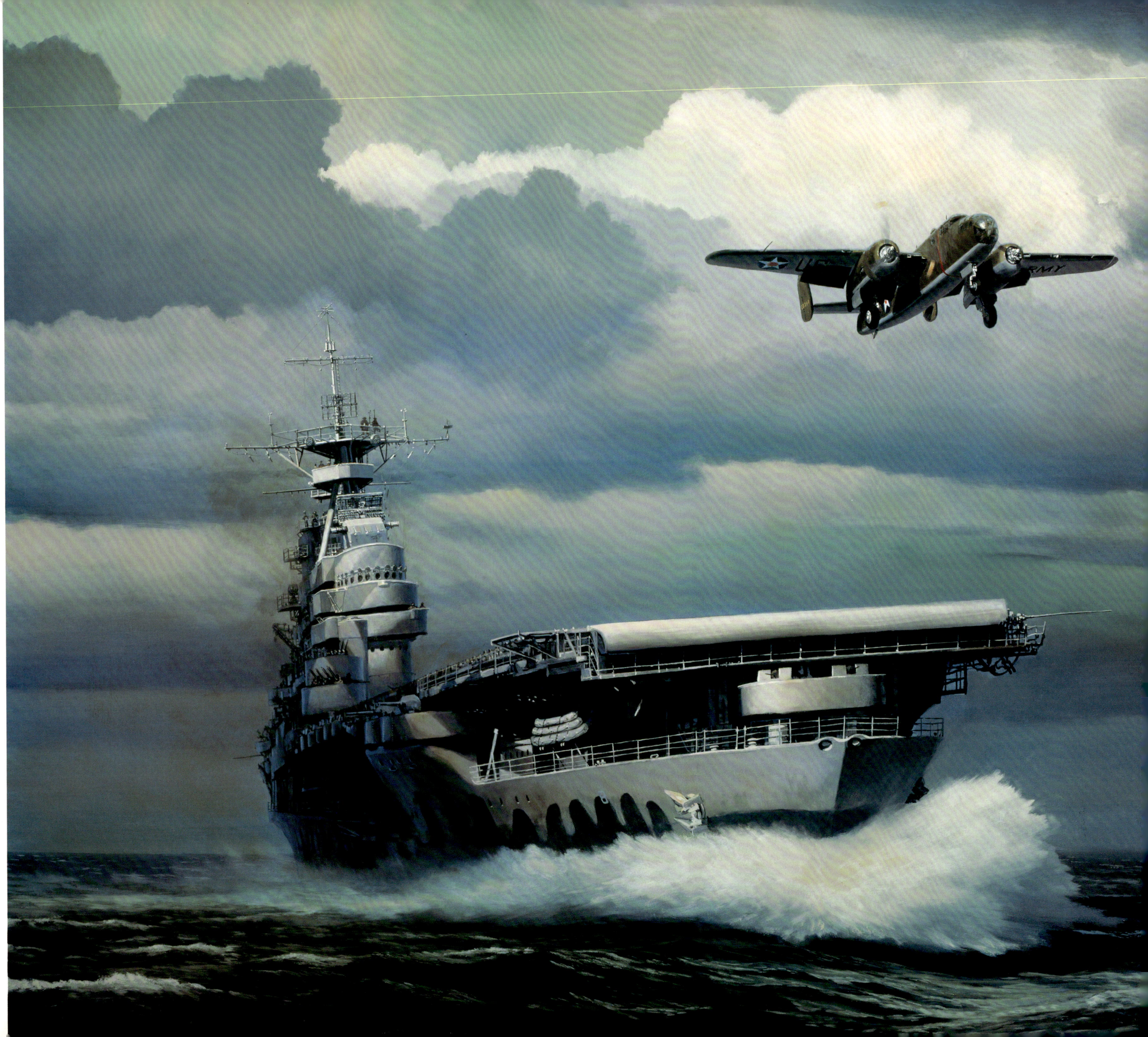

## Rising Into the Storm

Jimmy Doolittle's April 18, 1942 raid on the home islands of Japan is well known. Yet, Doolittle had been a man of firsts: in air racing, aeronautical studies, aerobatics, and instrument flight. His accomplishments, shared with his wife, Josephine "Joe" Daniels, tended to overshadow his persuasive leadership that was the key to his successes.

One major aspect of the April raid is less known. Without the well-planned cooperation of the U.S. Navy's leaders in the Pacific, it would not have happened. Admiral Nimitz, Commander in Chief Pacific, approved Admiral Halsey's plan to deliver the Raiders' planes, an approval masked in the secrecy required. Captain Mitscher took the *Hornet*, with the B-25s aboard, and seven supporting ships from San Francisco to the rendezvous point while Halsey in the *Enterprise*, with seven supporting ships, left from Hawaii. Task Force 16 converged and headed west with two submarines, *Trout* and *Thresher*, stationed on patrol and directed to report enemy forces that might pose a threat.

As the planned aircraft launch neared, tension mounted. The Air Force men struggled to maintain their planes while sailors increased their vigilance. At 3:00 a.m. on April 18, *Enterprise* signaled the spotting of two ships. Task Force 16 altered course to avoid detection. At 4:11, it turned back to the west. *Enterprise* launched eleven Wildcats and Dauntlesses as dawn patrols, and at 6:00 a.m. one patrolling aircraft reported spotting a small surface ship. Halsey ordered another course change and, at 7:38 a.m., a second small patrol vessel was sighted by a lookout on *Hornet* and an intercepted Japanese radio message confirmed the Task Force had been sighted.

Halsey dispatched the cruiser *Nashville* to sink the vessel and, simultaneously, ordered Mitscher to get Doolittle and his men on their way. The first B-25 was airborne forty minutes later. When the last departed, Task Force 16 reversed course and sailed east at full speed. *Hornet's* Navy aircraft were brought to the flight deck and patrols were launched continuously. Minor action ensued, but the ships returned safely to Pearl Harbor. It was a superb example of inter-service cooperation.

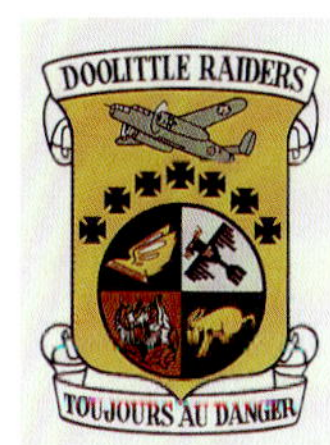


The elements of the Doolittle Raider crest represent the crews who took part in the Tokyo Raid. The seven Maltese Crosses come from the standard of the 17th Bomber Group, from which three of the squadrons were selected. The 34th Squadron is represented by the Thunderbird; the 95th Squadron by the Kicking Mule; and the 37th Squadron by the Tiger's Head. The winged Helmet represents the 89th Reconnaissance Squadron. A B-25 Bomber sits above, and the motto below reads "Ever into peril."

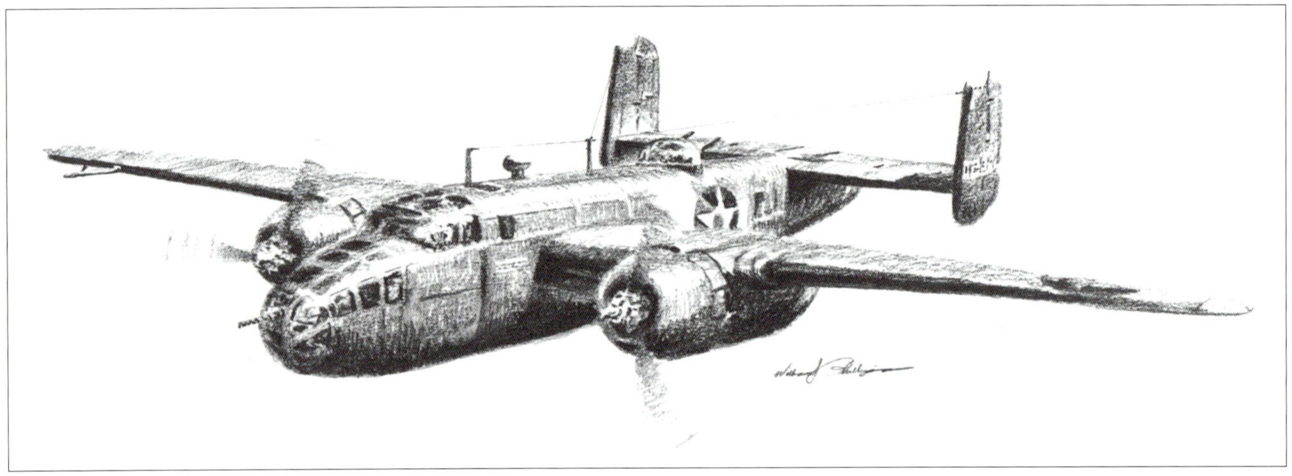

The B-25B "Mitchell"

## The Giant Begins to Stir

After departing the *Hornet* on April 18, 1942 and piloting the first of the sixteen B-25 Mitchells toward industrial targets in Japan, Lt. Col. Jimmy Doolittle neared Tokyo, and began dropping his bombs. Trailing Doolittle, and flying independent courses plotted by each crew's navigator, the other fifteen bombers—thirteen over Tokyo—followed suit. One aircraft was forced to jettison its load; the others damaged or destroyed targets including a tank depot, armory, oil tanks, steel and gas works, and small factories. All sixteen aircraft escaped unscathed from Japanese skies.

Landing in mainland China, however, proved challenging to all and fatal to some. Thick fog and a major storm forced eleven crews to bail out in total darkness and four crash-landed. The sixteenth and only aircraft to survive the raid, landed intact in the U.S.S.R. only to be interred by the Russians. In China, two men died in ditching their plane, and eight were captured by Japanese. (Three were executed and five were sentenced to brutal imprisonment; four later came home.) Survivors in China were treated well and helped toward safety on American bases, often at great risk to the Chinese, many of who were executed by the Japanese for assisting the U.S. crews.

Doolittle, assessing the modest damage done, the loss of some of the flyers and the loss of all of the aircraft, considered the raid ineffective. However, he vastly underestimated the major boost to morale. In that, the raid was a victory! Truly, the giant had begun to stir. In appreciation, a grateful President awarded him the Medal of Honor, promoted him to Brigadier General and decorated each of the remaining crewmembers.

Doolittle's raid was important to the artist, whose father played in the movie, *Thirty Seconds Over Tokyo*. Today, more than a half-century later, one of the Raiders, Tom Griffin, has said, "When we gathered in China, our airplanes wrecked and no assessment of the damage we inflicted, we thought of ourselves as failures. Now, after fifty-plus years of hearing accolades, we are beginning to believe that we won World War II single-handedly."

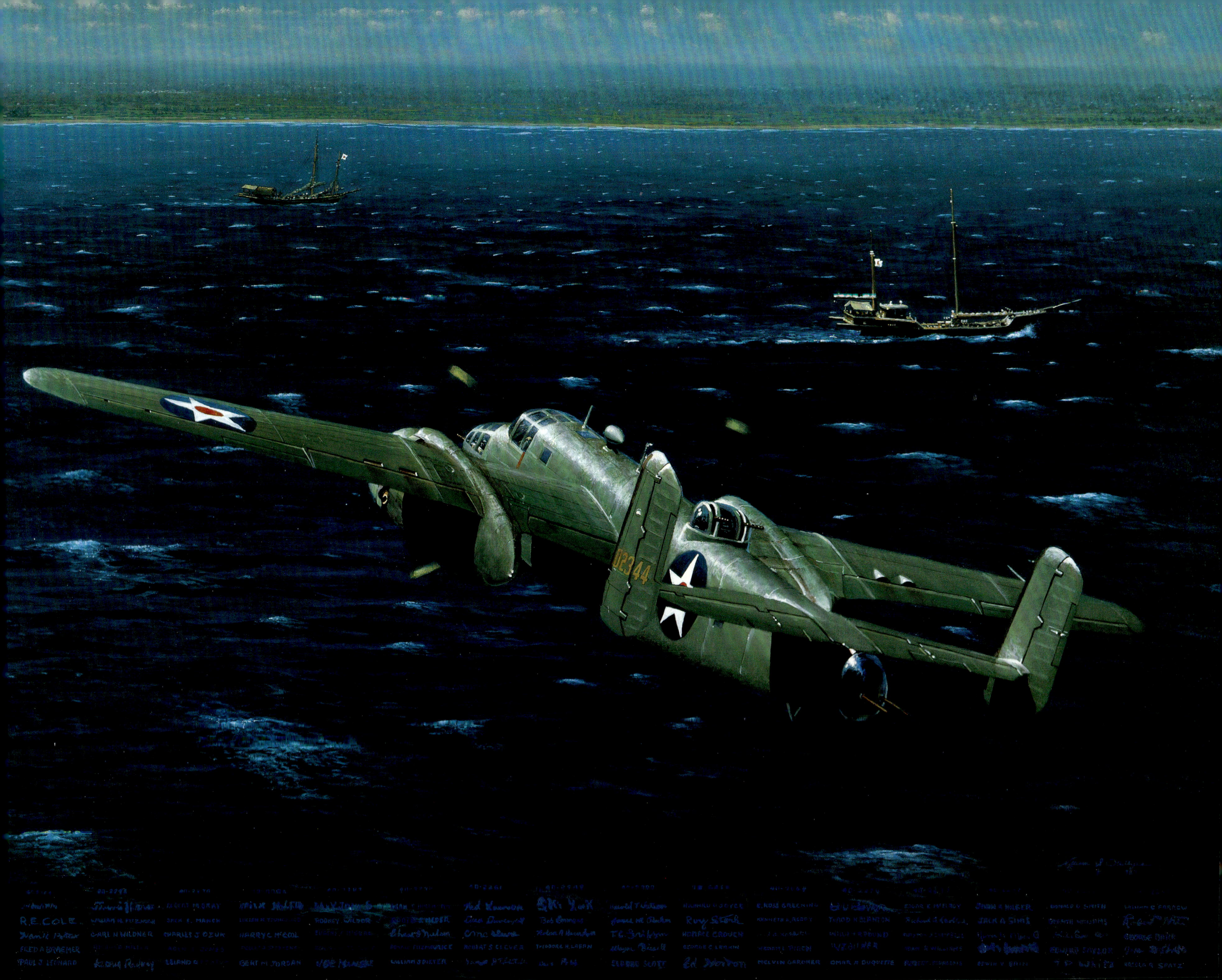
02344
R.E.COLE

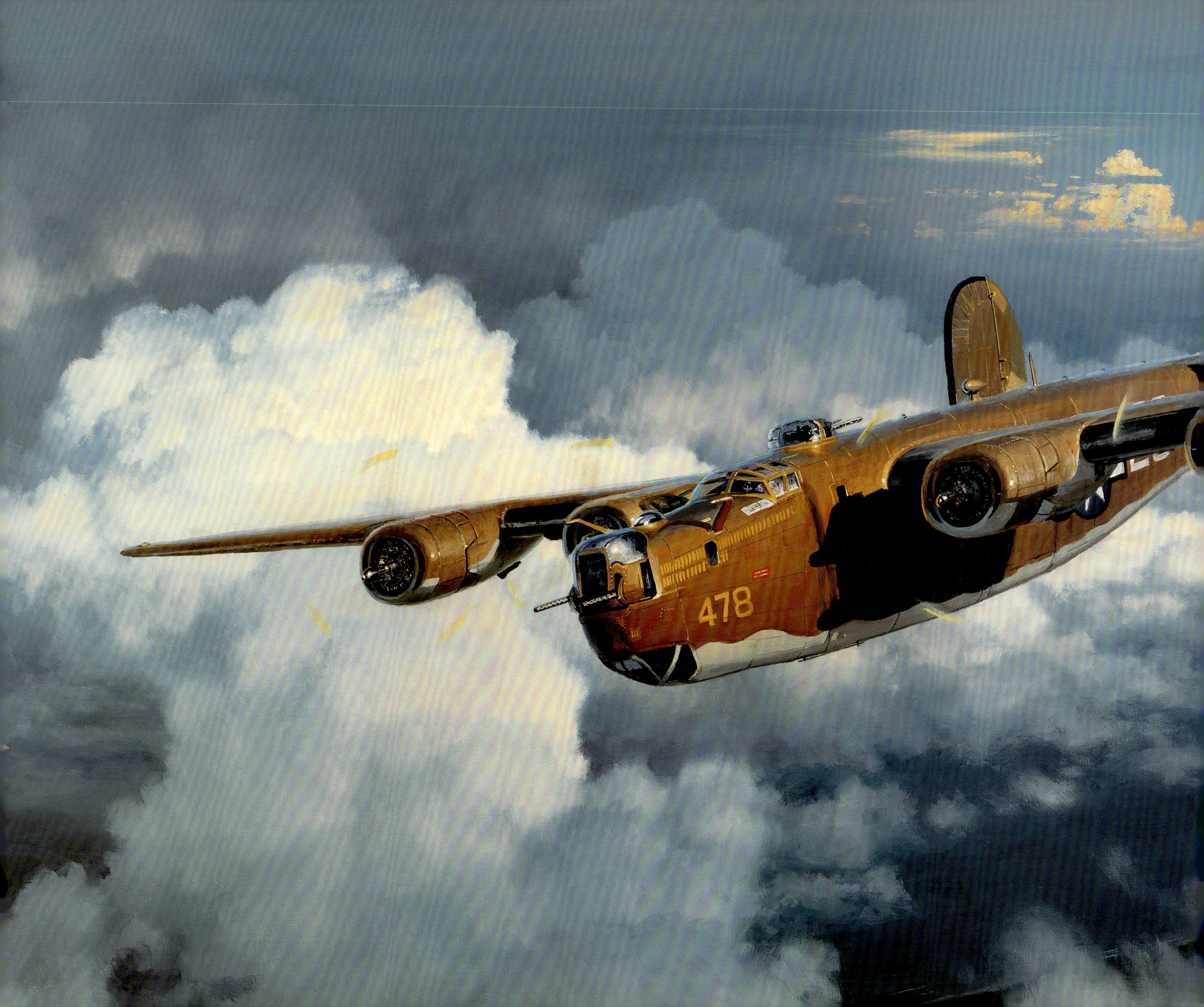
478

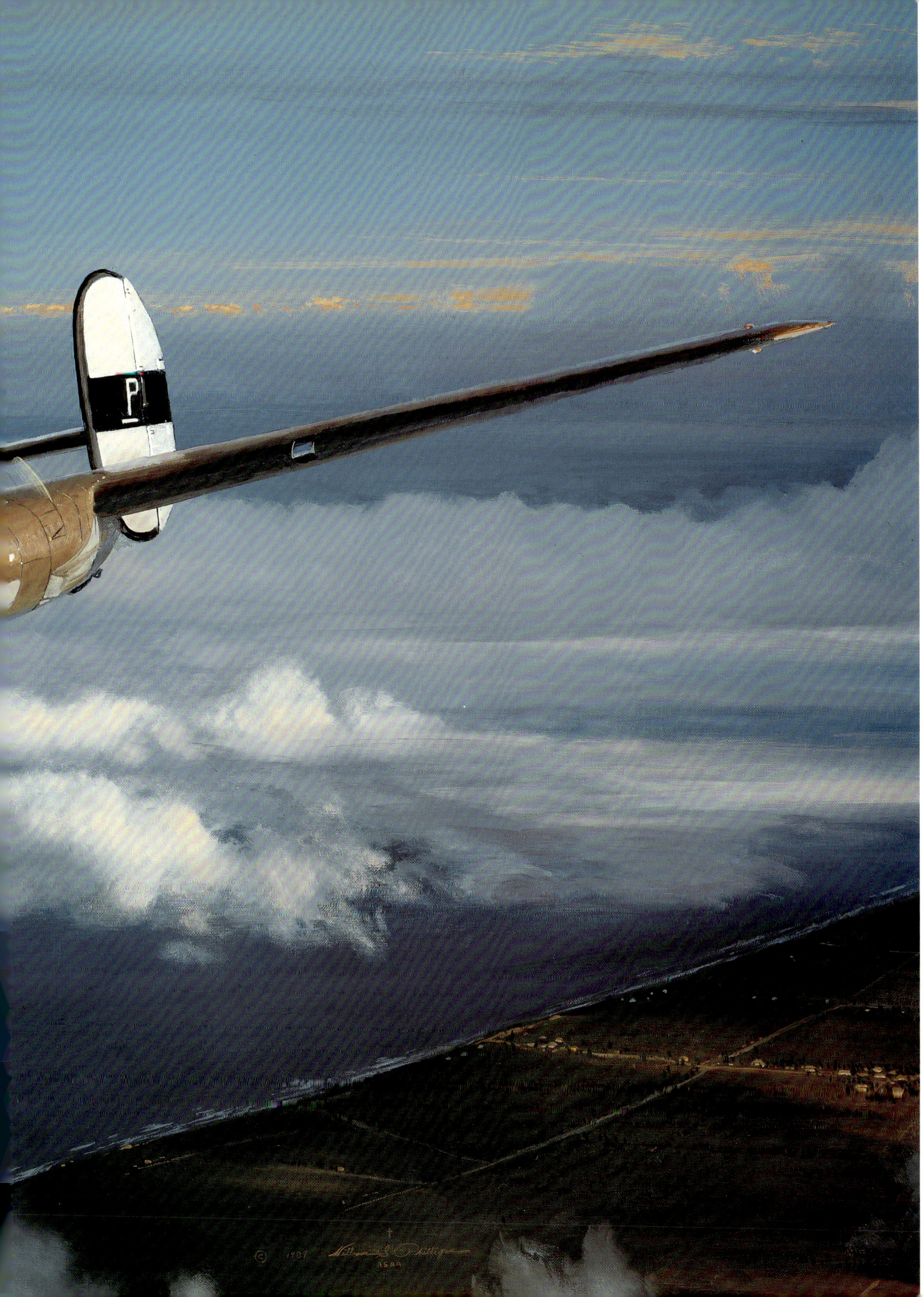

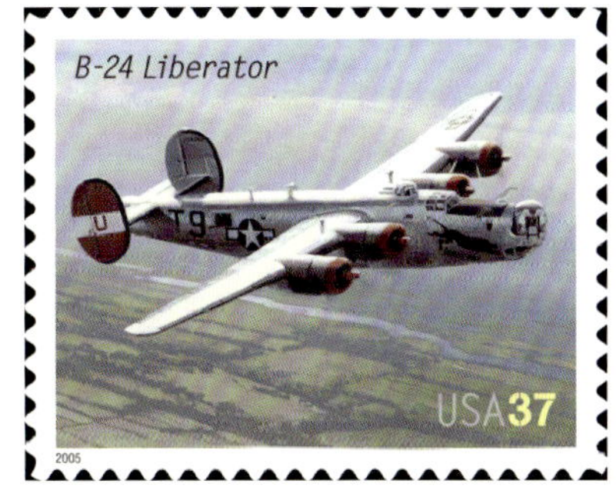


## Consolidated B-24 Liberator

U.S. Postal Service, Advances in American Aviation series, 2005

More B-24 Liberators were produced than any other U.S. warplane in history. Used by the US Army Air Force, the U.S. Navy, Britain's Royal Air Force, the Royal Canadian Air Force and others, this heavy bomber entered its primary wartime service against Japanese forces in the Pacific in January 1942.

## No Empty Bunks Tonight

Sometimes, good fortune smiled on crewmembers on an 8th Air Force bomber. All aboard counted a successful mission and plane and crew, out of the fire and fury of enemy skies, returned to England unscathed. They were exhausted, but flooded with relief to be spared having to dwell on the loss of a crewmember or friend. They could down a pint, a pie, and take some well-deserved rest before risking it all again.

Home, to the B-24 Liberator of the 392nd Bombardment Group, was Station #118 at Wendling in East Anglia, seemingly miles, and years, from training. In the nineteen months the 392nd was engaged against Nazi Germany, their B-24s flew 285 combat missions. In a horrendous loss of life, 184 aircraft and 835 young men of the Group were killed in action or in the line of duty. In February 1944, only six months after going to war, the 392nd was awarded the Distinguished Unit Citation for destroying an aircraft factory in Gotha, Germany; but it was small consolation.

As the fourth oldest B-24 Group in the Mighty 8th Air Force, the wartime record of the 392nd Bomb Group was a proud one indeed. The Consolidated B-24 Liberator had a proud record as well, as a strategic bomber, a transport, a tanker and a maritime patrol plane. It was the most widely produced aircraft of WWII with a total of 18,188 built. In combat, B-24s served well, edging out B-17s in speed, range, and bomb load.

Liberators made great contributions to the Allied victory. The skies are welcoming as these B-24s descend into East Anglia. Beneath the clouds, visibility is excellent and, in the Nissen huts, there would be no empty bunks tonight.

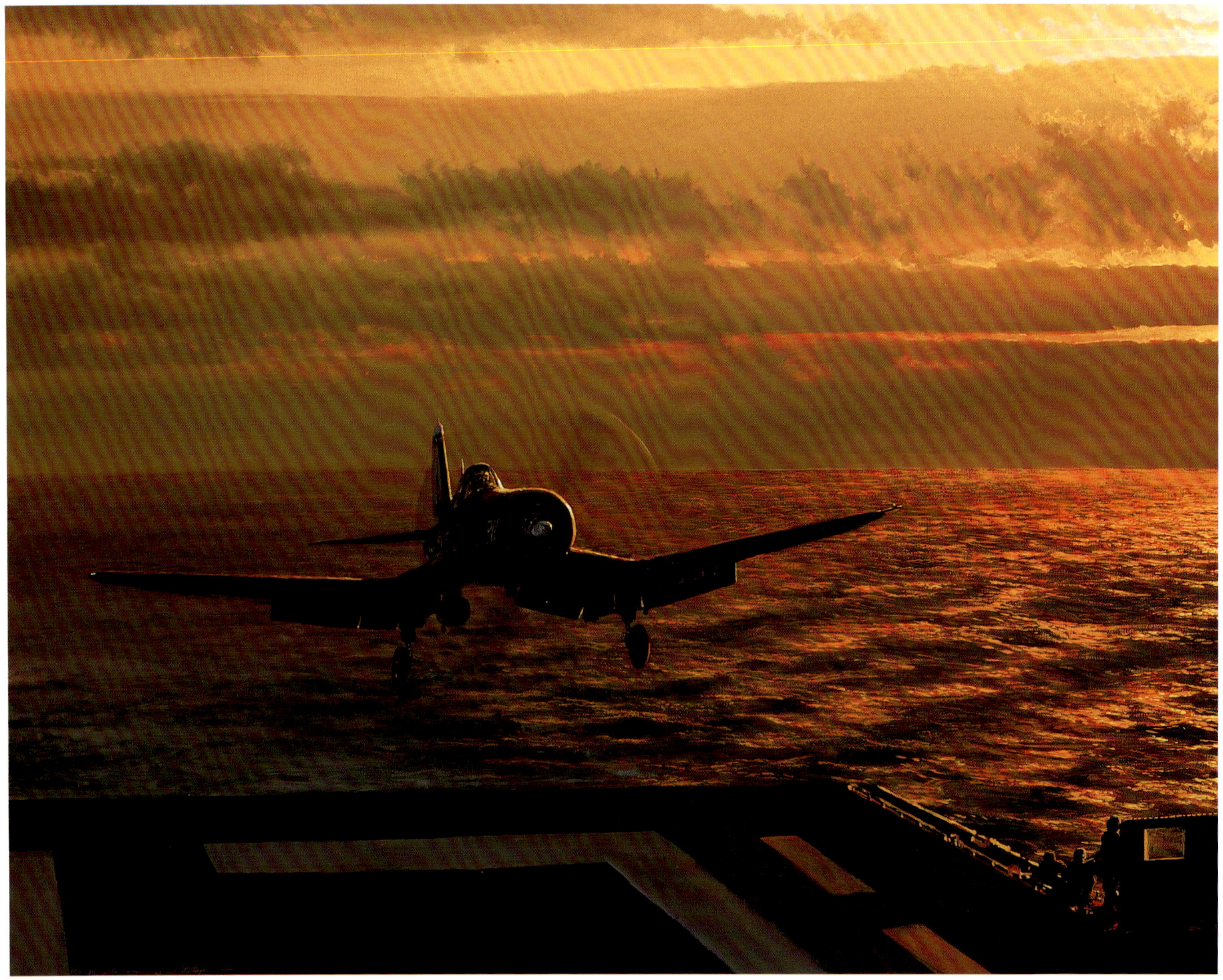

## Corsair Recovery

As the sun sank into the Sea of Japan, a lone Chance-Vought F4U-4 Corsair returns home to the USS *Valley Forge* (CV-45); another combat mission over North Korea is soon to end but first, recovery. His safety depends upon the Landing Signal Officer (LSO), whose task is exacting. LSOs are selected from a field of fully carrier-qualified, experienced, and outstanding pilots. They must appreciate the geometry involved in maneuvering the aircraft.

David McCampbell, WWII's top Navy ace, served as LSO as did two of the Navy's current Blue Angels. Today, LSO training is conducted at the Naval Air Station in Oceana, Virginia, and utilizes sophisticated simulation devices.

## And Now the Trap

In another of the artist's double-entendre painting titles, "trap" represents two views. For the Grumman F6F Hellcats, it meant the arrested landing on the carrier *Hornet*—the mission over, a meal and a rest were at hand. For Japan, the trap resembled a snare.

American and Allied forces were closing in. The Philippines had been re-captured, Iwo Jima had been conquered, and Okinawa lay under major attack. Successes that had spawned an overconfident "victory disease" in the spring of 1942 were being meticulously eroded. Japan's leaders refused to surrender; it was culturally unthinkable to a nation of Samurai warriors.

For now, the F6Fs are coming home. Their mission, perhaps a routine patrol, had been boring, as their drop tanks attest. They would have been jettisoned if the fighters had engaged in combat. But, the pressure of trapping the third wire on a rolling deck of the carrier creates a tension known to the very few. Although second nature to the experienced, details come unbidden—altitude, airspeed, gear, separation, fuel, flaps, directional control. Challenge excites aviators in carrier operations: their lives, their aircraft and their ships depend on accuracy, every trap.

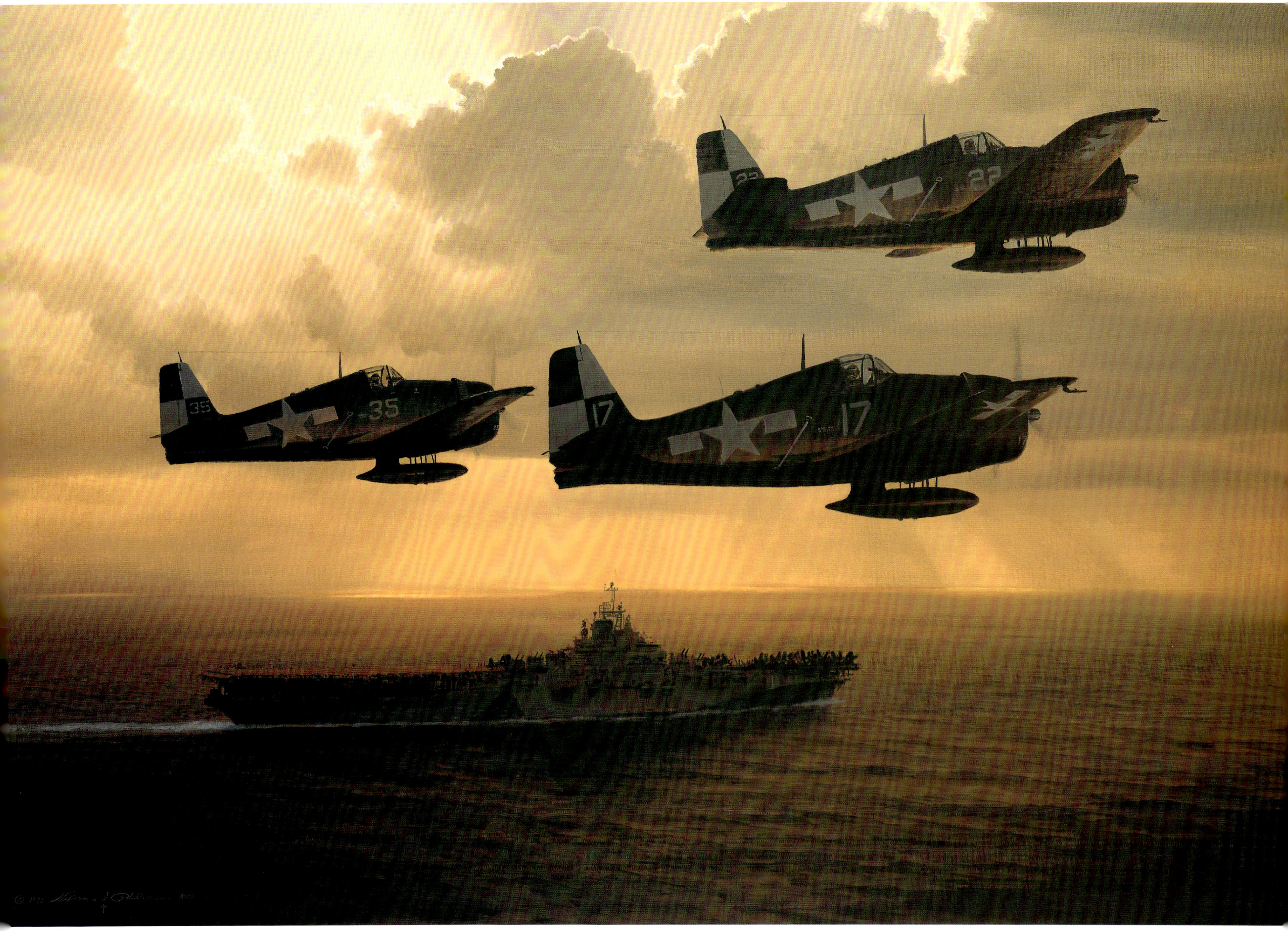

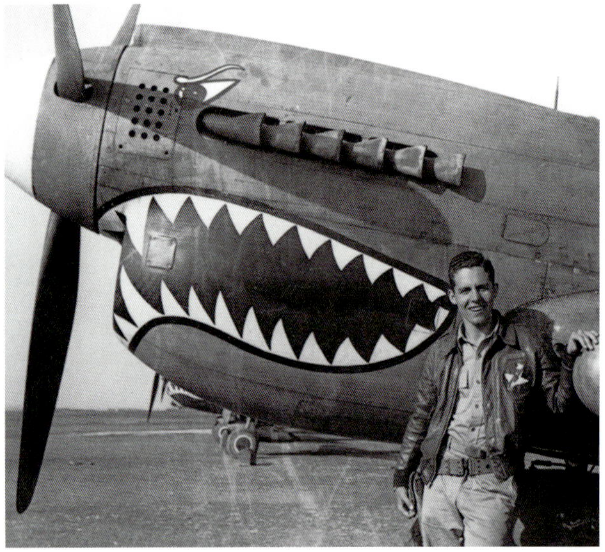

Don Lopez in 14th AF days beside *Lope's Hope.*

## Into the Teeth of the Tiger

Engaged in a game of aerial chicken on his first mission on December 12, 1943, the 75th Fighter Squadron's 1st Lt. Don Lopez, bared his P-40's teeth. In retaliation for a previous bombing of their Hengvang, China airfield, Lopez and eleven others took off to forestall a return assault. When intercoms crackled with warning shouts of incoming Japanese Oscars, he jettisoned his belly tank and dove into the fray. He fired on one, drawing the Oscar pilot straight toward him. Exchanging hits, the two held the head-on attack until the very last moment when the Oscar rolled into a vertical right bank. Both planes rocked in collision. The P-40's left wingtip was sheared, but the Oscar lost a major portion of the left wing and spun into a fatal dive. Ever the storyteller, artist Phillips captures this aerial combat moment in time.

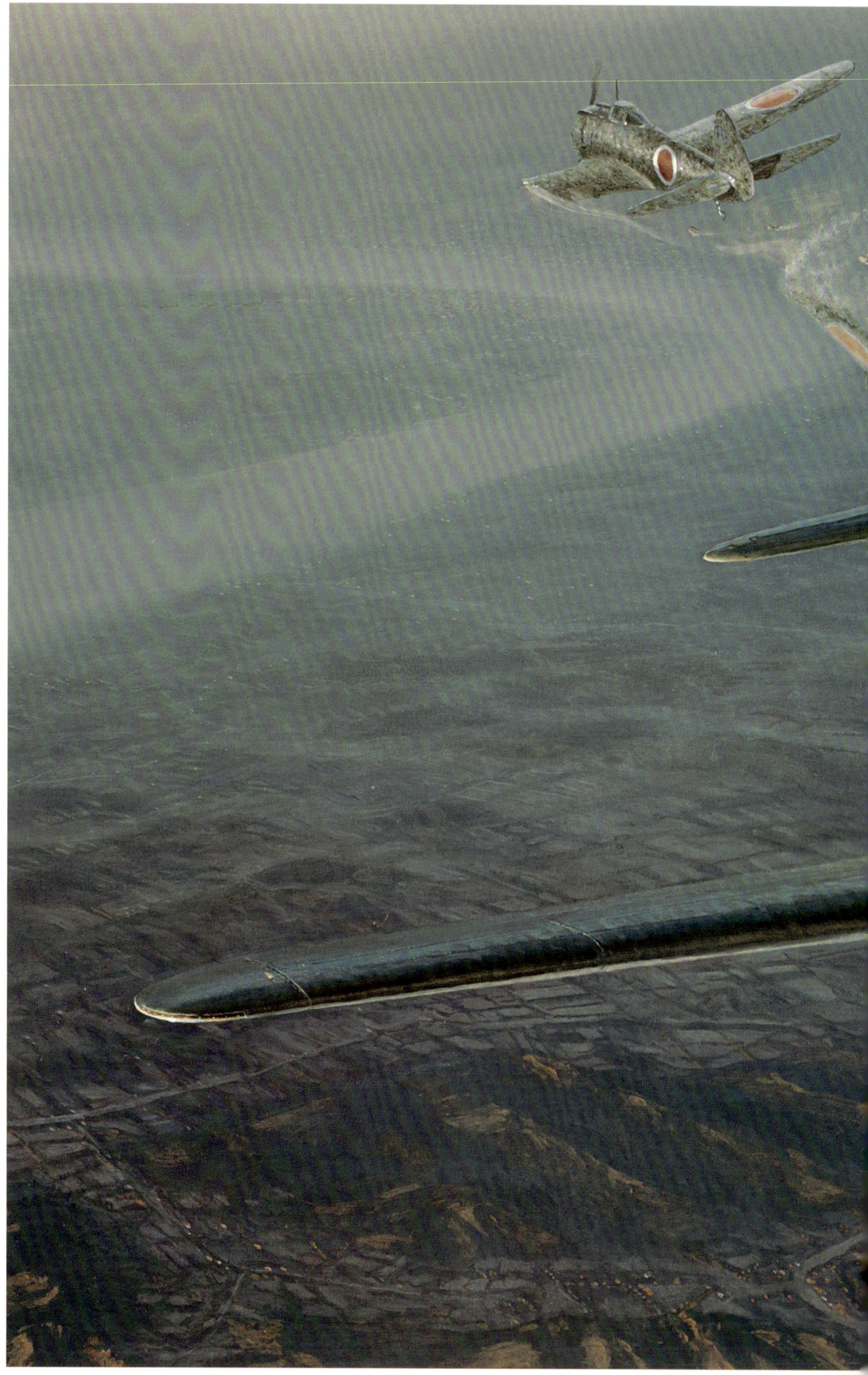

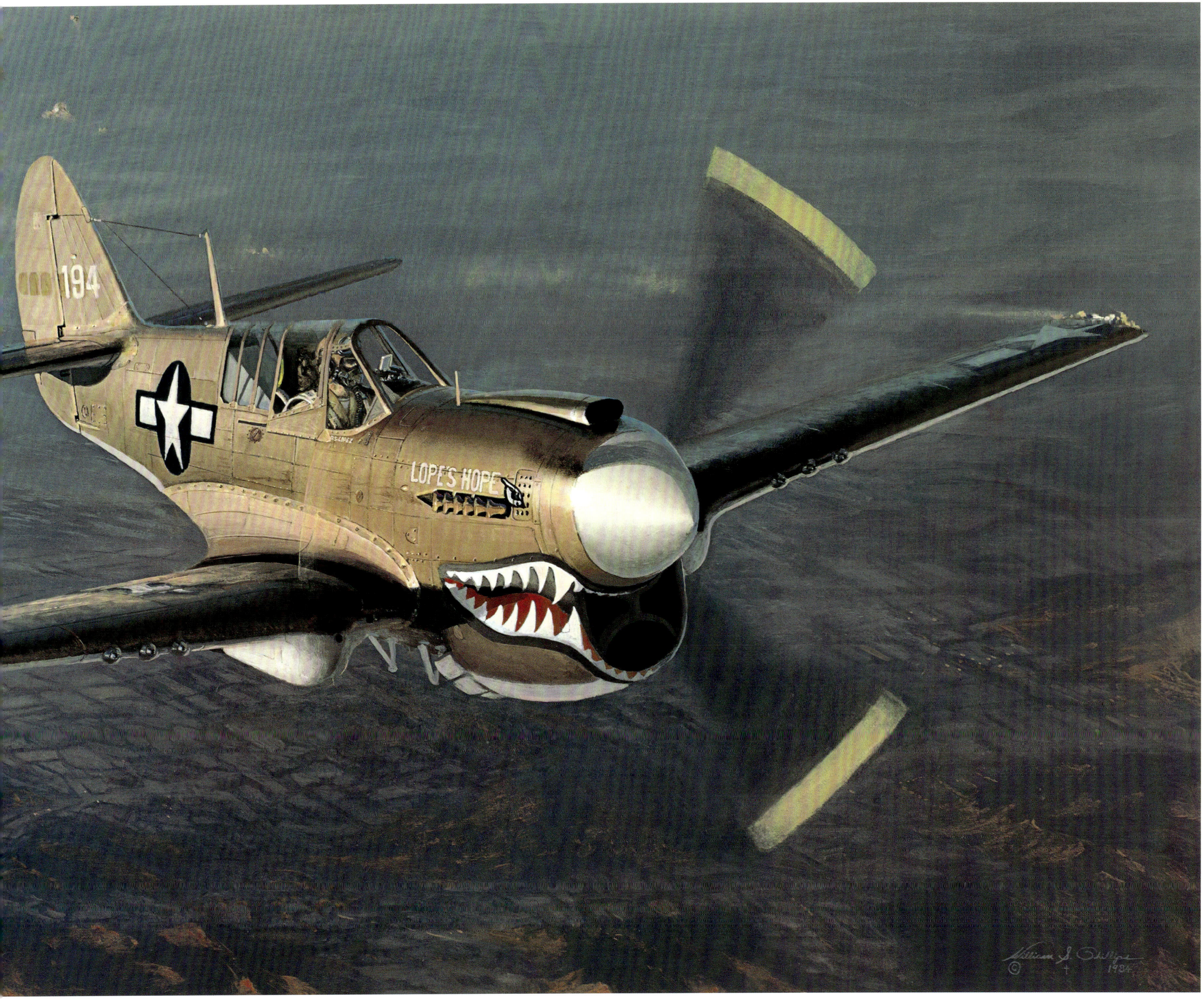
194
LOPE'S HOPE

## The Hunter Becomes the Hunted

Target: Berlin. It was March 6, 1944 and B-17s and B-24s of the Mighty 8th Air Force's 1st, 2nd and 3rd Air Divisions streamed in a ninety-four-mile wave to attack Berlin. Among the massed bombers, the 100th Bomb Group (H) manned 20 B-17s flying at 21,000 feet, 200 miles west of Berlin. There, twenty-one FW-190s bored in with deadly accuracy, downing fifteen Flying Fortresses into the fields below. Half of the crewmembers aboard the fallen bombers perished; the rest were imprisoned.

After the fray, one Focke-Wulf pilot, Oberleutnant Wolfgang Kretschmer, discovered he was alone. With reckless abandon, Kretschmer turned to attack the remaining B-17s and he, in turn, became the target of eight escorting P-47 Thunderbolts of the 56th Fighter Group, the "Wolf Pack" led by Colonel Hub Zemke. Fighters dove and shot him down. Though injured and badly burned, Oberleutnant Kretschmer survived. It was the worst single day of air warfare for the 8th Air Force. Sixty-nine U.S. heavy bombers and eleven escort fighters were lost.

The 100th Bomber Group flew to England in May 1943 and a ground echelon of approximately 1,500 shipped out aboard the converted *Queen Elizabeth*. All were stationed at #139, Thorpe Abbots, England, for twenty-two months between the 100th's first combat mission on June 25, 1943 and its final on April 20, 1945. There, flight crews and ground support personnel performed as vital teams to keep the aircraft flying. On 300 missions, they flew 8,600 sorties. Their losses were not the highest in the Mighty Eighth, yet they became known as "The Bloody Hundredth" with 229 losses in action and in accidents. Nonetheless, the Group had an impressive record of never going "off operations." As Eighth Commander, Lt. General Jimmy Doolittle, once said, "The 100th was always ready."

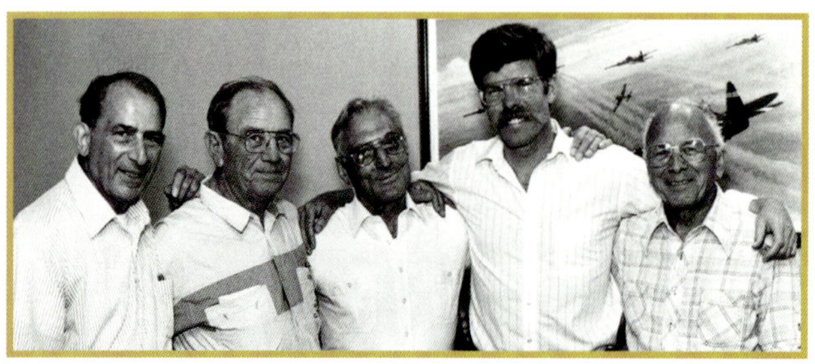

Four survivors of the March 6, 1944 mission are pictured here with the artist: B-17 top turret gunner and German POW, Harold Stearns; 56th Fighter Group Commanding Officer and leader of the Wolf Pack, Hub Zemke; Oberleutnant in the Luftwaffe, Wolfgang Kretschmer; artist Bill Phillips; and 100th Bomb Group pilot, Robert Shoens.

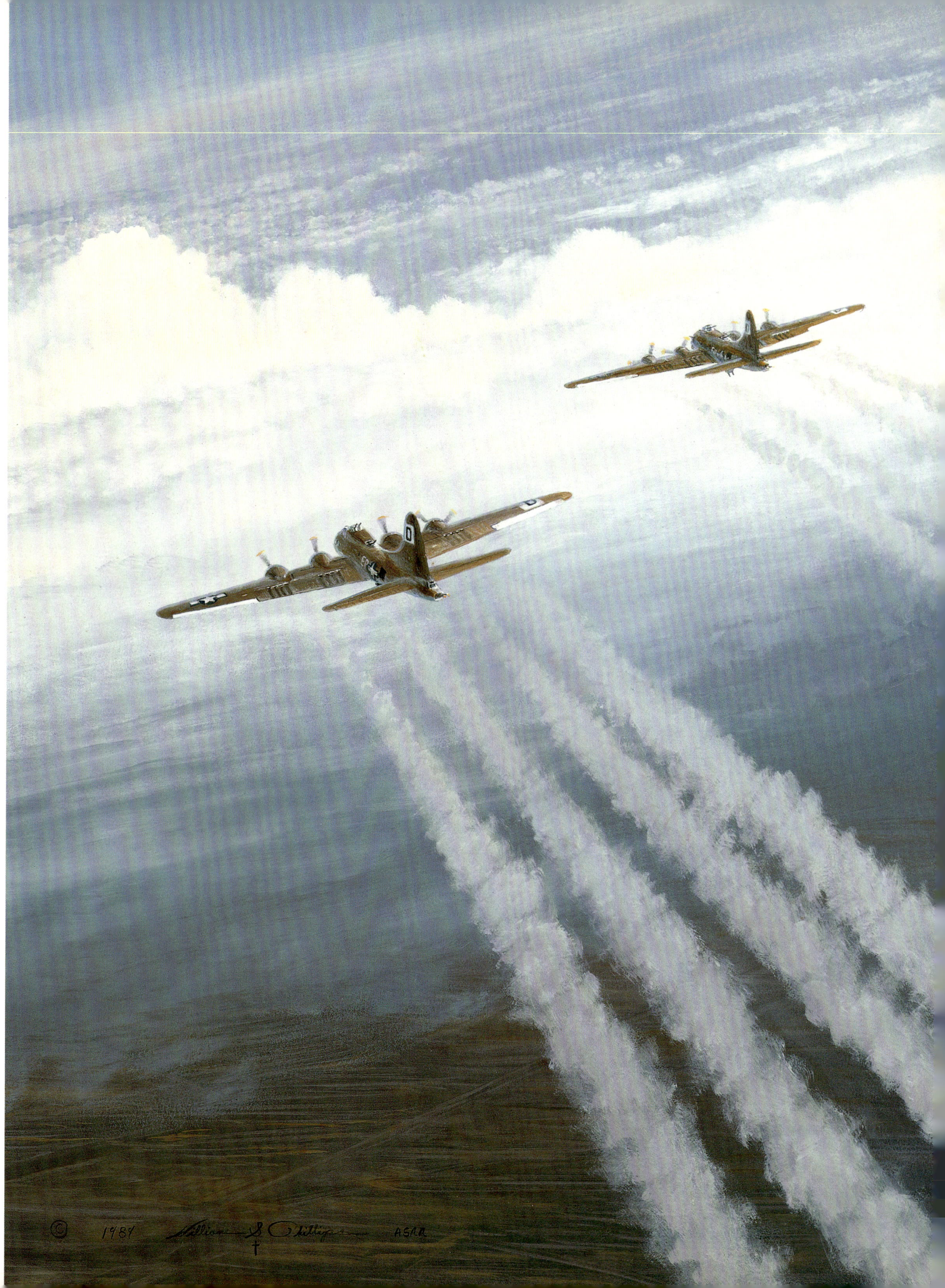

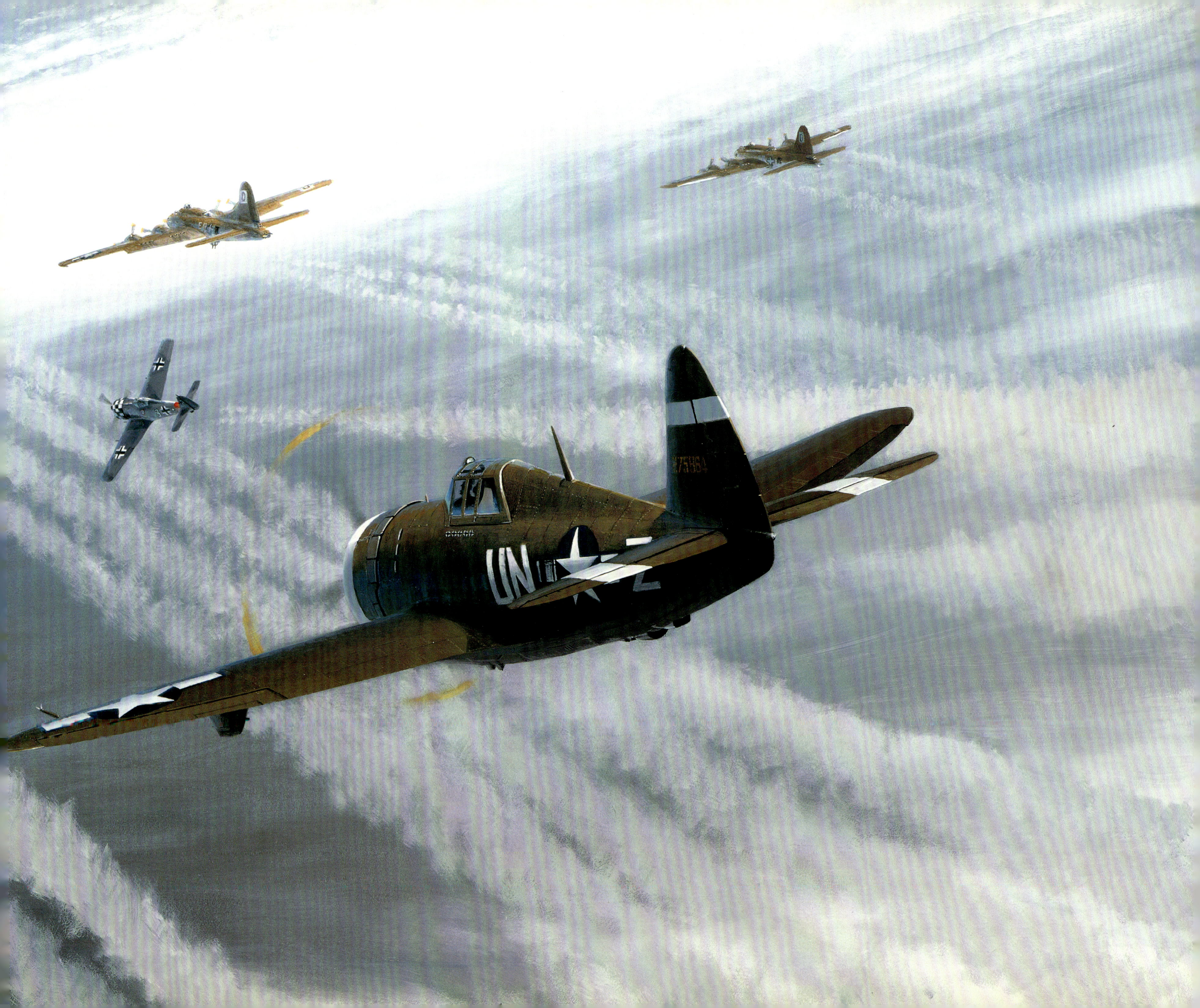

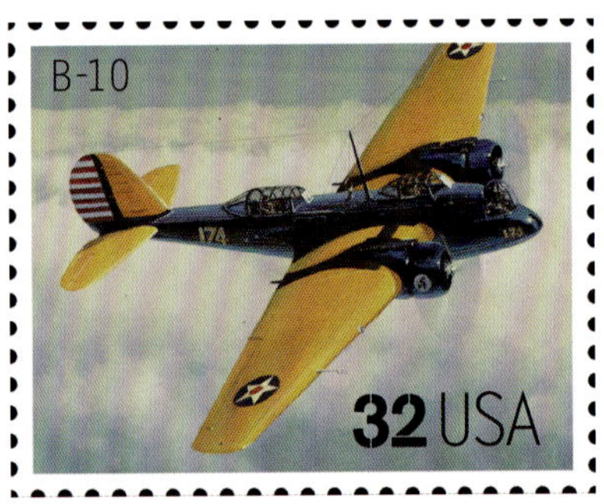


## Martin B-10

U.S. Postal Service, Classic American Aircraft series, 1997

Martin's B-10, America's first monoplane bomber, was built for, and flown by, the U.S. Army Accessions Command in the mid-1930s. Twin-engine B-10s provided an internal bomb bay for one ton of weapons, an enclosed crew flight deck and an auto-pilot. The B-10 heralded four-engine B-17s and B-24s, the strategic mainstays of World War II.

## Fifty Miles Out

*Dina Might*, a Boeing B-29 Superfortress, is headed toward Japan accompanied by two additional B-29s. Her crew, from the 421st Bomb Squadron, 504th Bomb Group, 313th Bombardment Wing (VH), was "going to hell," as her mine-dropping missions were described. Departing from the U.S. Navy Seabee-built base at North Field on Tinian Island in the Marianas, the bombers faced a 1,200 mile flight, each carrying 12,000 pounds of half- and one-ton mines.

Targeting the narrow Shimonoseki Strait between Honshu and Kyushu; the Inland Sea; as well as the harbors of Hiroshima, Kure, Tokyo, Nagoya, Tokuyama, Aki and Noda; mines were to be strategically and accurately placed in western approaches. Mine-laying missions were dangerous because the mines were dropped from a low altitude of 5,000 feet, and because the Japanese Navy searchlights and ship-based anti-aircraft weapons were more accurate than their land-based weapons. For a bomber to be caught in the Imperial Japanese Navy (IJN) searchlights during a mine run, it had to take everything thrown up at it. Yet, nearly 13,000 acoustic and magnetic mines were placed during March and August 1945, bringing Japanese coastal shipping to a standstill.

In May, Japanese merchant vessels were ordered to break through and eighty-five were sunk. Mining efforts were so successful that a postwar Strategic Bombing Survey gave the B-29s credit for 9.3% of total Japanese shipping losses during World War II. Coupled with effective incendiary raids and low-altitude missions against targets like oil-production facilities, mines were devastatingly effective.

On Tinian, sailors of the invaluable Construction Battalions—the Seabees—developed a special comradeship with B-29 aircrews. Many Seabees adopted a B-29 by painting their logo on the aircraft nose. B-29 crews carried the logo proudly as they personally represented Seabees in the bombing raids. The crew of the adopted plane was grateful to the Seabees for improving the quality of life through providing better quarters, washing machines, mattresses, ice cream, and cold beer. For crews like that of *Dina Might*, it was a win-win situation.

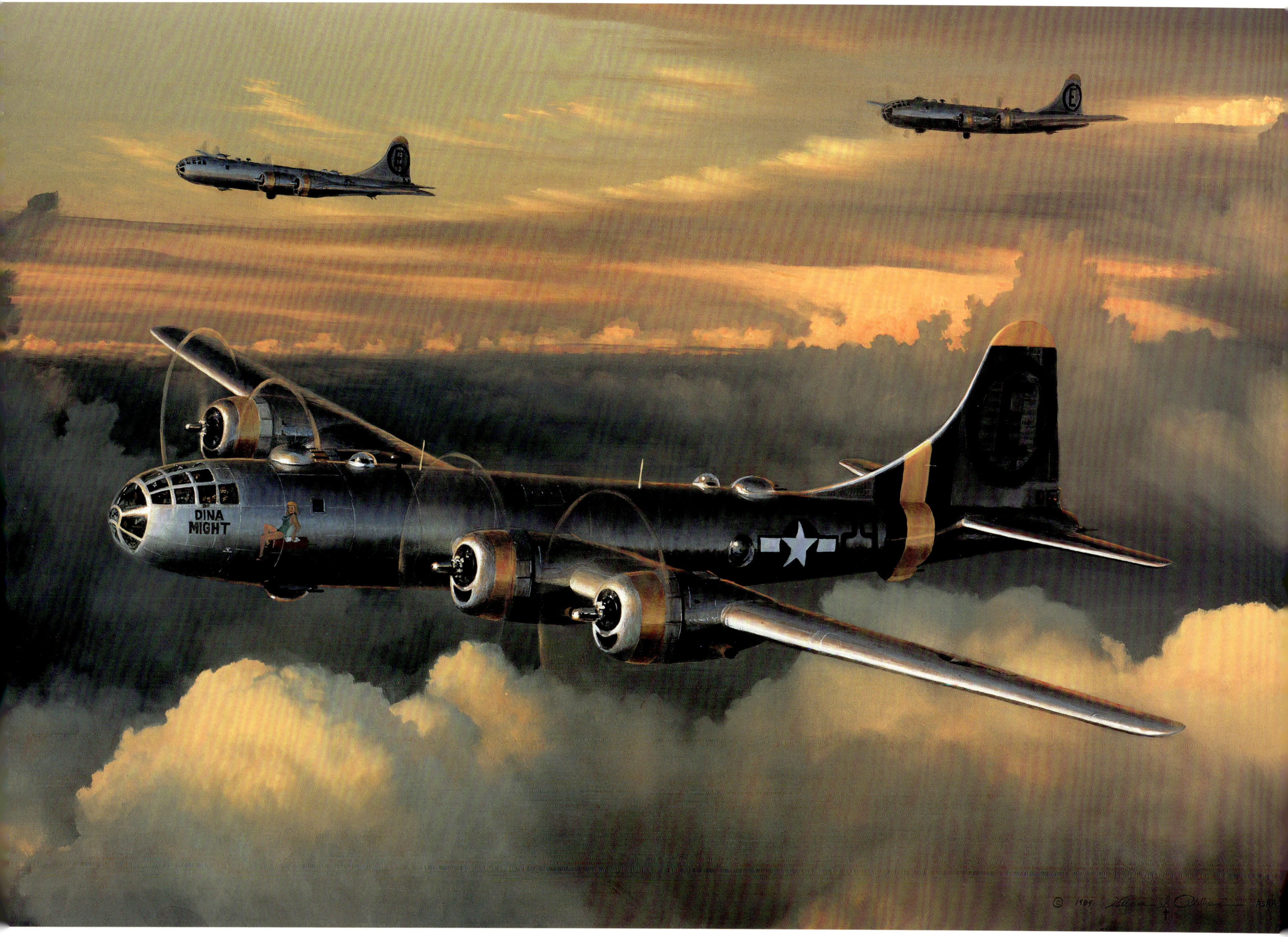
DINA
MIGHT

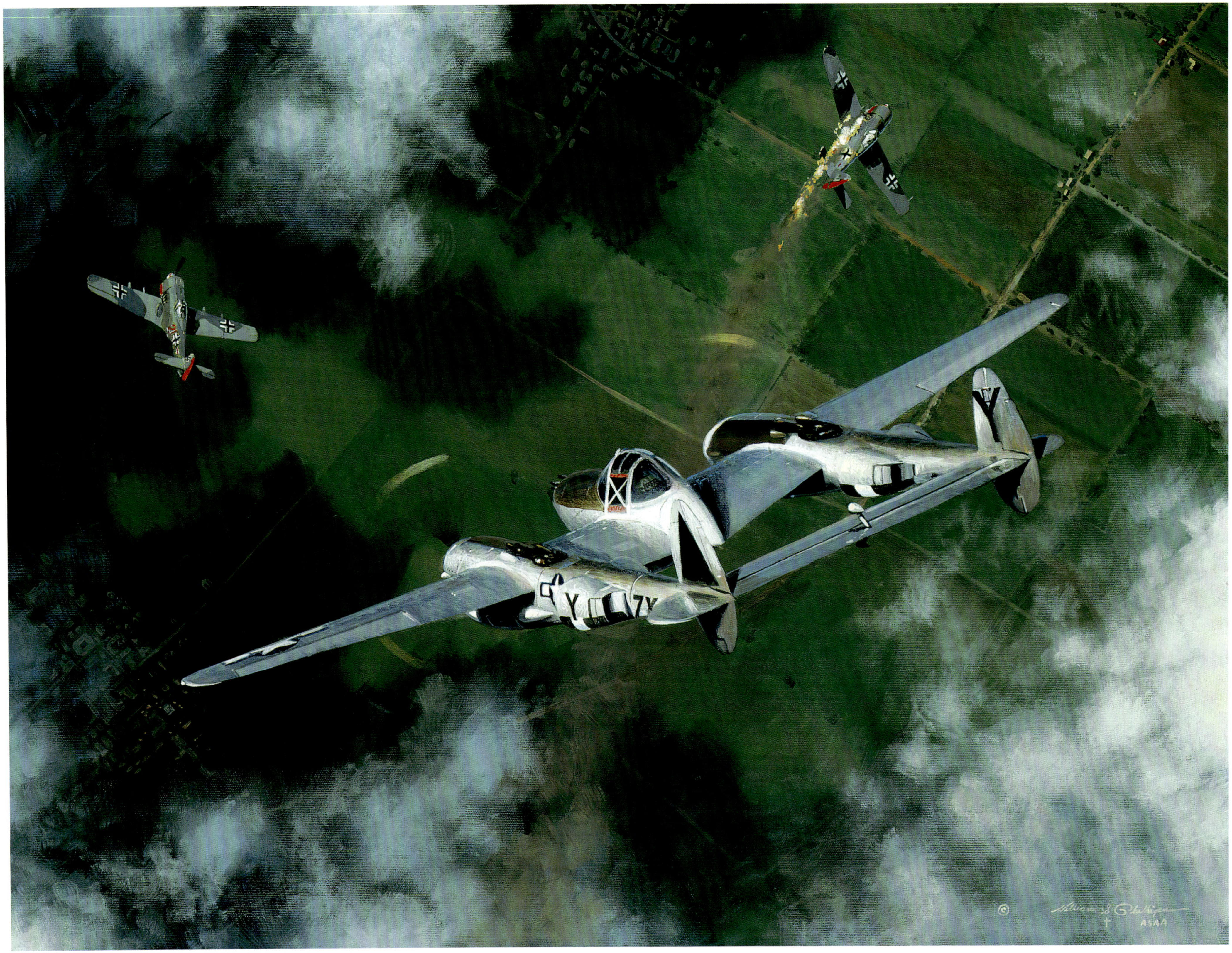

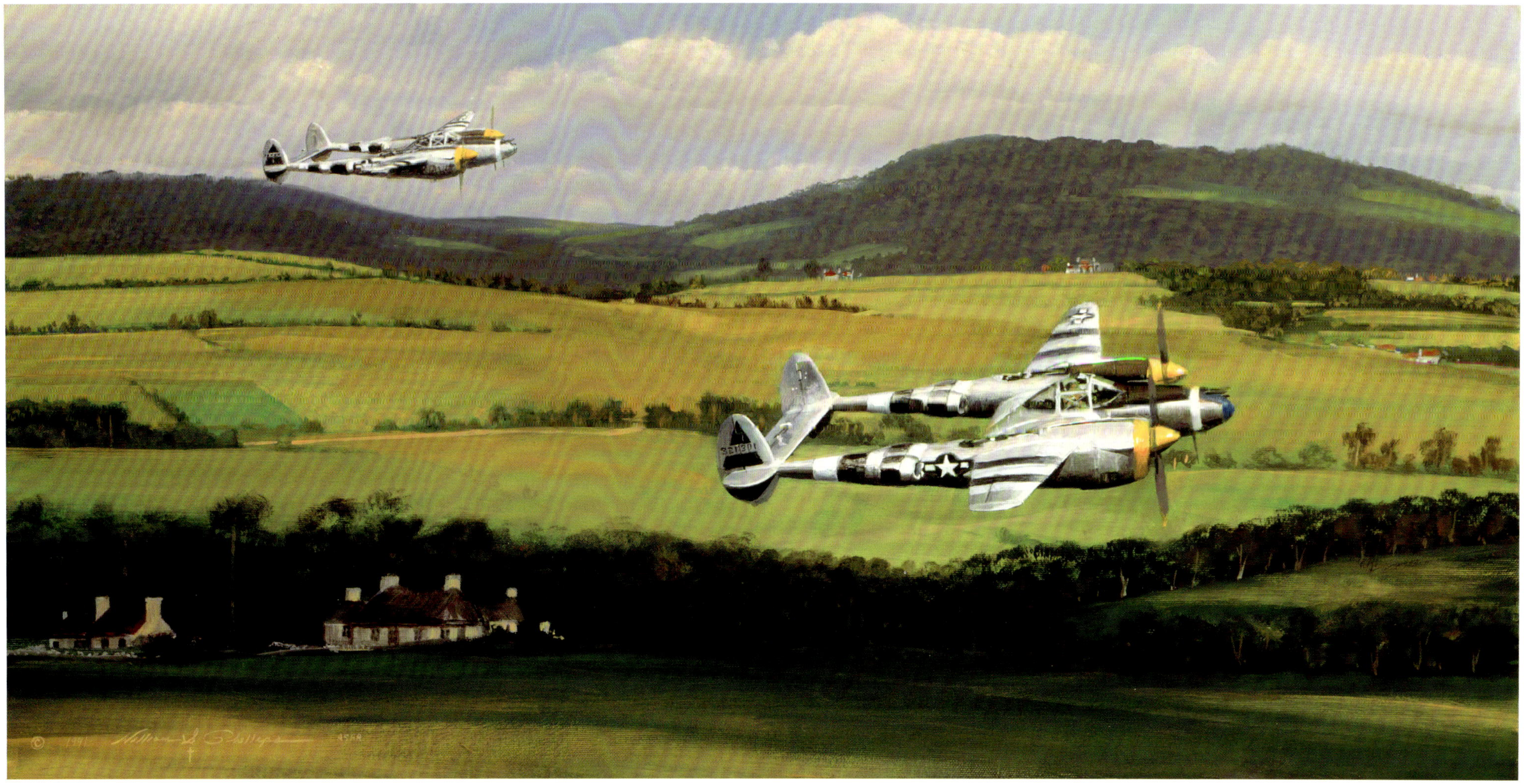

## Splitting a Pair

One down and a Ninth Air Force P-38 pilot, 429th FS, 474th FG, moves toward claiming a second Luftwaffe FW-190 in a single outing. The Ninth Air Force moved from North Africa to England to build a Tactical Air Armada for the invasion of Europe. By D-Day, it produced the largest air force ever assembled under one command: 250,000 personnel, with 3,500 aircraft in 1,500 units. Soon after the invasion, Ninth engineers built airstrips in Normandy for aircraft in support of ground troops. Between March 1944, and April 1945, the 429th FS moved from England to bases in France, Belgium and Germany. The Ninth was on the move toward defeating Fascism.

## Pass Down the Valley

Easily identifiable as part of the D-Day invasion on June 6, 1944, two P-38s wear the bold invasion stripes painted on all participating Allied aircraft. Assigned to the 55th Fighter Squadron, 20th Fighter Group, they make a pass through a pastoral French valley. They contributed to the 312 missions logged and the 212 aerial victories achieved. Based at Kingscliffe in England, the 55th, whose lineage traces to the First World War, began flying the Lockheed P-38 Lightning in the European Theater of Operations on December 28, 1943 and continued in the speedy fighter until July 21, 1944 when they moved into the North American P-51 Mustang.

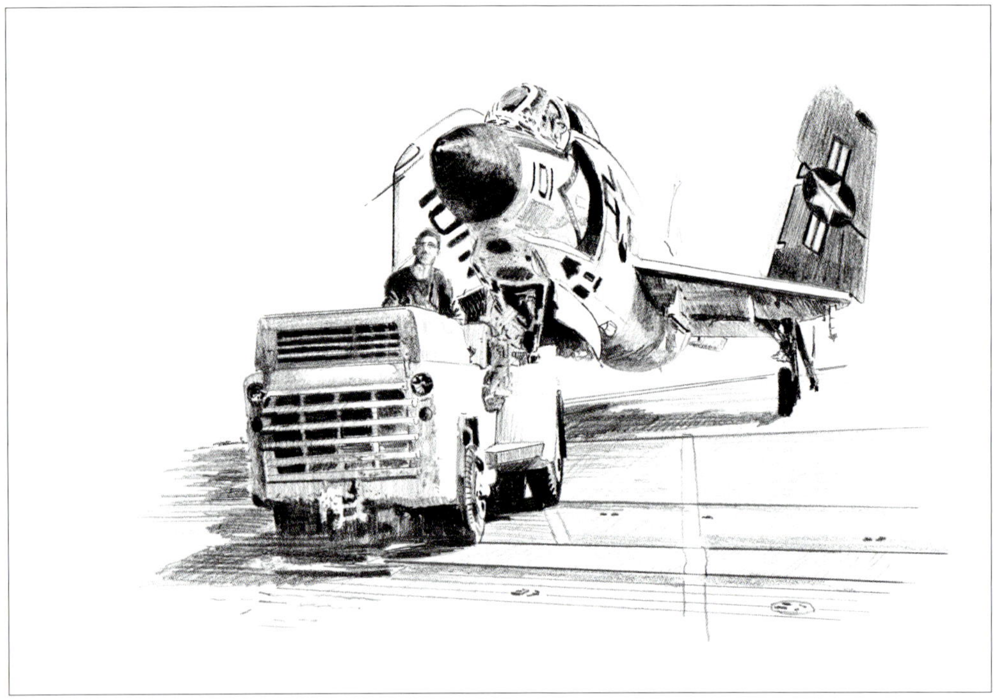


*Towing the Demon*
A Navy deck crewman and tractor aboard ship respot an F3H Demon in the 1950s.

## In the Beginning

Arrival at Naval Air Station Pensacola precedes six weeks of preflight training covering engineering, aerodynamics, navigation, flight regulations, aviation physiology, and water survival. At its conclusion, Navy and Marine pilot trainees take a breath before launching into primary flight. Also included in primary training are those from Air Force, Coast Guard and foreign countries. A syllabus includes sixty-six flight hours in the Raytheon (Beechcraft) T-34C, Navy's primary trainer, a turboprop development of the T-34A, (the civilian Bonanza) and the T-34B, Navy's Primary Trainer until 1978. The "C," with max cruise at 250 mph and service ceiling above 30,000 feet, makes for a great beginning.

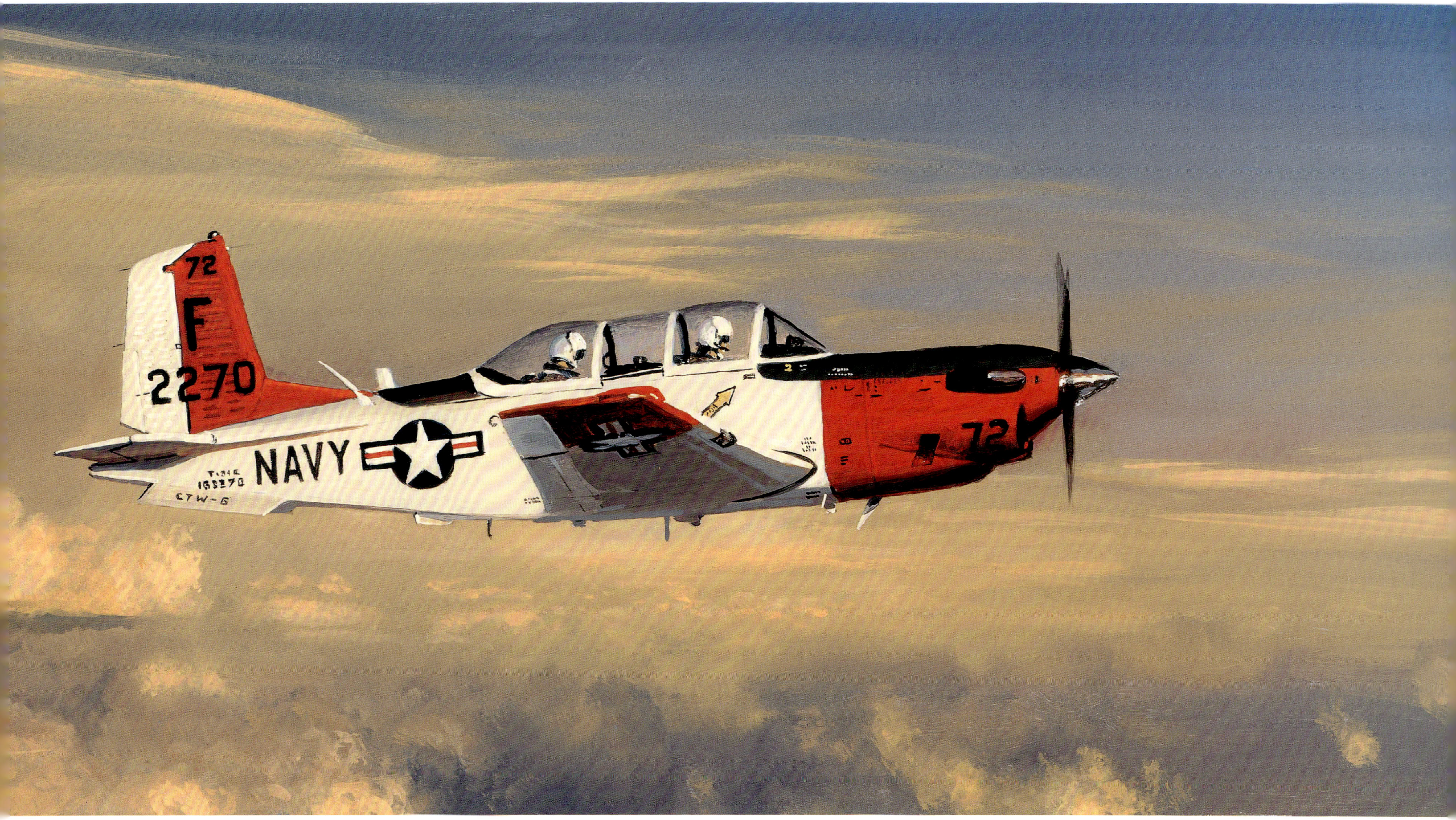
72
F
2270
NAVY
72
CTW-6

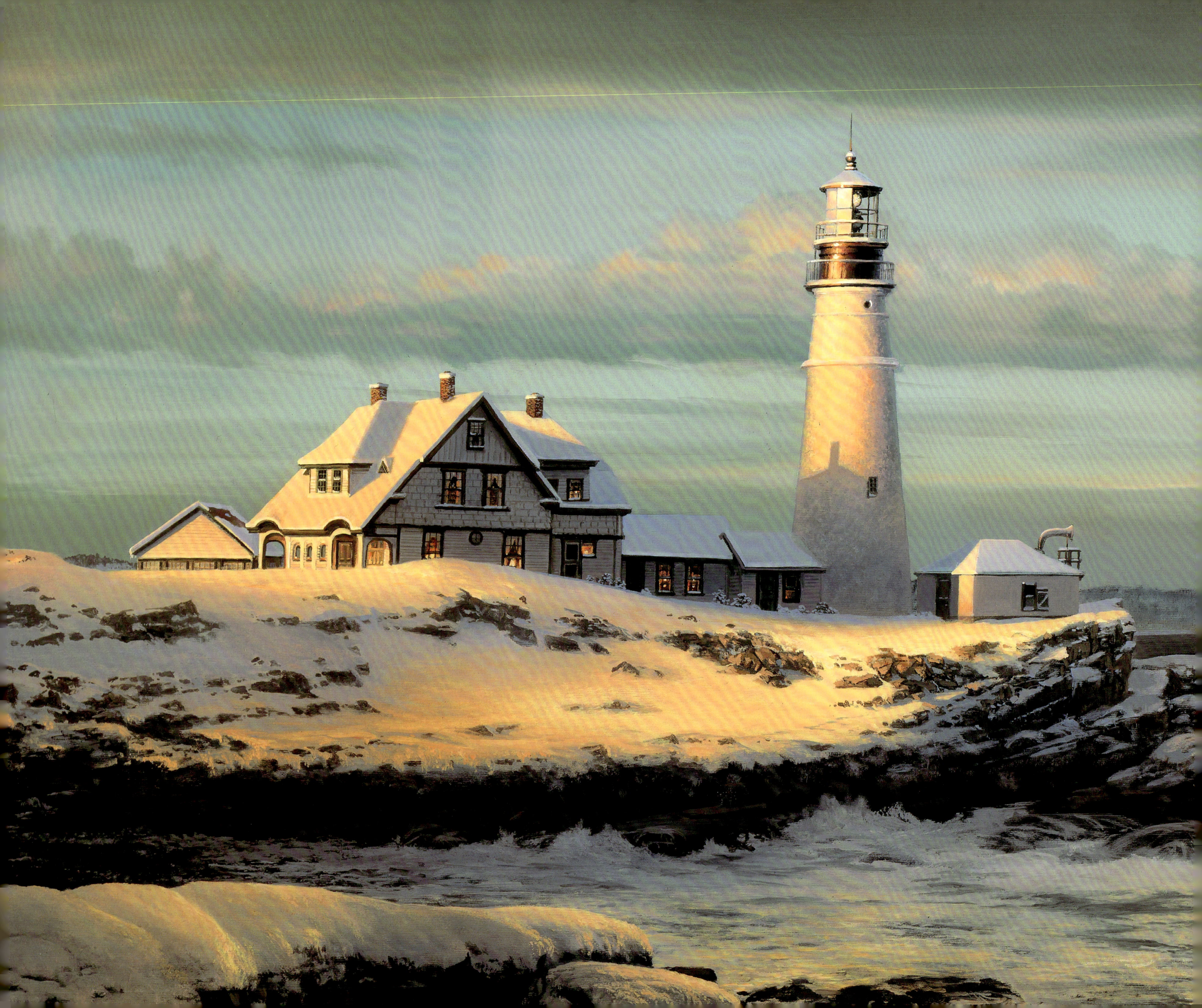

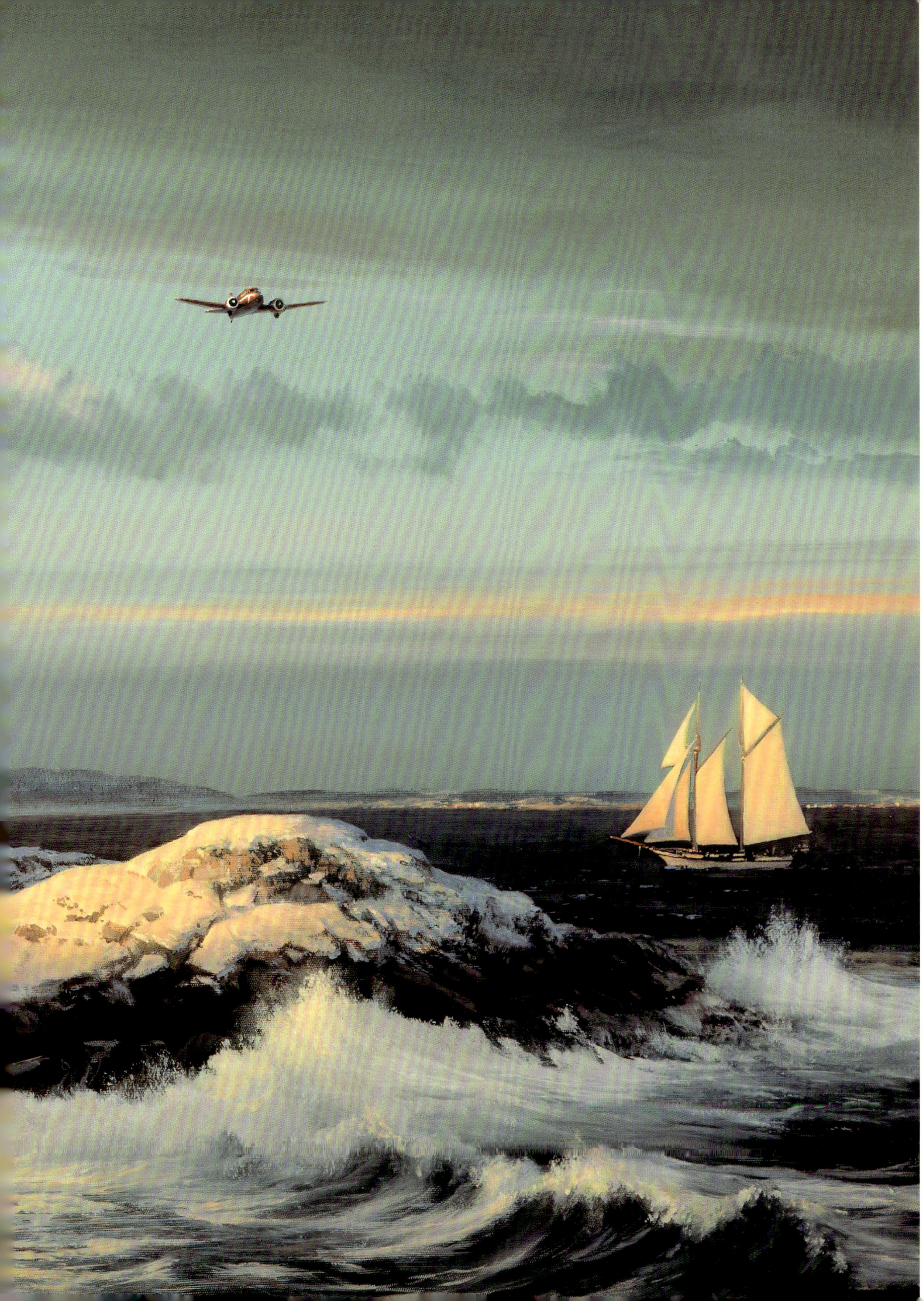

## The Lightkeeper's Gift

This painting of picturesque Portland Head Light near Cape Elizabeth, Maine pays homage to a tender story, a true tale that has risen to the realm of myth. The story begins on a wintry, cold evening, one week before Christmas in 1929. A pilot, Captain Bill Wincapaw, had been delivering supplies and mail throughout Maine's Penobscot Bay and, as he finished his rounds and planned to head home, he was overtaken by a strong, sudden squall and enveloped in blinding whiteout conditions.

When able to maintain ground reference, Wincapaw knew the coves, islands, inlets, and small bays of the uneven coast of Penobscot Bay. Hopelessly disoriented in foul weather, and with his plane running low on fuel, Wincapaw stared into the mist and suddenly glimpsed a light. Dropping toward it as a moth is drawn to a flame, he was relieved and thankful to recognize the Dyce Head Lighthouse of Castine, Maine. He oriented himself and turned his airplane to a heading for home. Following the lights gleaming from six other lighthouses, he credited each of the lighthouse keepers for saving his life.

As Christmas Eve neared, he carefully wrapped gifts, and retraced his flight of the week before from lighthouse to lighthouse, dropping a gift for each. Named the Flying Santa, Wincapaw continued his tradition and, as years passed, a local historian, Edward Snow, carried it on. For forty-three years, Snow played the role of Flying Santa, expanding his territory and venturing to lighthouses as distant as Michigan and California. The Friends of the Flying Santa, a non-profit organization, now heads up the Christmas tradition and counts twenty-nine lighthouses as beneficiaries of its gifts.

In Northport, Michigan, the U.S. Coast Guard adopted the role, using the story to augment flight training while seeing that children benefit from the aerial delivery of surprise gifts. Artist Bill Phillips painted the Cessna UC-78 Bobcat, a twin-engine aircraft made popular by the radio program, *Sky King*, to dramatize the Flying Santa, and to capture a moment when gifts are given and received.

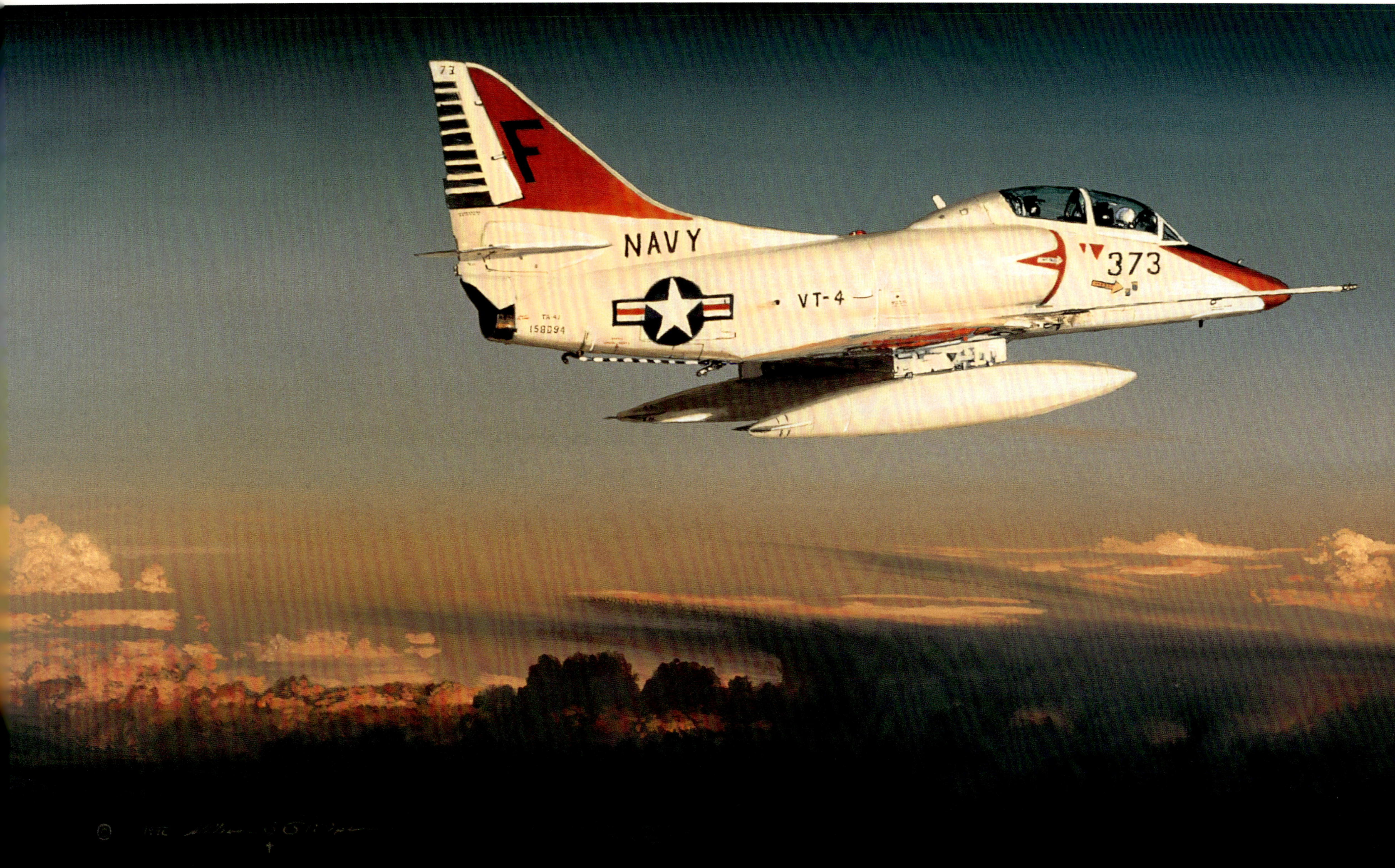
NAVY
VT-4
373
158094

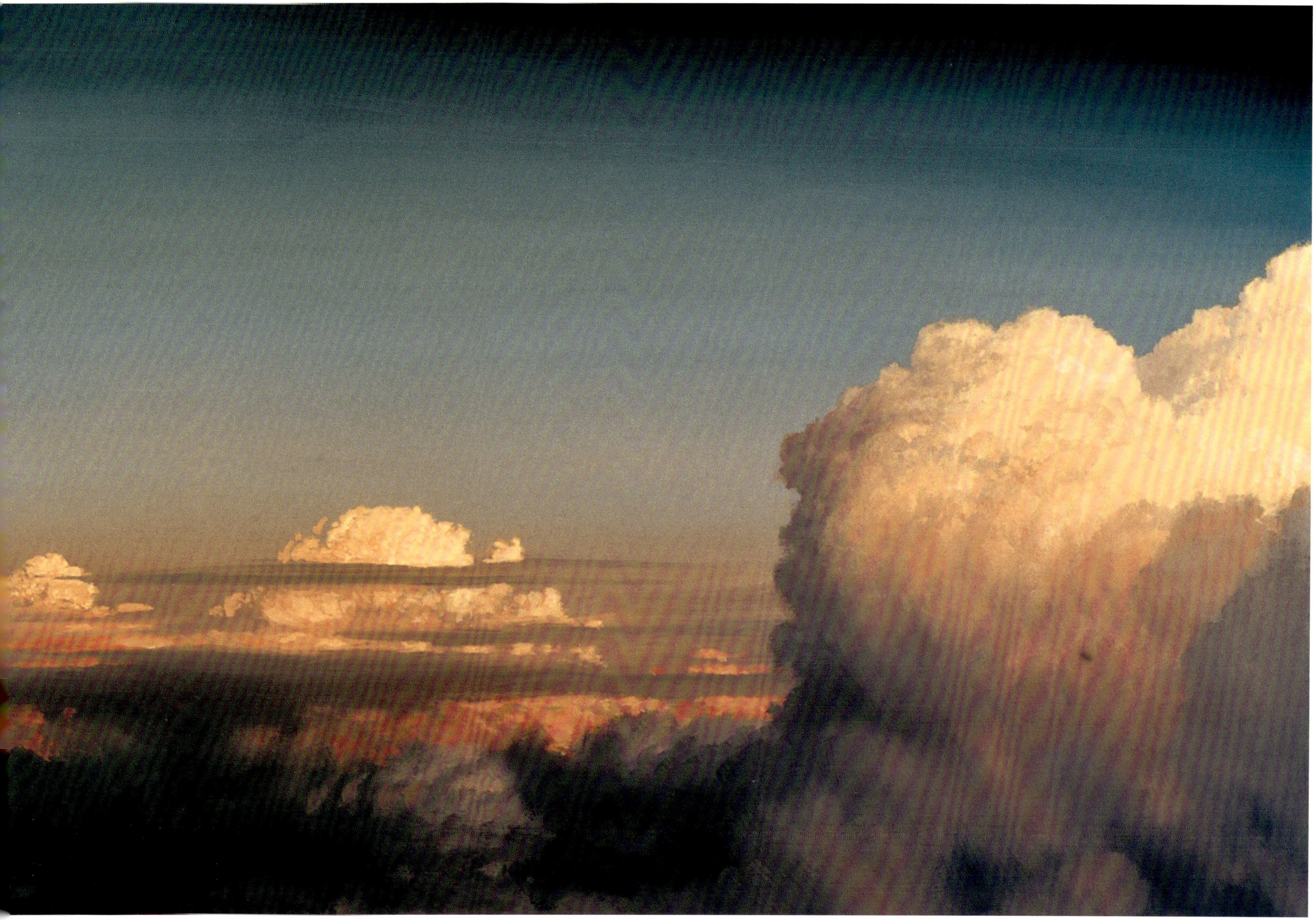

OVERLEAF:

## Solo Cross Country

Alone, perhaps for the first time in this advanced trainer, the pilot has plenty to occupy his thoughts, especially with the darkening clouds building ahead. Is there a clear path between the clouds? How's the fuel supply? Should a request to divert, or for a higher altitude, be made now? Certainly the cumulus clouds will bloom into mature thunderstorms as they do virtually every summer day over the Gulf of Mexico.

Conditions are perfect for a storm—vertical updrafts, plentiful water vapor, and warm temperatures cooling as the vapor rises. This jet, the McDonnell Douglas (Boeing) TA-4J Skyhawk, is strong, but the penetration of mature thunderstorms should be avoided. A storm can reach 30,000 feet and its dark interiors could conceal turbulence, wind shear, pelting rain, hail, icing, and violent up- and down-drafts that could develop into microbursts. Having made it this far in the U.S. Navy's demanding flight training program, the pilot doesn't want to foul up on this big test.

The Skyhawk is a two-seat trainer version of an airplane with a proud history in Navy and Marine Corps service. It served in combat in Vietnam, the Blue Angels for a dozen years, was the aggressor aircraft in air-to-air combat training, and has been an advanced trainer for almost thirty years, providing a safe but challenging platform for students close to pinning on their gold wings.

Students can't dwell on history since more immediate demands take their full attention: managing the fuel, navigating over water, transmitting essential radio calls, swiveling to ensure the skies are clear of other aircraft, considering alternate airfields, weighing the possibility of a carrier landing, maintaining altitude, airspeed and control, and monitoring the storm. Fully aware that the other seat is empty, the pilot is busy and alone—truly solo.

## Last Sortie of the Day

Naval Flight Training, dating from 1914, has participated fully in the dramatic technological change of the first century of powered flight. More than 100,000 Naval aviators have been awarded the coveted "Wings of Gold" and have served their country. Candidates for Naval Flight Training are college graduates. Each is launched from one of five bases—two in Texas, two in Florida and one in Mississippi—sites blessed with generally favorable flying weather and opportunities to get airborne over water. In Primary Training, new students begin with the T-34 Mentor or the T-6A Texan II.

Following the successful completion of Primary, trainees are given five "pipeline" choices. The pipeline selected—jet, E2/C2, Maritime Prop, Helicopter, E6 Tacamo (command and control aircraft)—determines where Intermediate and Advanced training will be launched. After eighteen months, wings are bestowed upon successful graduates.

Naval Air Training Command also graduates Naval Flight Officers (NFOs) who go on to perform multiple roles in strike aircraft like the F/A-18 Super Hornet, in electronic countermeasures, navigation or intercept operations.

In the painting, two U.S. Navy advanced flight training pilots turn their T-2 Buckeyes for home base and a well-earned "cool one," against a brilliant sunset over the Gulf of Mexico. Built by North American Rockwell, the Buckeye began service with the Naval Air Training Command in the early 1960s. Over 11,000 student pilots have been introduced to high-altitude, high-speed flight in one of its five variants and many have made their first traps in this reliable trainer.

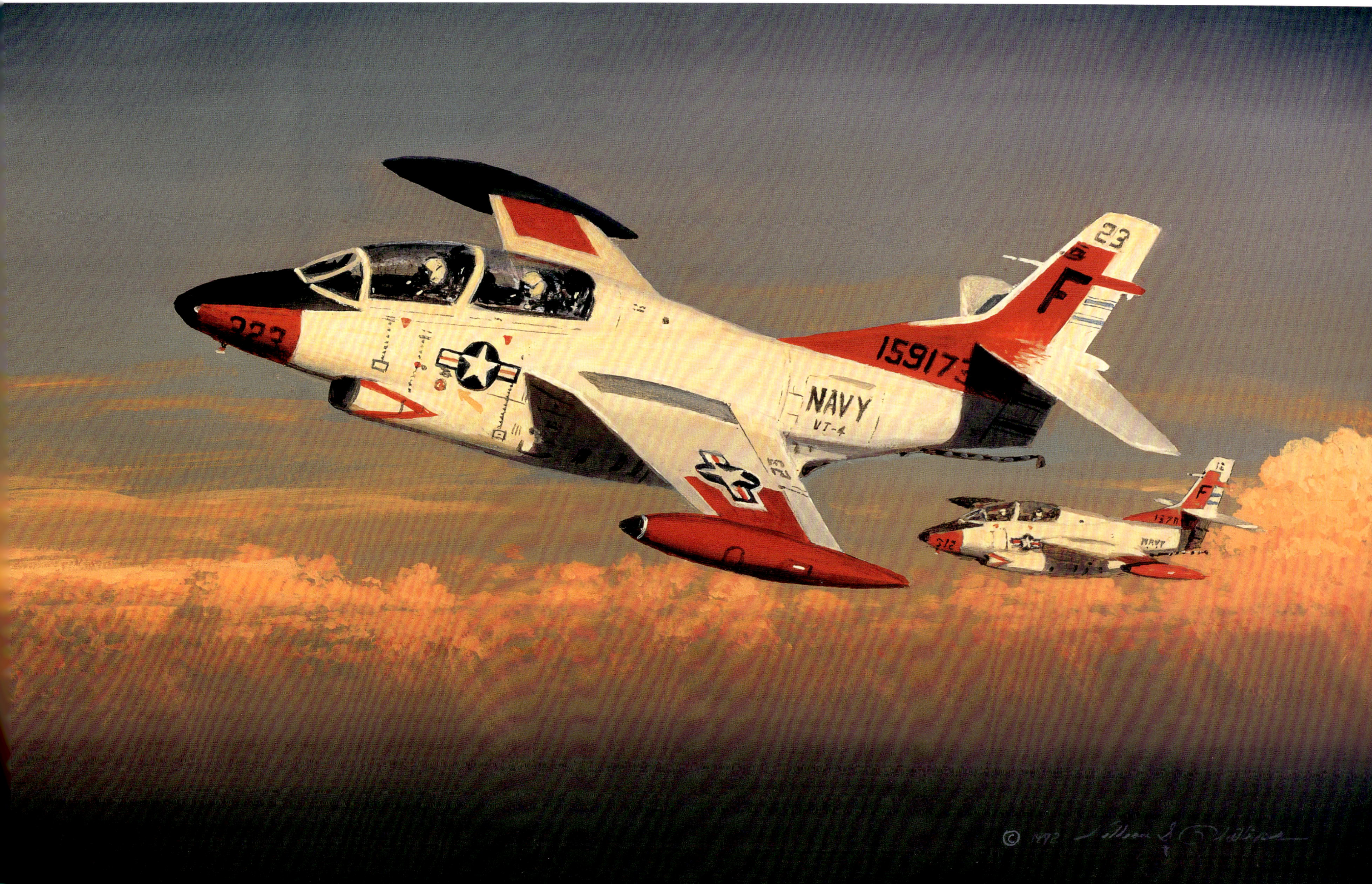
23
F
159173
NAVY
VT-4

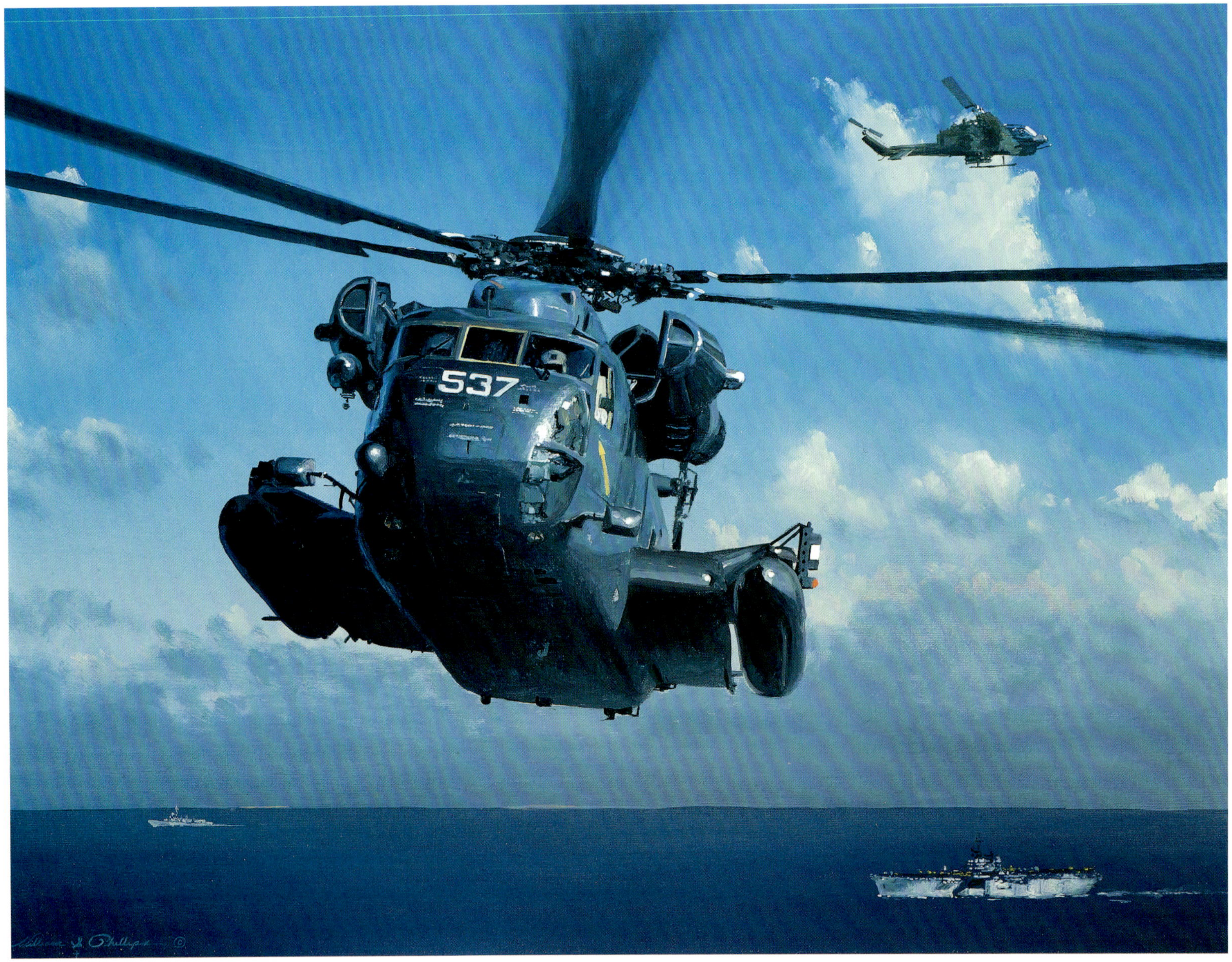


## Heading For the Mine Danger Zone

Prior to the Iraqi invasion of Kuwait, there was protracted war between Iran and Iraq. Waters of the Gulf, the Straits of Hormuz and the Gulf of Oman became the site of what the artist refers to as "Gulf War One." The U.S. Navy entered the fray to safeguard international oil tankers and part of their operation was detection and destruction of mines at sea.

The vehicle that performed minesweeping missions was the MH-53E Sea Dragon, a modification of Sikorsky's CH-53E developed for heavy vertical lift. In 1986, the Sea Dragon became the Navy's primary Airborne Mine Counter-Measures (AMCM) helicopter. With three GE turboshaft engines and a seventh main rotor blade, the MH-53 performed aerial refueling, search and rescue, and cargo transport operations. Its offensive punch in the mine danger zone, however, came from its ability to tow a mind-sweeping sled, a side-scan sonar, and a mechanical minesweeping system.

# Pedal to the Metal Over Smith Ranch

The X-15 research program was aviation's most significant research effort following the development of the first powered flight by the Wrights. Knowledge gained from each of the X-15's 199 flights, combined with technology transfer, contributed to every aspect of the U.S. Space Program. North American Aviation was selected to produce a rocket-powered aircraft to conduct scientific research, and the first of its three X-15s rolled out on October 15, 1958. Ten years later, the last of 199 flights landed at Edwards Air Force Base. In that decade, records for altitude, 67 miles; and speed, Mach 6.7, were set. But more importantly, data gathered on hypersonic aerodynamics offered insights into intense heating, flight outside the atmosphere, re-entry, stability, control at extremes of speed and altitude, and the human factors involved. Artist Bill Phillips depicts the release of an X-15 from its NB-52 over Smith Ranch, Nevada, with the F-104 flying chase.

OVERLEAF:

# The Long Green Line

In November, 1965, U.S. Army's Bell UH-1 Iroquois, "Huey" helicopters, delivered troops and supplies in the central highlands of South Vietnam as reinforcement for the battle in the Ia Drang Valley. The battle—the first between U.S. soldiers and North Vietnamese regulars—lasted three days. Troopers were landed safely from the helicopters, but were immediately surrounded by approximately 3,000 enemy soldiers, arriving in waves. Strong leadership, air support, and artillery fire prevented the badly outnumbered Americans from being overwhelmed. When the battle was over, 2,000 North Vietnamese and 79 Americans were dead.

With the distinctive sound of its Pratt and Whitney engine-powered rotor, the "Huey" is familiar to all who served in Vietnam, including the artist. More than 5,000 were introduced during the Vietnam conflict; they performed gallantly in the air assault mission depicted in the painting, in medevac, command and control, transport roles, and as gun ships taking the fight to the enemy.

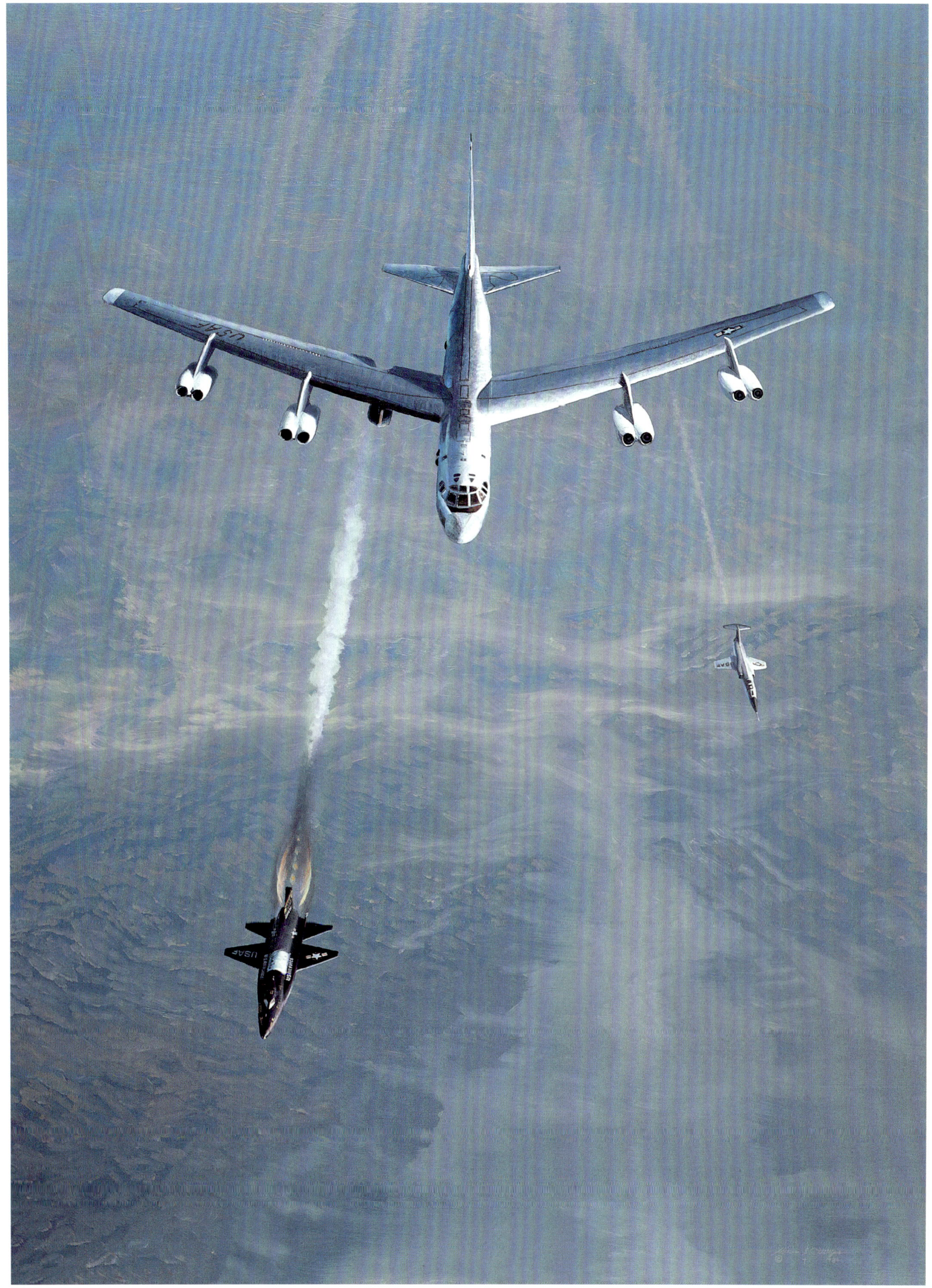

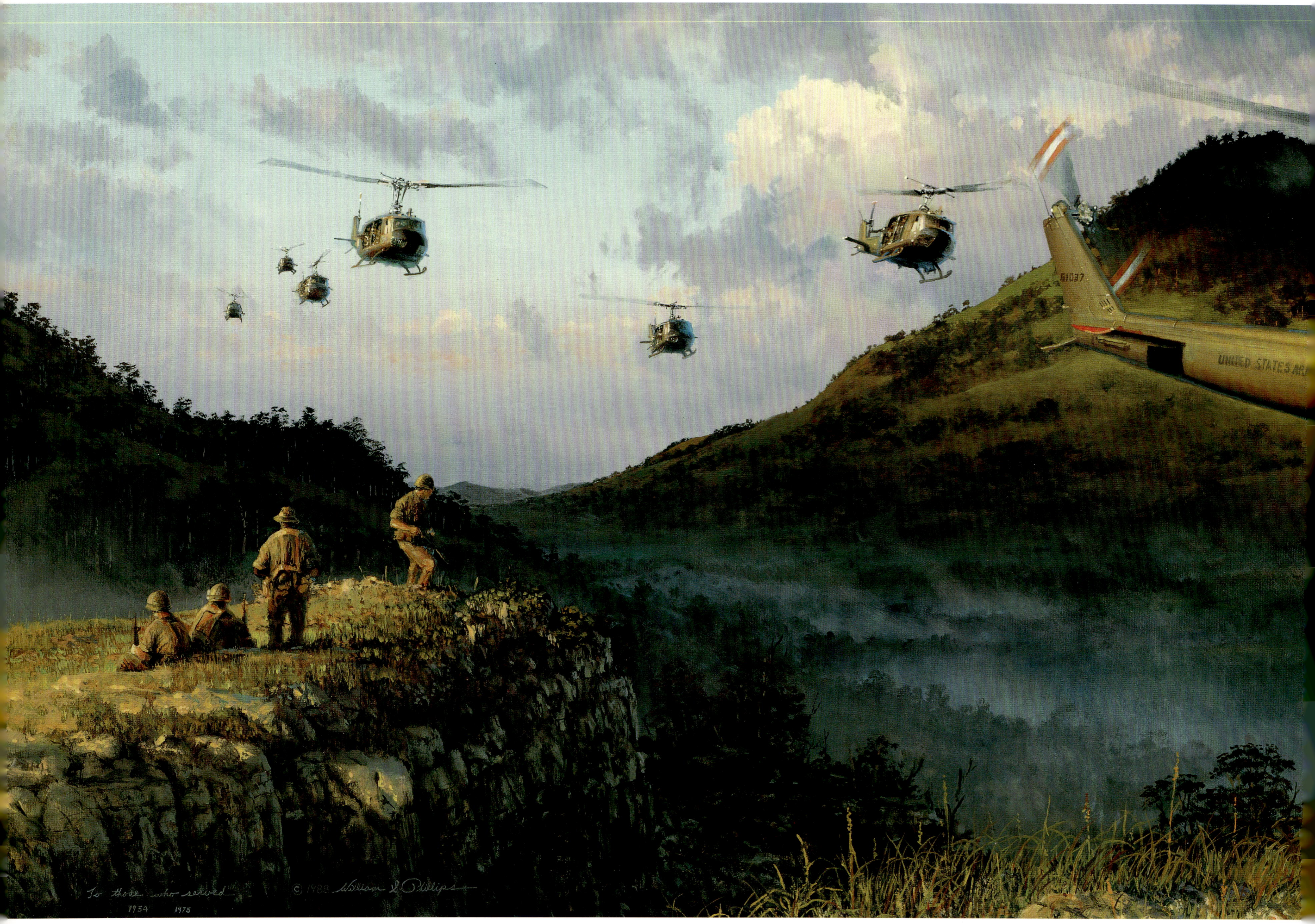
UNITED STATES AR
To those who served
1954 1975
© 1988 William S Phillips

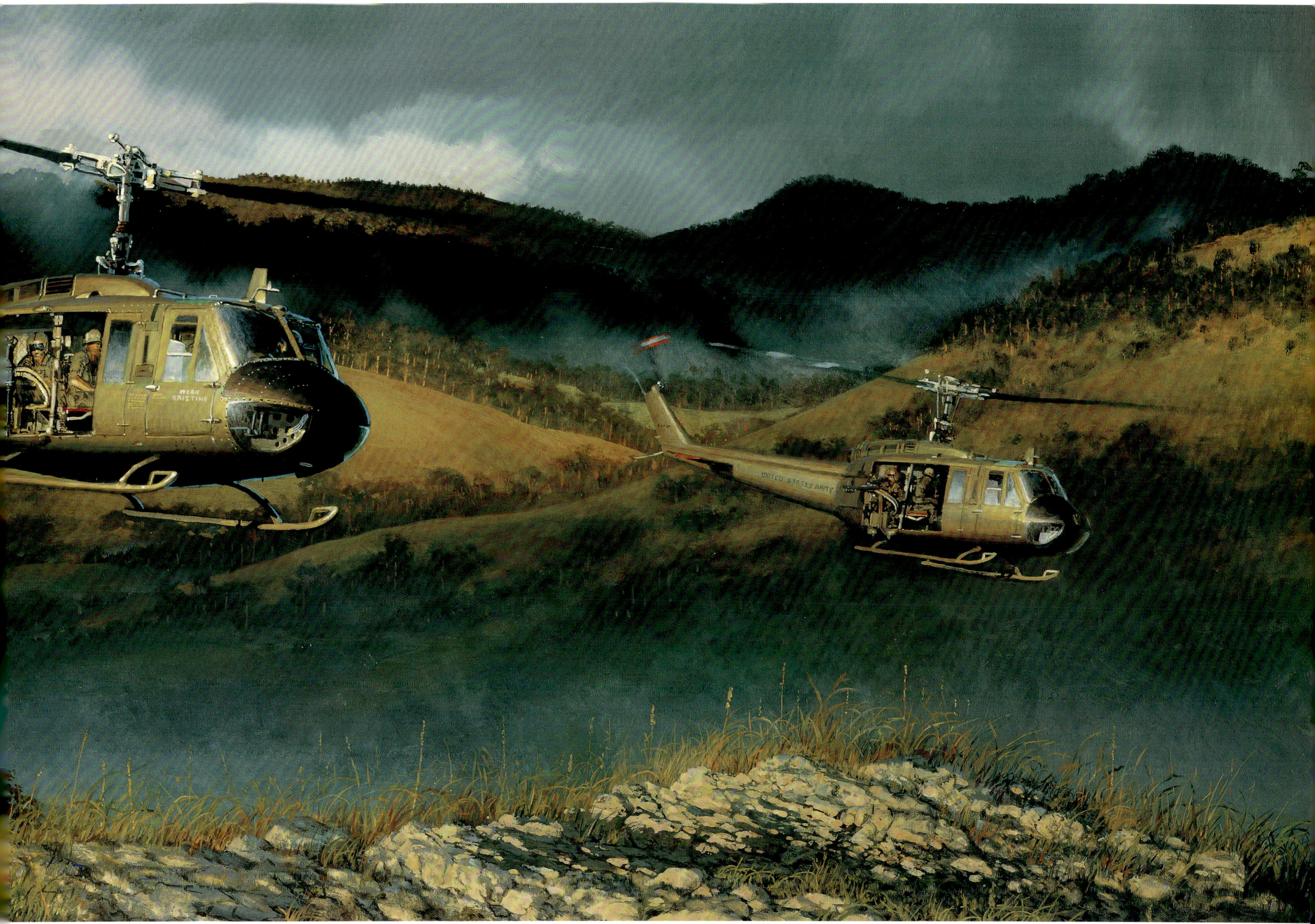

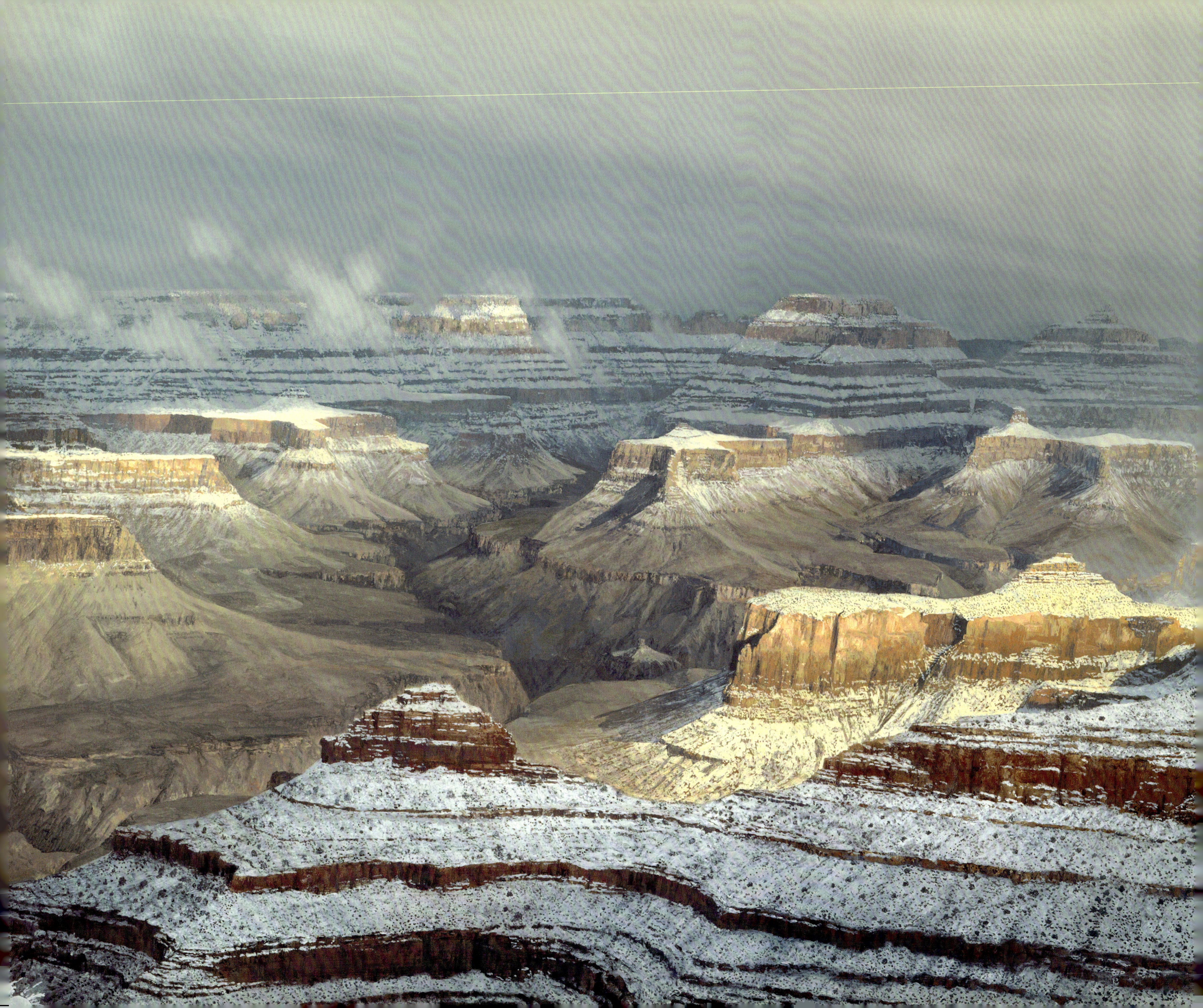

# ARTISTRY

## INTO THE SUNLIT SPLENDOR

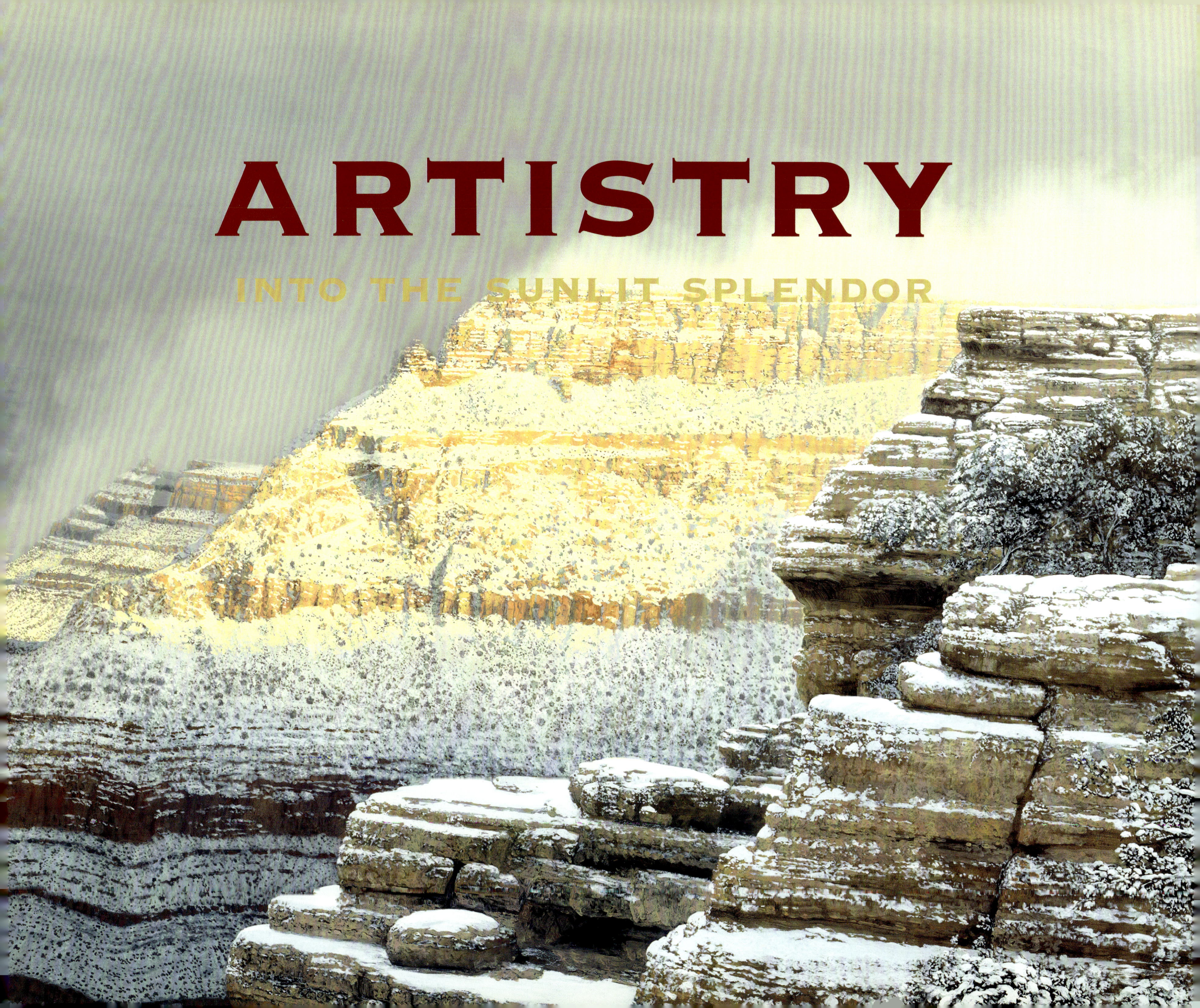

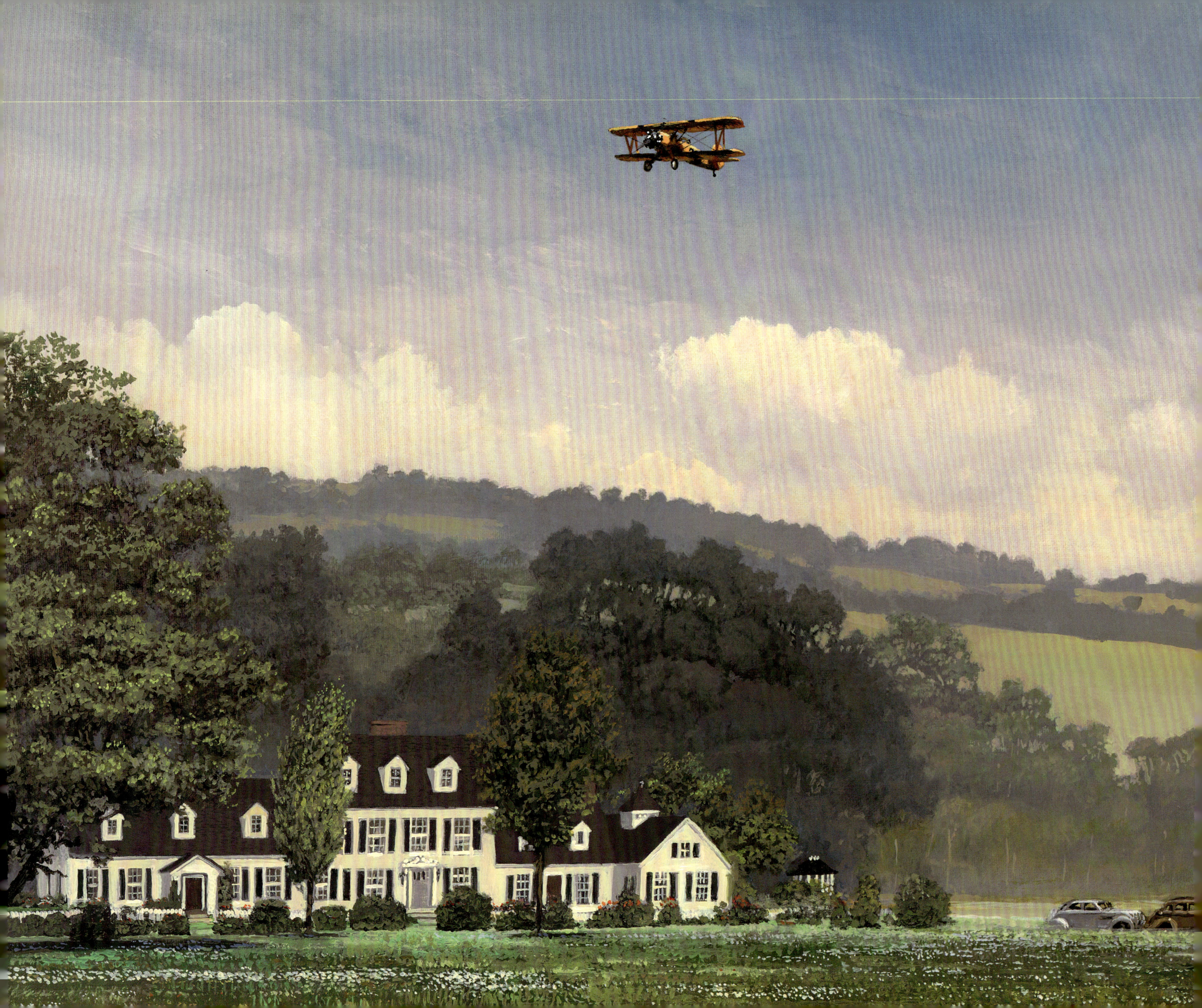

*Every artist dips his brush in his own soul, and paints his own nature into every picture.*

Henry Ward Beecher

Artists are skilled in creating works of aesthetic value. Artistry carries that description further to mean the craftsmanship in synthesizing form, composition and perspective into works of compelling beauty. "I have taken every opportunity to see and value artistry throughout my life," said Bill, "and these impressions influence my paintings. As I observe an aircraft being maneuvered skillfully by a pilot, I experience a display of something beautiful. The impact of this experience is so exciting that I am compelled to capture the essence in my own way, with my own artistry, for others to experience.

"Painting fills every corner of my life. When airborne in a commercial airliner, I prefer a window seat and spend most of the flight composing future paintings while observing the sky in all its moods. From the air or from the ground, I am captivated by ever-changing sights and sensations: the elegant grace of flight, the movement of wind, light and shadow playing over ancient rocks or the startling, magical forces of lightning imbedded in massive evening thunderstorms."

For thousands of years, man's aerial view of earth was limited to what could be seen from the top of the nearest mountain. Until the invention of flying machines—balloons in the eighteenth century and powered craft in the twentieth century—landscapes mirrored an artist's largely horizontal view. The advent of powered flight introduced dramatic challenges to painters. Suddenly, artists had to depict vast distances between subject and background, lit primarily by the sun, yet influenced by reflected and artificial light. The atmosphere, far from empty space, bore gases, dust, and reflective particles. As airborne machines achieved great speeds and record-breaking altitudes, artists who sought to paint the world of flight had to master a new perspective of the three-dimensional expanses through which aircraft travel.

"When one thinks of an artist, the image often includes studios, easels and canvas," said Bill. "But, at times, research for a painting can take an artist far from the studio and be quite exhilarating. Before documenting 'Top Gun' missions at Miramar Naval Air Station, I was required to complete the Navy's water survival school, physiological training, and courses in flight safety. I underwent grueling training and testing before I could climb aboard the VF-126 TA-4 Skyhawk to capture the 'Top Gun' experience."

Bill's research ranges from high flight to low level tactics. He rode in an F-15 during air combat training maneuvers at altitude, and careened close to the desert of northern Nevada on a RF-4 practice mission. He has spent weeks on naval vessels at sea, and traversed the rocky promontories of his beloved West. He has witnessed nature's magnificence and his artistry reflects the integrity of one who has seen this majesty and spent countless hours flying and photographing the sky and the atmosphere.

The appreciation of subtle gradations in light as it is reflected from clouds and other surfaces is crucial in aviation art. Bill Phillips' understanding of light infuses each painting with an element of drama. "When I flew with the Blue Angels over the Gulf of Mexico," said Bill, "I saw how sunlight reflected from aircraft flying close to the water and observed how reflections shift within shadows on metal. In Jordan, even at high altitudes, dust particles suspended in hot desert winds tinted the clouds, as well as the landscape below, with shades that were almost 'otherworldly.' The scene had to be seen from above before its delicate hues could be effectively translated to canvas. Only after much experience and observation did I learn the subtleties required to make a painting come to life, without sacrificing authenticity and without making the work overly complex. My goal is to offer enough leeway for the viewer to fill and imagine."

In 1987, Bill was the subject of a one-man show at the Smithsonian's National Air and Space Museum. As the show was being hung, Bill had the chance to wander quietly among his paintings and reflect. "As I looked at the paintings displayed on the walls of the Smithsonian's gallery, I was touched by the experience we call 'art,' and by the fact that artists are never really finished, no matter how old or experienced they have become. There is always something more to learn, some new challenge to understand. There is a vitality that comes from each experience, and the rewards play out across the canvas and express to others the deep and emotional bond you have with your subject. I was overwhelmed by the realization that I was a witness to history and in some small way had become a part of it."

For many years, aviation dominated Bill's choice of subject for his art, but in 1991 he created a project that would evolve into his very successful Phillips Bay series. A charming, fictional town set in the not-too-distant past, Phillips Bay embodies the American values of family, service, and community that the artist cherishes. The portrayals of twentieth century American life through this series are yet another expression of Bill's artistry. With each painting, Bill captures a lyrical quality while holding fast to historic foundations—a dance that layers history, architecture, and inevitably a peek at aviation (since no Phillips Bay painting is without a distant plane in the sky).

When invited to speak about Phillips Bay, the artist recounts the remarkable tale of how Phillips

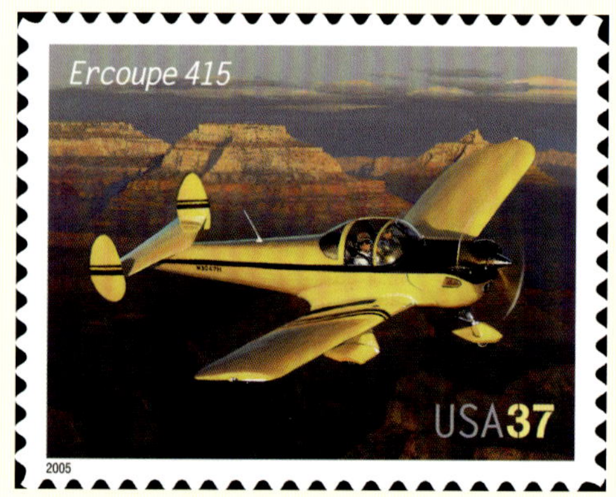


**Ercoupe 415**
U.S. Postal Service, Advances in American Aviation series, 2005

Designed by Fred Weick, one of America's aviation pioneers, this craft was meant to become "Everyman's Airplane." A low-wing, two-seat, tricycle gear light aircraft, the Ercoupe had coordinated controls, rendering rudder pedals unnecessary. It was flown and taxied much like a car. Certified in 1938, 112 Ercoupes were built before WW II's impact halted production.

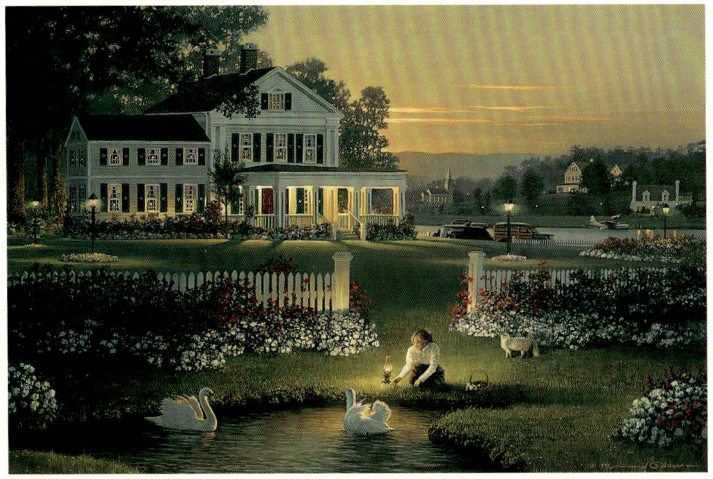

**Evening Song**

In a nostalgic return to a peaceful half-century ago, the shape of hearts abound. They decorate the shutters at each lit and welcoming window; they form the shape of the pond, and the arched wings of a pair of swans.

Bay came to be (Captain John Bittle, blown off course during a passage to the colonies, found safety in a small harbor) and the families and businesses that make up the small town. The Homefront Series, among the most well known paintings, portray a scene in Phillips Bay during each year of World War II. At the end of Bill's presentation, there is invariably an audience member who asks for directions to Phillips Bay, despite Bill's frequent reminders that it is fictional.

Bill's ongoing love affair with the Grand Canyon began with a postcard sent to him by his father. Bill has hiked, rafted and painted its majesty for more than two decades, always challenged by its grandeur. In all seasons, he has sketched and documented the canyon's changeable moods, its colors, light and shadows, through unimaginable heat at its depths to blizzards on its rim. Bill's depictions of the textures, shapes, and lines of the multi-hued canyon have been enhanced by numerous flights over and through this remarkable natural wonder. Experience has strengthened his talents as a landscapist, which, in turn, has enriched his aviation art.

In 2004, when the U.S. Park Service offered Bill the chance to be Artist in Residence at the North Rim, he accepted gratefully. The resultant paintings depict his impressions of the geological features of the Canyon's North Rim and his artistry in capturing, in Bill's words, "the essence of this most beautiful of canyons." Bill arranged to be involved with the daily activities of the dedicated men and women who steward the National Park. "I was afforded the opportunity to ride along on law enforcement patrols and observe fire management activities," he said. "Being in residence, I took advantage of approaching my work deliberately. Time slowed as I observed things so often missed while hiking or rafting. I was afforded the luxury of time to sit for hours and observe not only the grand vistas before me, but also the small dramas at my feet. One day, near the edge of the rocks at Cape Final, I noticed several feathers floating on the rainwater captured in one of the many depressions on the rocky surface. Looking further, I discovered more feathers caught in a small bramble just short of the canyon's edge and one small drop of blood on a rock nearby. I began to realize that within the immensity of the whole, there are countless small dramas that, unless we take the time to observe, escape us entirely. These small details are what artistry is really all about."

## Low Pass for the Home Folks

Early in World War II, honoring a tradition among all fighter pilots, two eager Luftwaffe pilots buzz the farmers harvesting crops near Garmisch-Partenkirchen nestled below the snow-dusted Bavarian Alps. The gutsy lads careen over the wheat field in their Messerschmitt Bf 109s, dropping below church steeples to offer a raucous aerial salute. Had they been reported, their squadron commander would have growled, "Das ist verboten!" At the same time, recalling his own youth, he would have admired their aggressiveness—a key characteristic in a fighter pilot.

Initially test flown in 1935, the all-metal Bf 109, designed by Willy Messerschmitt and Walter Rethel, had an enclosed cockpit and credited its speed to retractable gear, and the Rolls-Royce Kestrel V engine that powered it to a world speed record of 379.39 mph at the International Zurich Air Show in July 1937. Until the Battle of Britain in 1940 proved otherwise, the Bf 109 was considered the best fighter in the world. Like its premier rival in Europe, the Supermarine Spitfire, the Bf 109 was continually upgraded and, from its 1936 service in the Spanish Civil War until the end of WWII in 1945, more than 35,000 were produced.

These two young Luftwaffe pilots faced a long haul ahead, as it was the practice for the German Air Force to keep pilots in combat much longer than their foes. Several German pilots were credited with over 300 kills, which amounts to the accrual of an exhausting number of flight hours, if they were lucky and lived.

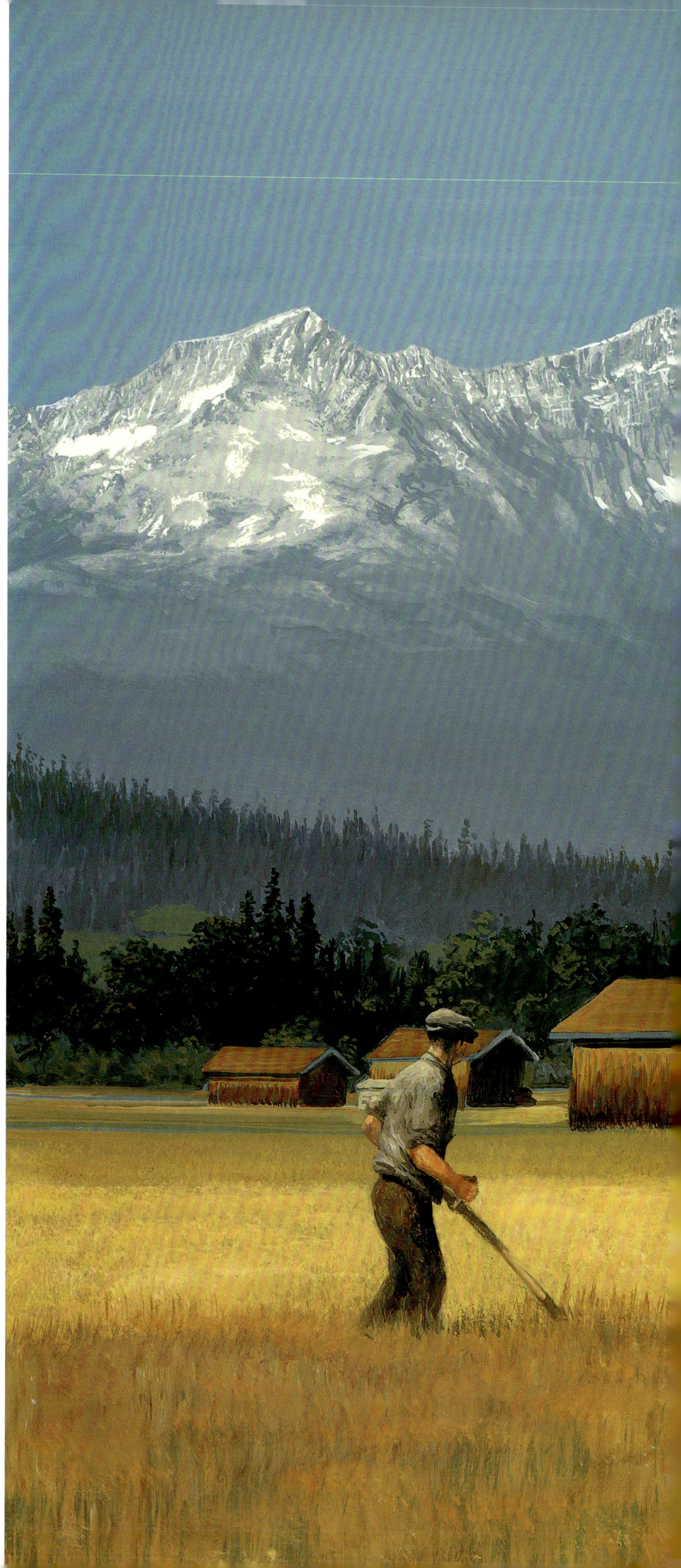

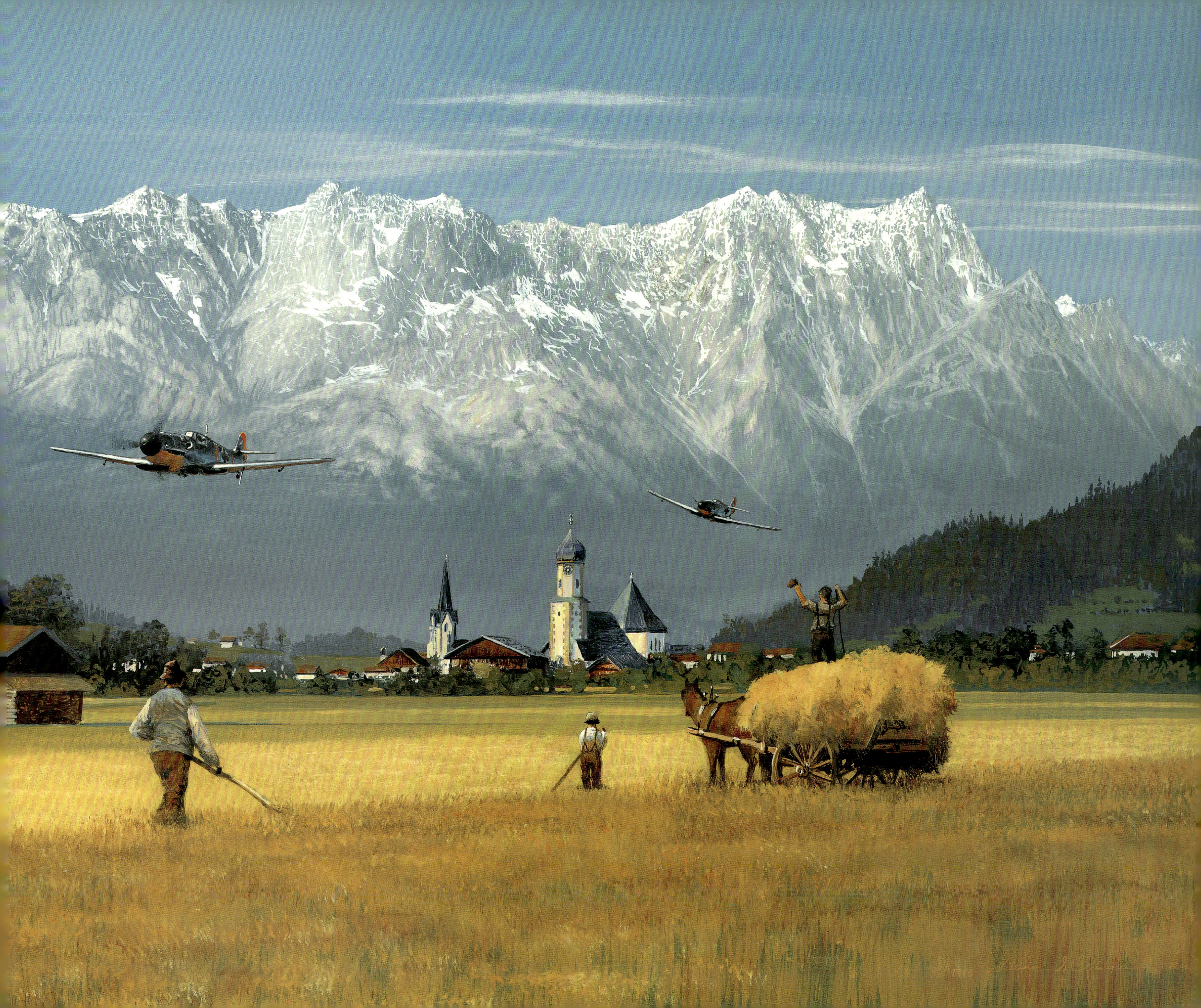

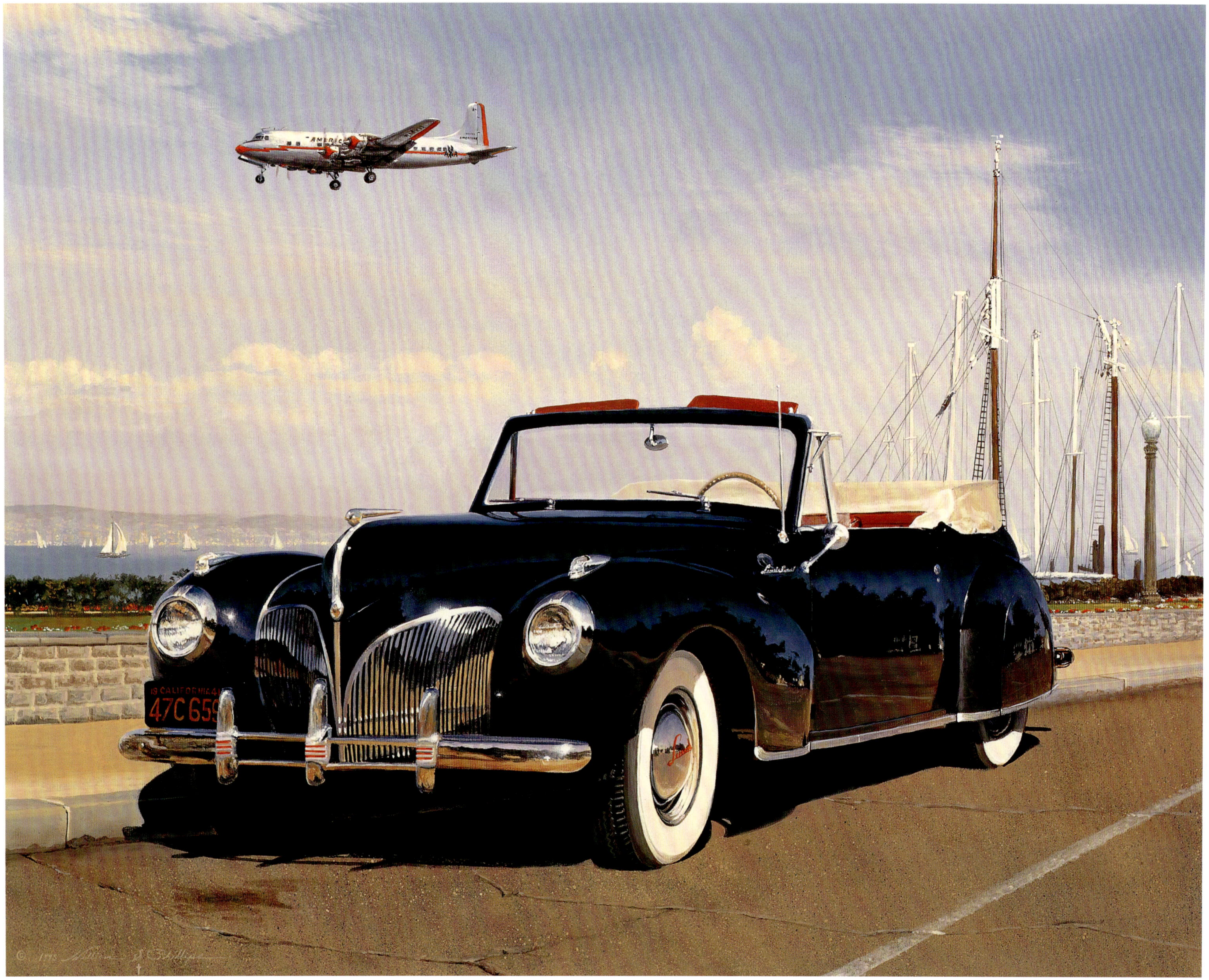

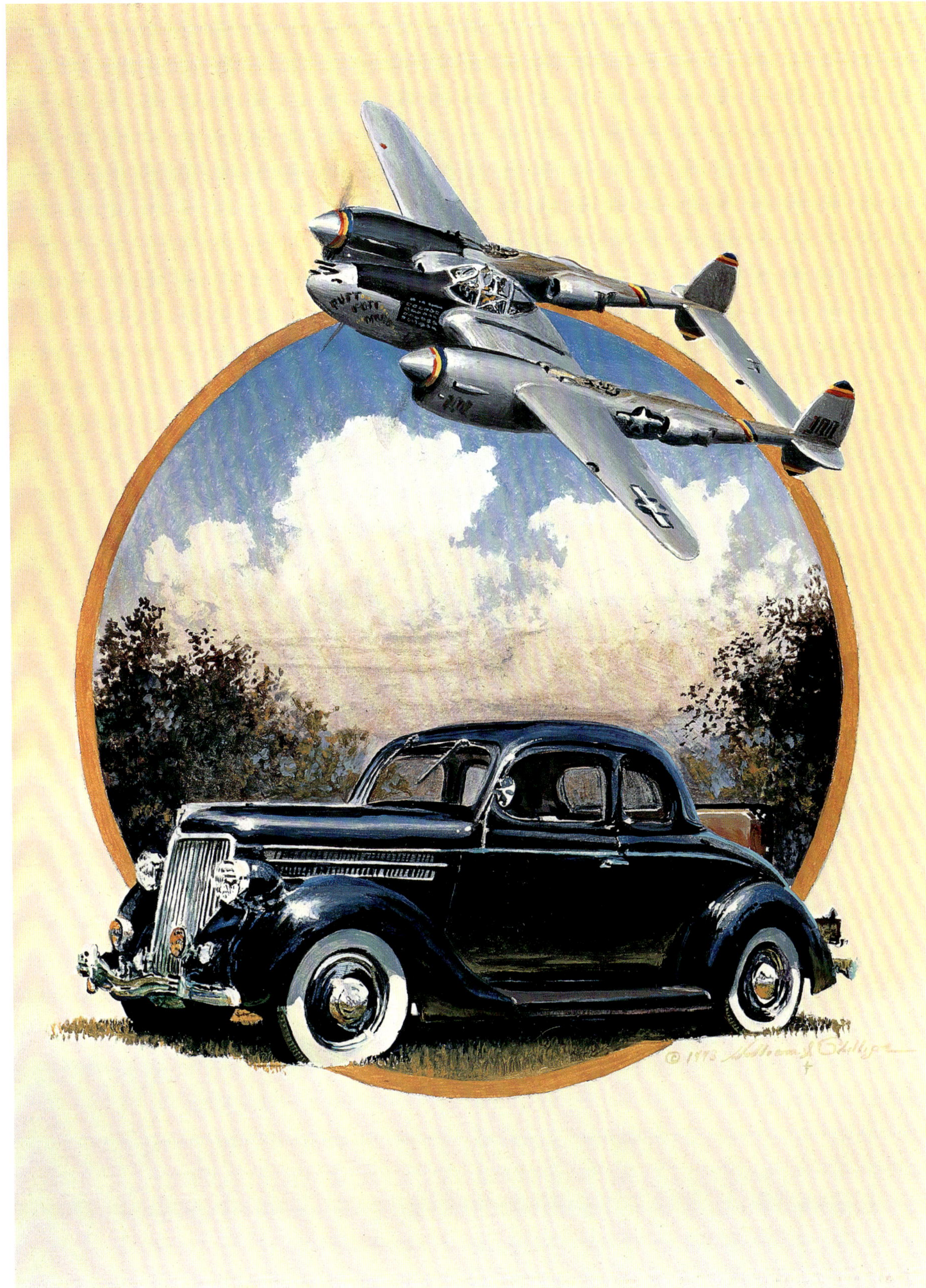

## Classics: P-38 and Car

This highly stylized painting combines two classics of the late 1930s including one of WWII's premier fighters, the twin-tail, twin-engine, sleek shape of the Lockheed P-38 Lightning. Dubbed *Putt Putt Maru,* this Lightning was flown in the Southwest Pacific by Col. Charles MacDonald, a twenty-seven-victory ace and Commander of the 475th Fighter Group, 5th USAAF. He worked with Charles Lindbergh to improve the range and performance of his aircraft. And the roadster . . . it's parked briefly during a Sunday afternoon drive in the country.

## Hank's Baby

(opposite)

Commissioned by a San Francisco Bay Area resident to portray three of his loves: planes, boats, and cars. The scene centers on Hank's third pride and joy—his 1941 Lincoln Continental Cabriolet convertible, one of the last of the pre-WWII automobiles. Behind the sleek convertible, sailboats rock gently in the waters of the marina where Hank and his wife moored their boat. Overhead, an American Airlines Douglas DC-7 commits to landing, entering its final approach with landing gear and flaps extended. The DC-7 was Douglas' largest and last piston-engine passenger plane, and the airliner on which Hank's wife served as a stewardess.

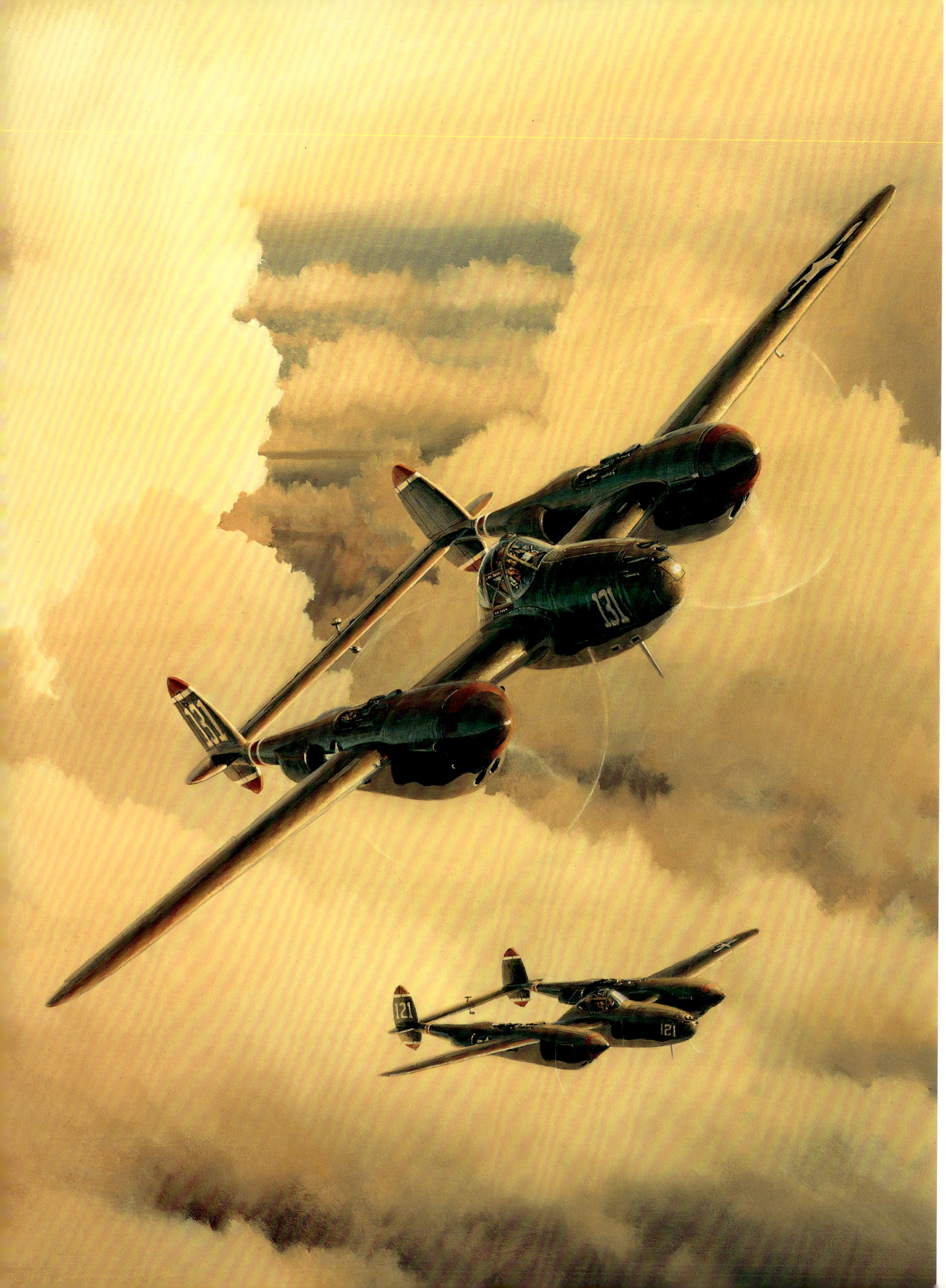

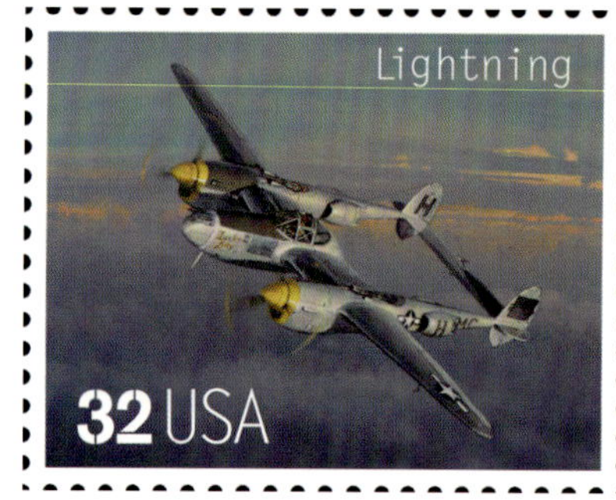


## P-38 Lightning

U.S. Postal Service, Classic American Aircraft series, 1997

Lockheed's P-38 was its first military airplane. Nearly 10,000 P-38s were produced, in eighteen variants. Known to the Germans as the "Forked Tailed Devil," a leading P-38 Ace called it, " . . . one of the great aircraft of WWII." And Japanese Ace Saburo Sakai said, "I was astonished to find an American aircraft that could outrun, outclimb, and outdive our Zero . . ."

## Thunder and Lightning

The last rays of sunlight lighten the sky to gold as two Lockheed P-38 Lightnings careen among building cumulonimbus. One of the two fighters, from the 431st Fighter Squadron, 475th Fighter Group, bears the Number 131 of C.O. Major Thomas McGuire, a Medal of Honor recipient and, with thirty-eight victories, the second leading ace of all time. He was honored posthumously in New Jersey's McGuire Air Force Base.

Adding again to the painting title's play on words, a *lightening* of the P-38 occurred when two external fuel tanks were jettisoned to facilitate tight turns in air-to-air combat. Unfortunately, McGuire's decision to hold onto his tanks proved fatal on January 7, 1945. While aiding his wingman, Captain Edwin Weaver, McGuire overstressed in banking too sharply at low altitude and crashed on Negros Island in the Southwest Pacific Theater.

In addition to serving as the combat plane flown by America's two top aces—Dick Bong and Tommy McGuire—the P-38 is credited with a number of firsts. It was the first fighter with two engines—Allison V-1710 liquid-cooled developing 1,425 hp; the first with counter-rotating propellers that negated torque; and the first with a central pod containing the armament and eliminating the need to synchronize firing with the eleven and one-half-foot props.

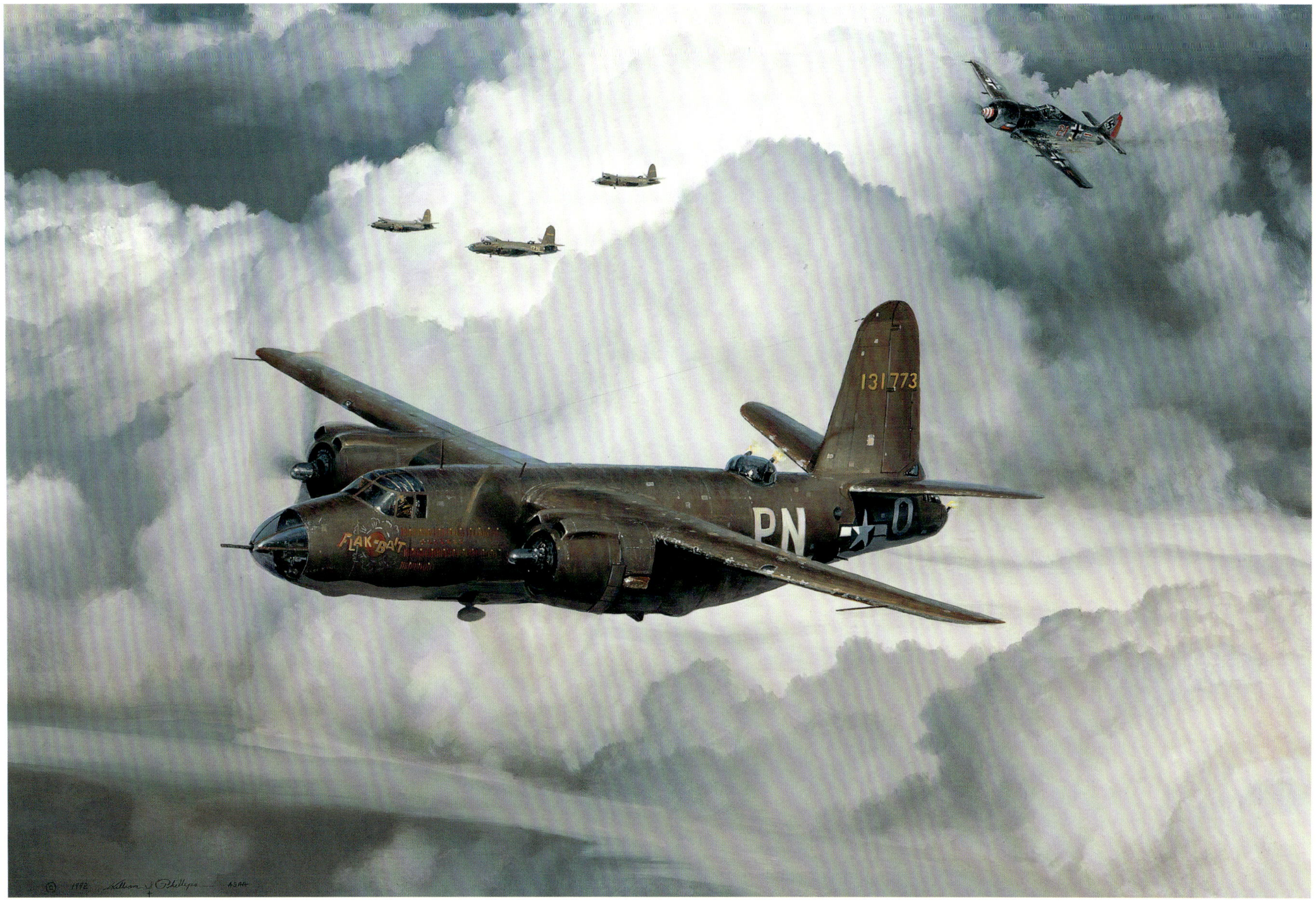


## The Bait Bites Back

*Flak Bait*, a Martin B-26B Marauder, was named for its pilot's dog, Flea Bait. This bomber flew more missions over Europe than any other WWII U.S. airplane. She completed, and survived, 202 operational missions but, not without damage. Over 1,000 patched flak holes bear witness to its service.

In artist Bill Phillips' painting, *Flak Bait* is dishing out some punishment of its own, as the German Focke-Wulf 190 will attest. More than 5,000 Martin Marauders were built and, while it initially experienced a high accident rate and developed a reputation as a "widow-maker," two aerobatic greats—Lt. Colonel Jimmy Doolittle and Major Vincent "Squeek" Burnett—were asked to demonstrate single-engine emergency procedures. When Doolittle and Burnett landed smoothly time and again, one propeller feathered, engine out, it was determined that pilot training was inadequate. Once in combat, the Marauder and its pilots had the lowest loss rate and greatest bombing accuracy of any U.S. WWII aircraft.

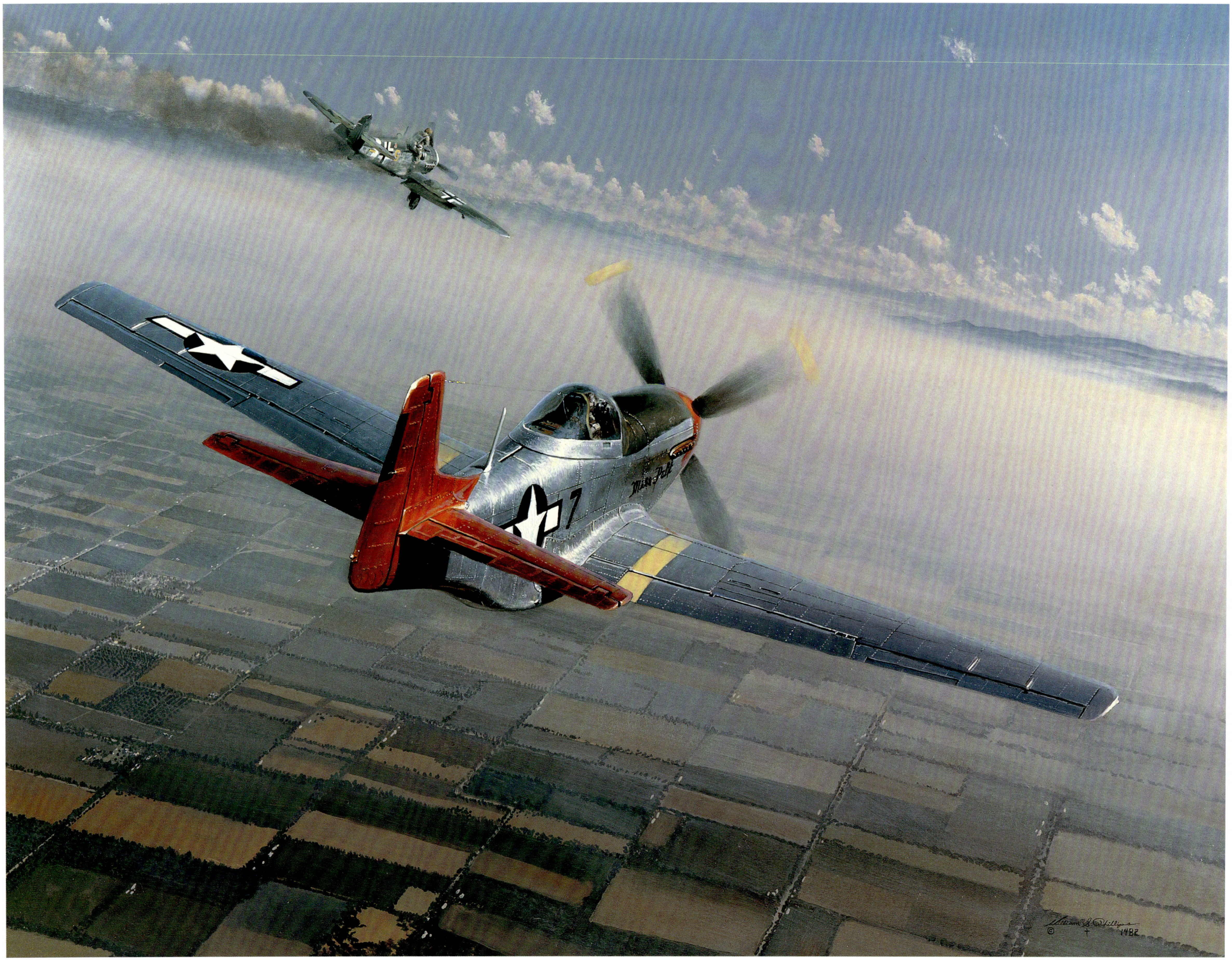
7
© 1982

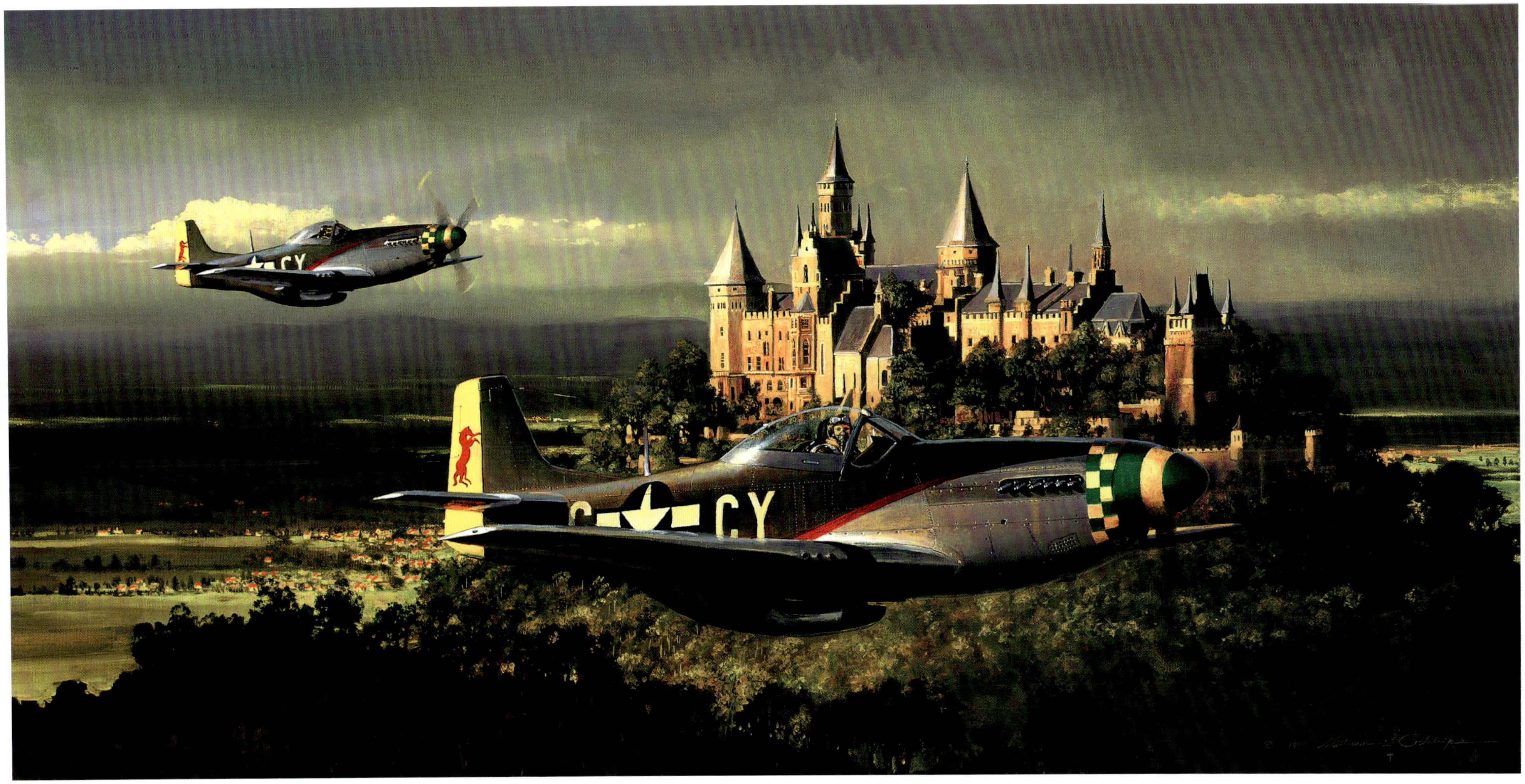

## Two Down, One to Go
(opposite)

Imagine that you are high in the skies over Italy on July 18, 1944 as the 100th Fighter Squadron, 332nd Fighter Group's Second Lieutenant Clarence "Lucky" Lester, counts his second victory of the day and his opponent, the pilot of the Messerschmitt Bf 109, prepares to bail out of his crippled aircraft. Lt. Lester's third victory came later on July 18th and another Bf 109 was dispatched.

It was celebration time for the pilots of the 332nd as the day's tally totaled eleven enemy fighters downed, to only three 332nd pilots lost. In what became true of their entire Tuskegee Airmen experience, and for their Group alone, no bombers for whom they flew cover were lost to enemy air actions. The "Red Tails" of the 332nd, and on this date, Lt. Lester of the famed WWII Tuskegee Airmen, were justly proud.

After a delayed start into combat, the first black American squadron flew P-40Ls out of North Africa in June 1943. Action from Sicily followed and, as the enemy was pushed north by the armies of U.S. General Patton and British General Montgomery, the Tuskegee Airmen, now a Group, moved into Italy. As the U.S. Army Air Corps' only black American unit, the Tuskegee Airmen fought—and won—two wars: one against Fascism and the other against prejudice.

## Victory Pass

In 1944, after D-Day, the 343rd Fighter Squadron of the 55th Fighter Group transitioned from the P-38 Lightning to North American P-51 Mustangs. With its extended range, the Mustang afforded the 343rd the capability of escorting 8th Air Force bombers to the target and back. The squadron saw action against ground targets during the Battle of the Bulge in the winter of 1944–1945. Two of the fighters are shown in loose trail formation streaking past a hilltop castle fortress that, as a symbol of Teutonic might, dominated the German countryside. Now, softened in the glow of low-cast sunlight and guarding peaceful sunlit fields, the castle symbolizes hope for a better future.

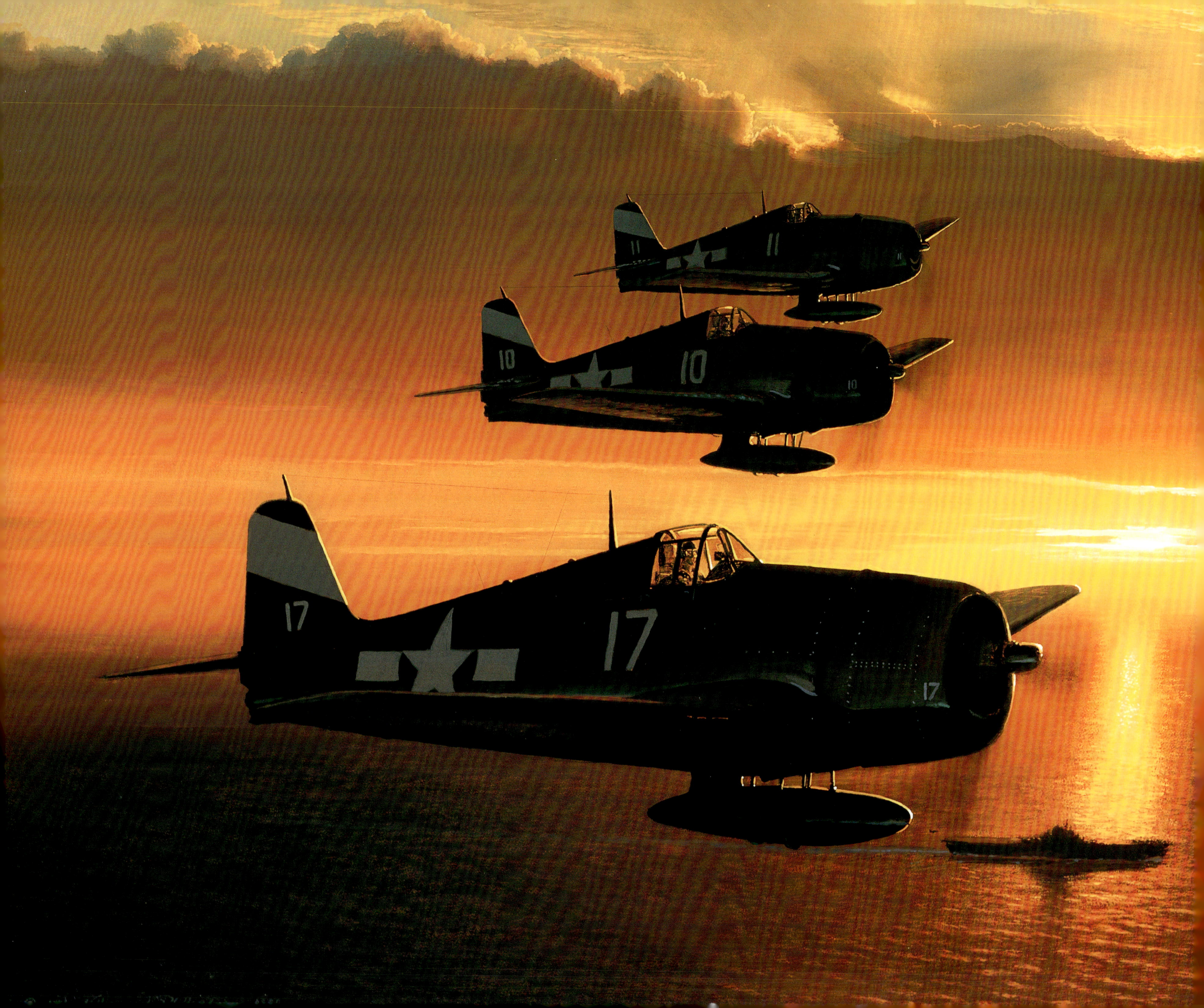
11
11
10
10
10
17
17
17

## Return to the Lady

With the dipping of the sun into the waters of the Pacific and a cloud deck threatening, a flight of Grumman F6F Hellcats sweeps back to their carrier, USS *Lexington* (CV-16). The fifth U.S. Navy ship to bear this proud name, its keel was laid down as USS *Cabot* in Massachusetts on July 15, 1941. She was renamed *Lexington* a year later, and commissioned in February 1943. She sailed for Pearl Harbor in August, raided Tarawa in September, Wake in October, and covered actions in the Marshalls and the Gilberts in November. On November 23–24, her aircraft downed twenty-nine enemy planes.

In December, a Japanese torpedo damaged the *Lexington*. She put in for repair and then reentered action, participating in Dutch New Guinea, the Marianas and, in March, played a major role in the "Great Marianas Turkey Shoot" in which 300 enemy aircraft were downed and a carrier and a destroyer were sunk. Despite taking a Kamikaze strike and subsequent major repair, *Lexington* continued gallantly.

Hellcats proved to be the primary shipboard fighters for the last two years of World War II and, with more than 12,000 produced, they accounted for the destruction of 6,000 Japanese aircraft. The extraordinary combination of rugged, capable Hellcats and fast, powerful, and defensible carriers, made for U.S. victory at sea.

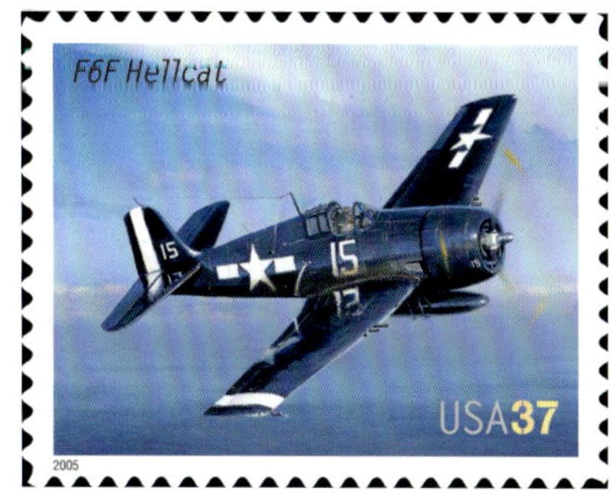


## Grumman F6F Hellcat

U.S. Postal Service, Advances in American Aviation series, 2005

A product of the Grumman "iron works," the Hellcat came to the Pacific Theater as a follow-on to the earlier F4F Wildcat. Faster, more maneuverable, well armed, and with a superior rate of climb, Hellcats immediately outclassed the opposition. F6Fs took starring roles in striking enemy warships and proved successful as search aircraft and fighter-bombers.

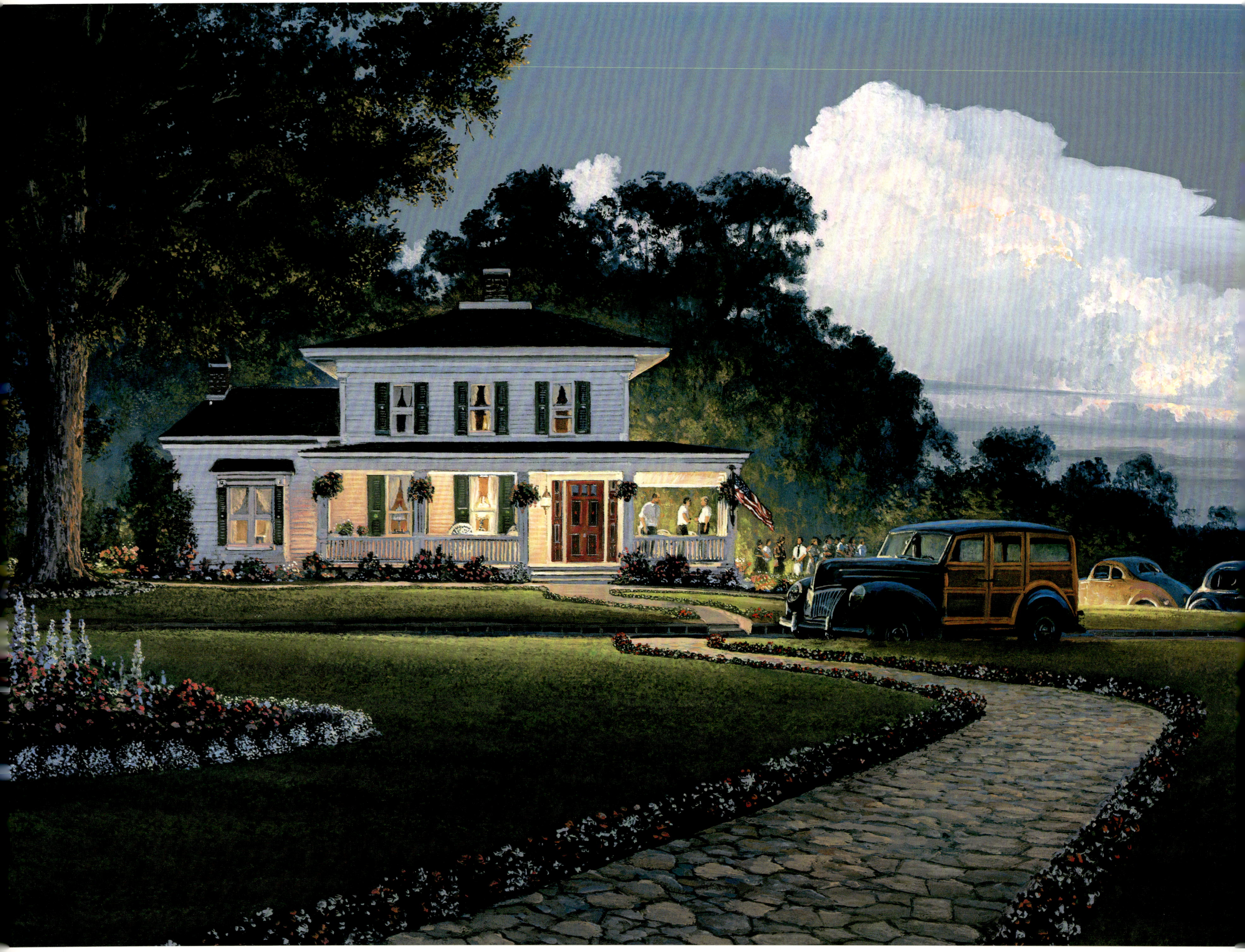

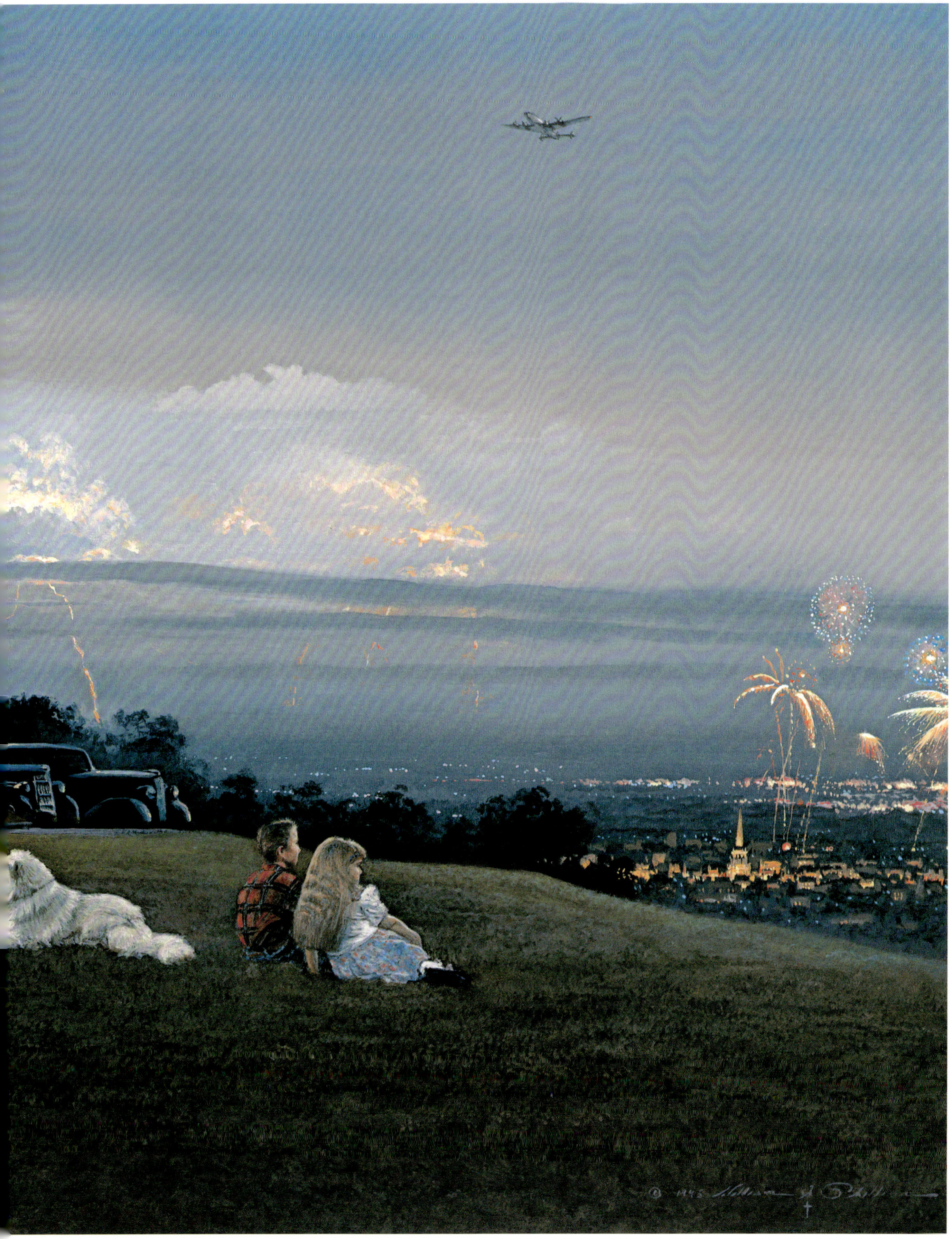

## Summer of '45 and All Creation Rejoiced

In this nostalgic and evocative painting, the artist invites us to revisit a summer evening that could have taken place in countless backyards sixty years ago. The devastating war in Europe has come to an end, but Pacific skies are still blackened with on-going battles. Do the distant storm clouds represent the passing fury of nature, or do they also suggest the dissipation of world war with thoughts of those still in harm's way? The lights of the home illuminate a homecoming party, home fires kept burning, and the light of hope for the future. Do the fireworks celebrate 169 years of independence, or do they joyously welcome our service men and women home? The Lockheed Constellation reflects its role as a military transport and also signals a return to civilian air travel. In the foreground, the children settled on the lawn with their white dog bring to mind peace and a hopeful future. With his appreciation of the power of symbolism, the artist invites the viewer's own interpretation of the summer of 1945.

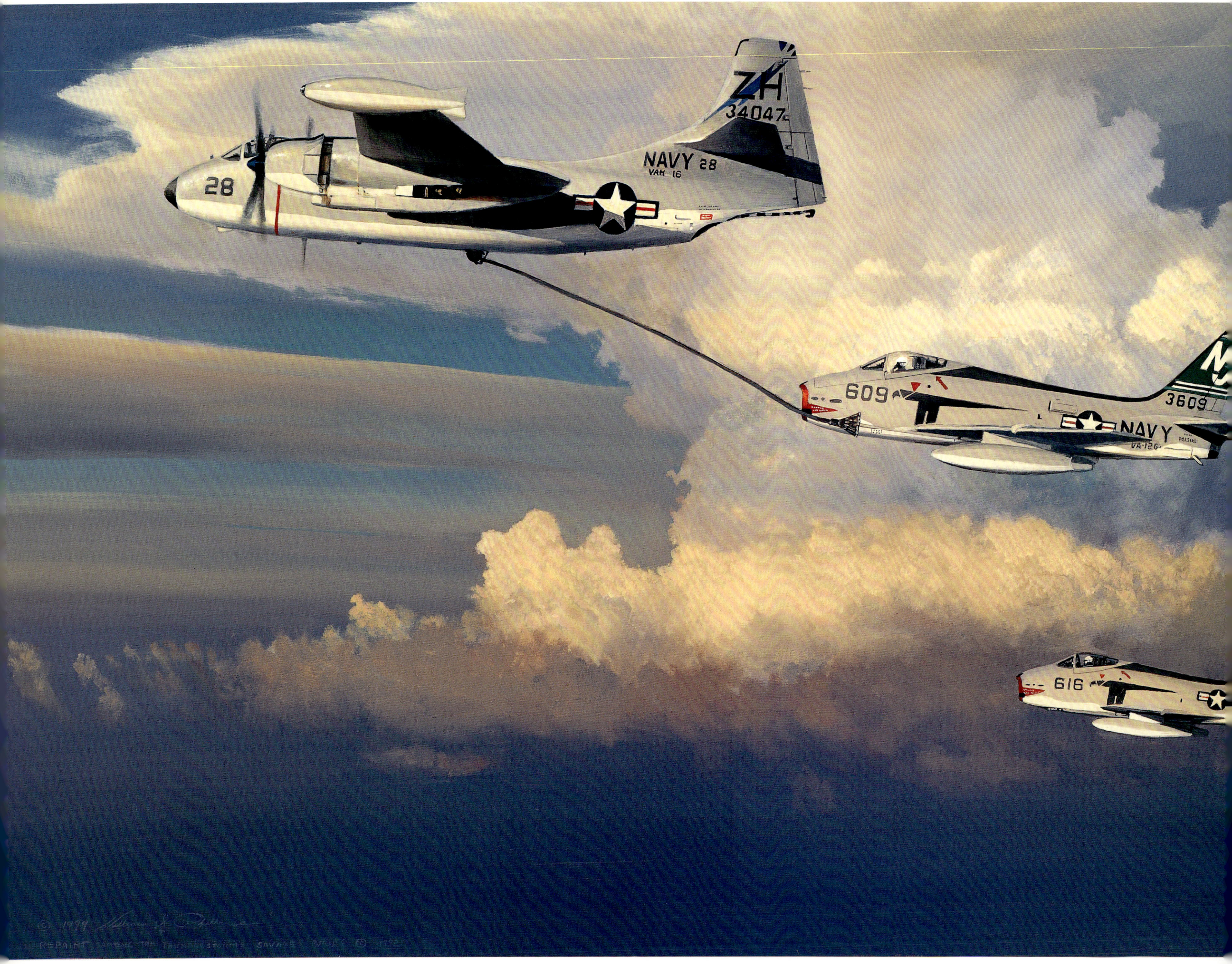
ZH
34047
NAVY 28
VAH 16
28
609
3609
NAVY
VA-126
616

## Among the Thunderstorms
## Savage Furies

Waiting in line for gas is common for ordinary mortals on the ground but it's now becoming commonplace at over 20,000 feet as military aircraft use aerial refueling to extend range. During an exercise at altitude with four aircraft in close proximity to one another, an encroaching thunderstorm promises danger, with its vicious turbulence, hail, heavy rain, and potent electricity. The danger is exacerbated for the FJ-4B Furies if they are a great distance from the carrier, with low-fuel lights warning their pilots, so the hazards must be met. Sidling up to the North American AJ Savage with its fuel supply, each Fury waits a turn at the drogue—a basket at the end of a flexible hose—and all crewmembers involved hope to accomplish refueling before the maturing thunderstorms disrupt them.

The concept of in-flight refueling, spawned by endurance flight record attempts in the 1920s, resurfaced with the advent of long-range strategic bombers and extended-reach fighter operations during the Cold War. Boeing's conversion of the Stratocruiser to the KC-97, and the B-707 civil transport to the KC-135 Stratotanker, met the needs for many years. The U.S. Air Force focused on the use of tanker-mounted refueling booms "flown" by the boom operators aboard the tanker into a receiver's receptacle. Refueling large helicopters called for the U.S. Navy's use of the probe and drogue method in which the receiving pilot engages the aircraft's probe with the drogue like the Savage is providing for the Fury.

This composite-powered, carrier-based AJ Savage was designed as an attack bomber equipped with two P&W R-2800 piston engines and an Allison J33 turbojet in the rear fuselage. Ten were converted to in-flight refueling tankers after 1959. The FJ-2 Fury was the Navy's first swept-wing fighter, more closely related to the North American F-86 Sabre than the FJ-1, a straight-wing model. The Navy was pushed to respond to the speed achieved by the MiG-15 that out-flew straight wing Navy fighters—Panther, Banshee, and FJ-1 Fury—and led to the design of the FJ-2. With its three variants, including the FJ-4B, a total of 1,100 were built by 1958.

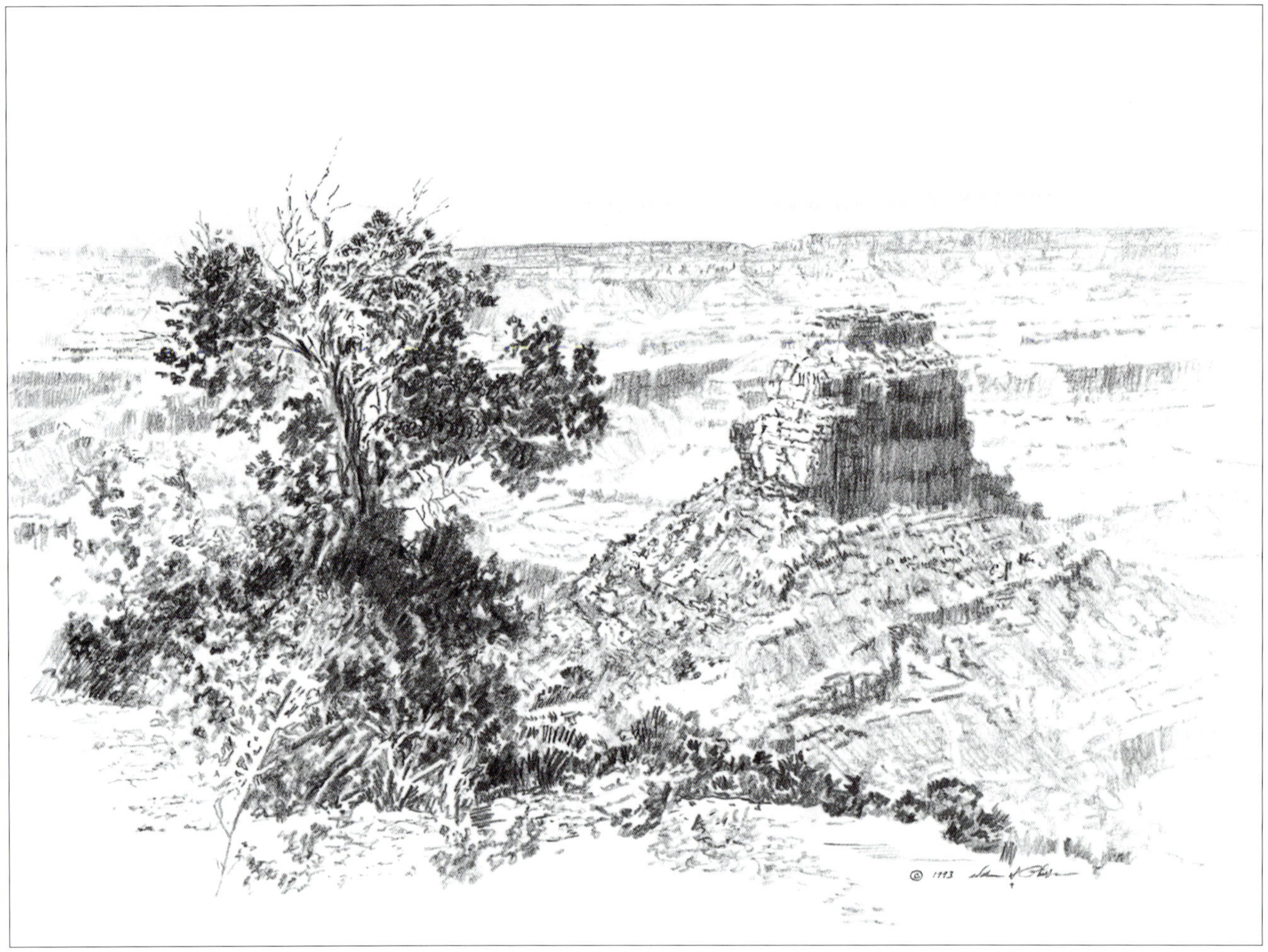

## Canyon Visitors '57

In the early 1950s, the F-100 Super Sabre, with its distinctive swept-wing, introduced supersonic speeds to the U.S. Air Force. Selected by its Demonstration Squadron, the Thunderbirds, the Super Sabre thrilled nineteen million fans with precision aerobatics. Here, in a stylized painting, it streaks over a sleek 1957 Ford Thunderbird against the background that symbolizes reverence for the Native American Thunderbird and unites man's machines with the natural beauty of the Grand Canyon.

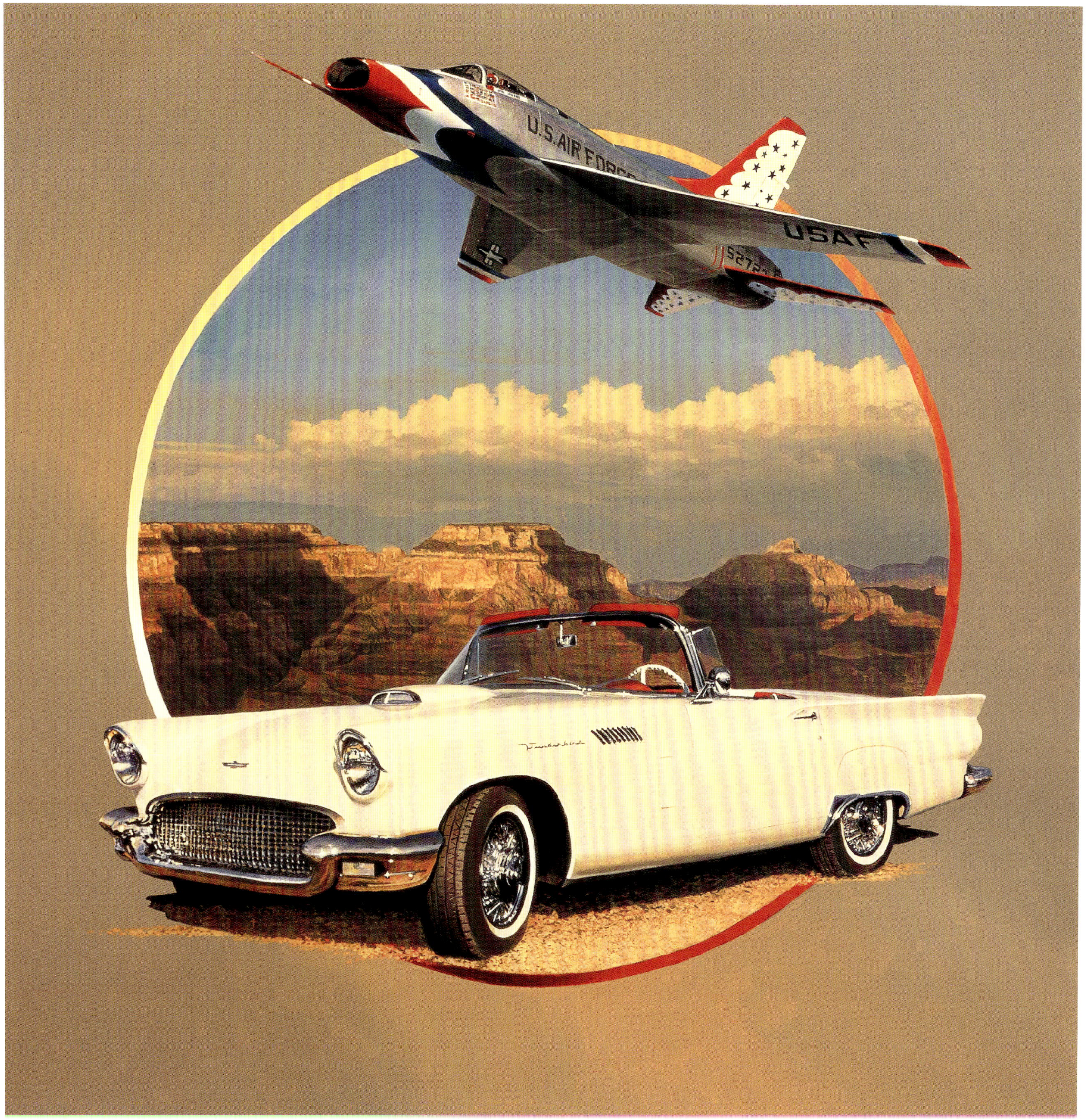
U.S. AIR FORCE
USAF

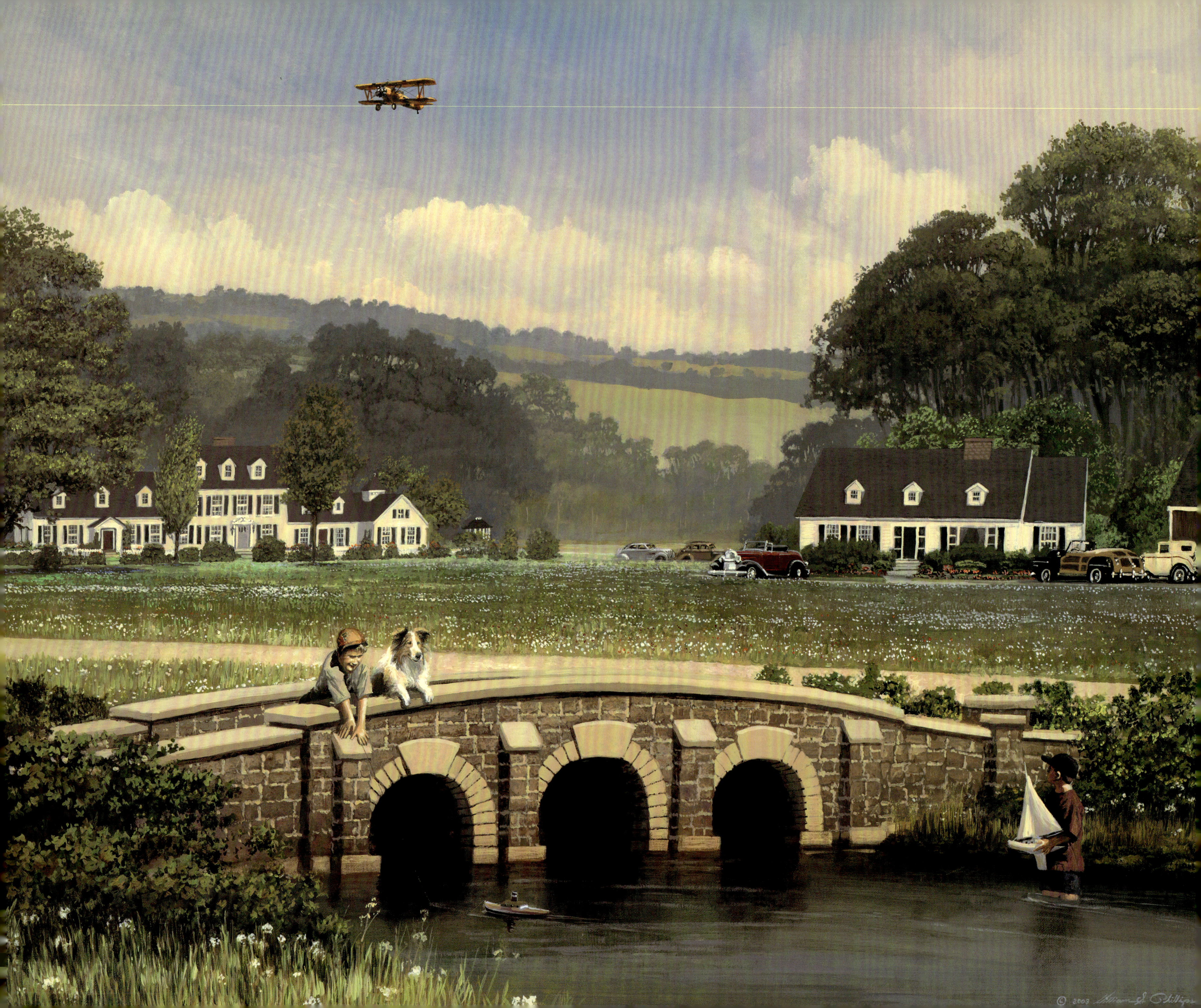

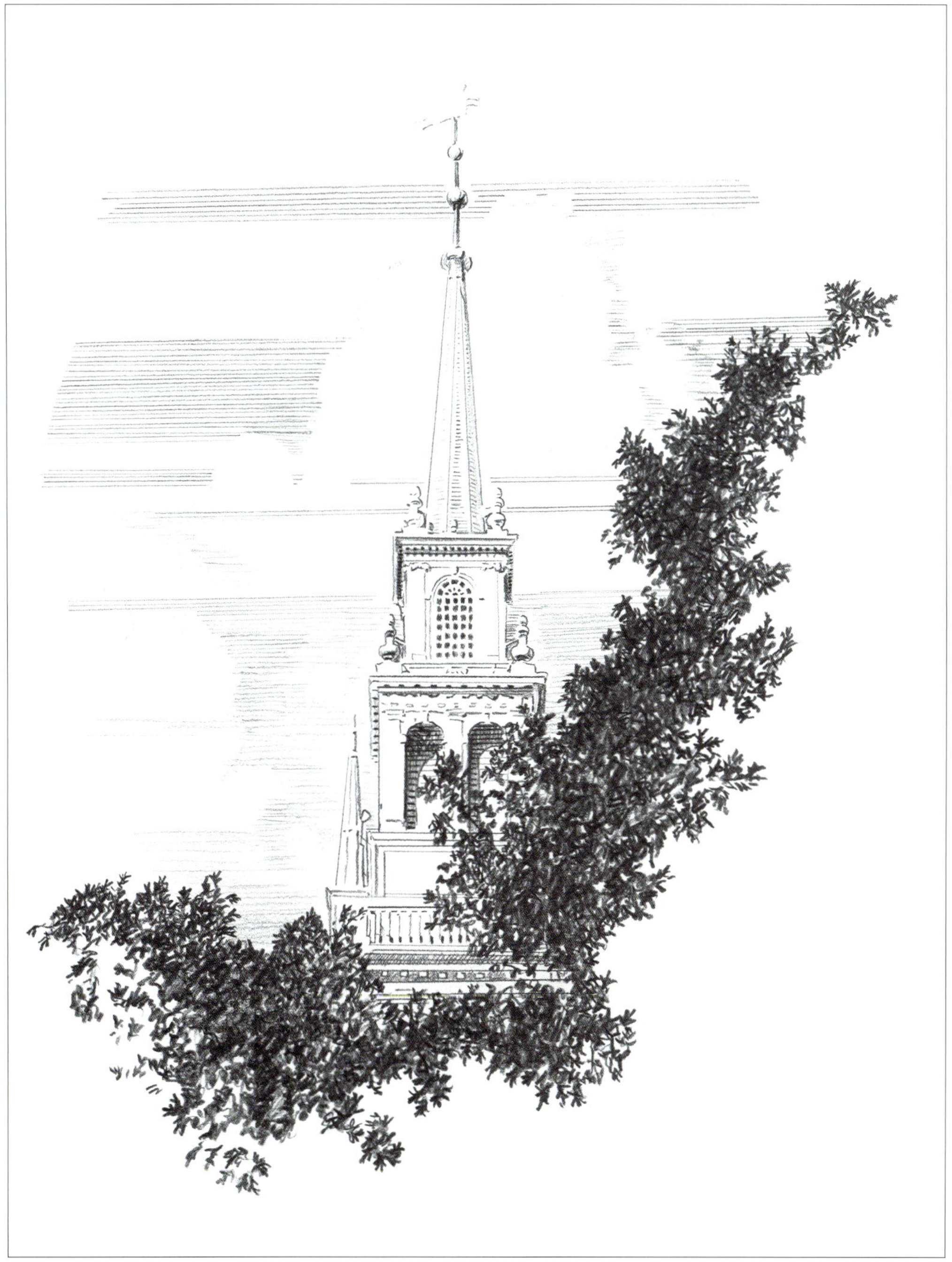

## Apple Creek Yacht Club

Apple Creek Valley, thirty minutes west of Phillips Bay, is at its best on a summer day. Two boys with their collie have temporarily stilled the voices of bullfrogs, while both their sailboats and imaginations float in the clear spring water. Vintage cars and the passing Stearman set the scene in time.

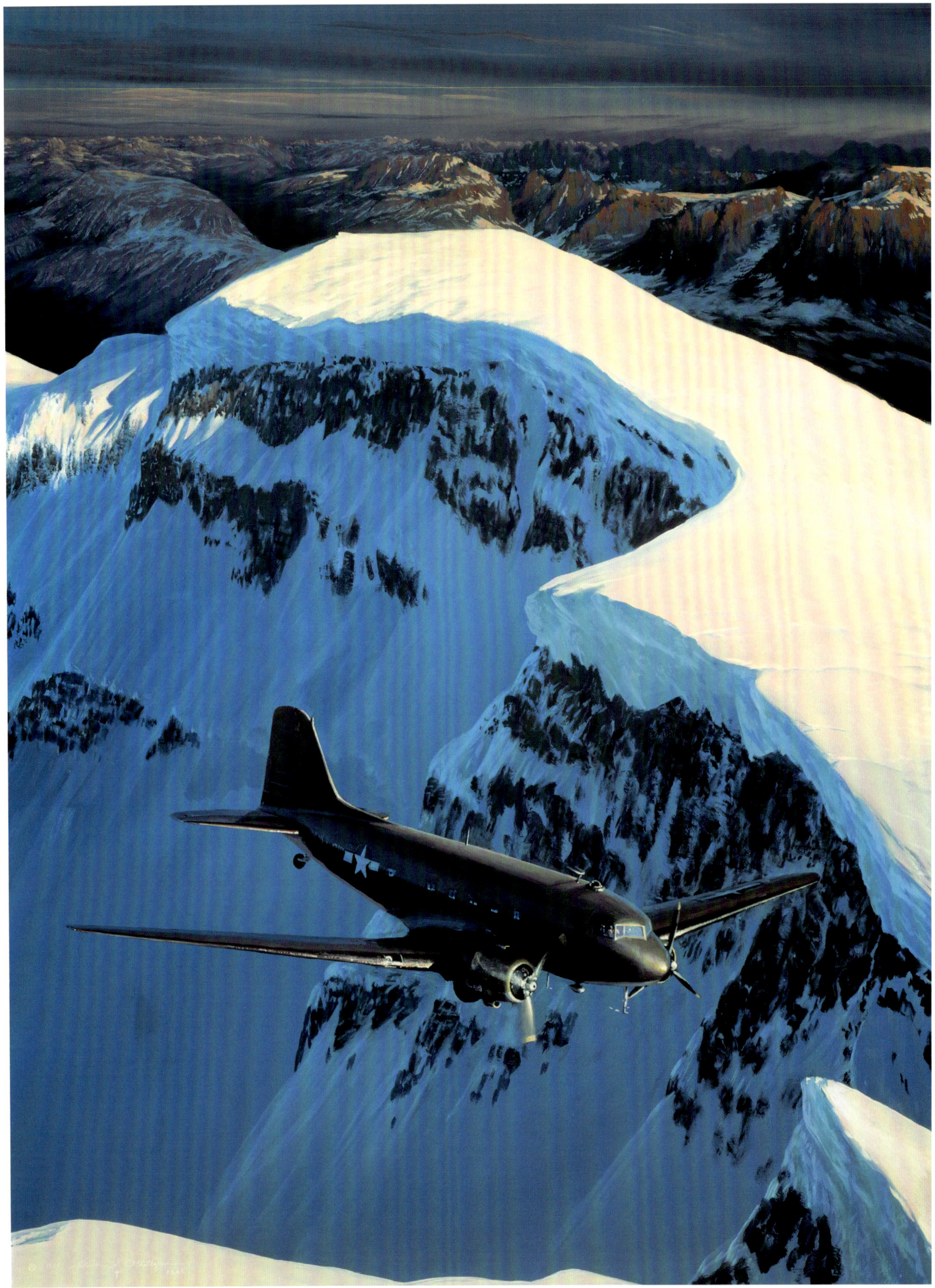

## High Anxiety

Crossing the "Hump," the Himalaya Mountains, which boasted nine of the world's ten highest peaks, made flying the distance between India and China among the most hazardous flying of World War II. Aircraft like this C-47 Skytrain operated heavily loaded, at or above their service ceilings. The weather was extremely changeable and plagued with high winds, severe turbulence and structural icing. Few navigation aids existed and emergency landing fields weren't to be found. When inclement weather wasn't a factor, pilots faced enemy fighter attacks.

Due primarily to weather-related accidents, 460 aircraft and 800 crewmembers were lost along what became known as the "aluminum trail." Losing an engine over the corrugated terrain was full justification for "high anxiety."

## Desert Resupply

This painting might have been a classic travel poster contrasting a modern transport aircraft with the ancient "ship of the desert," the camel of Jordan. Although threatened with obsolescence by technological advance, like the camel, the twin-engine DC-3 "Gooney Bird" flies on in many parts of the world where she provides vital services for those in need. In this case, the DC-3 might be delivering food and medical supplies for an ailing Jordanian child.

In the mid-1980s, Bill and Kristi Phillips were invited to visit Jordan. The artist was commissioned to produce paintings for the Royal Jordanian Air Force, and had the opportunity to fly in the Mirage F-1. He supped with the Bedouins and was given unlimited access to ancient archeological sites. The experiences inspired sixteen paintings.

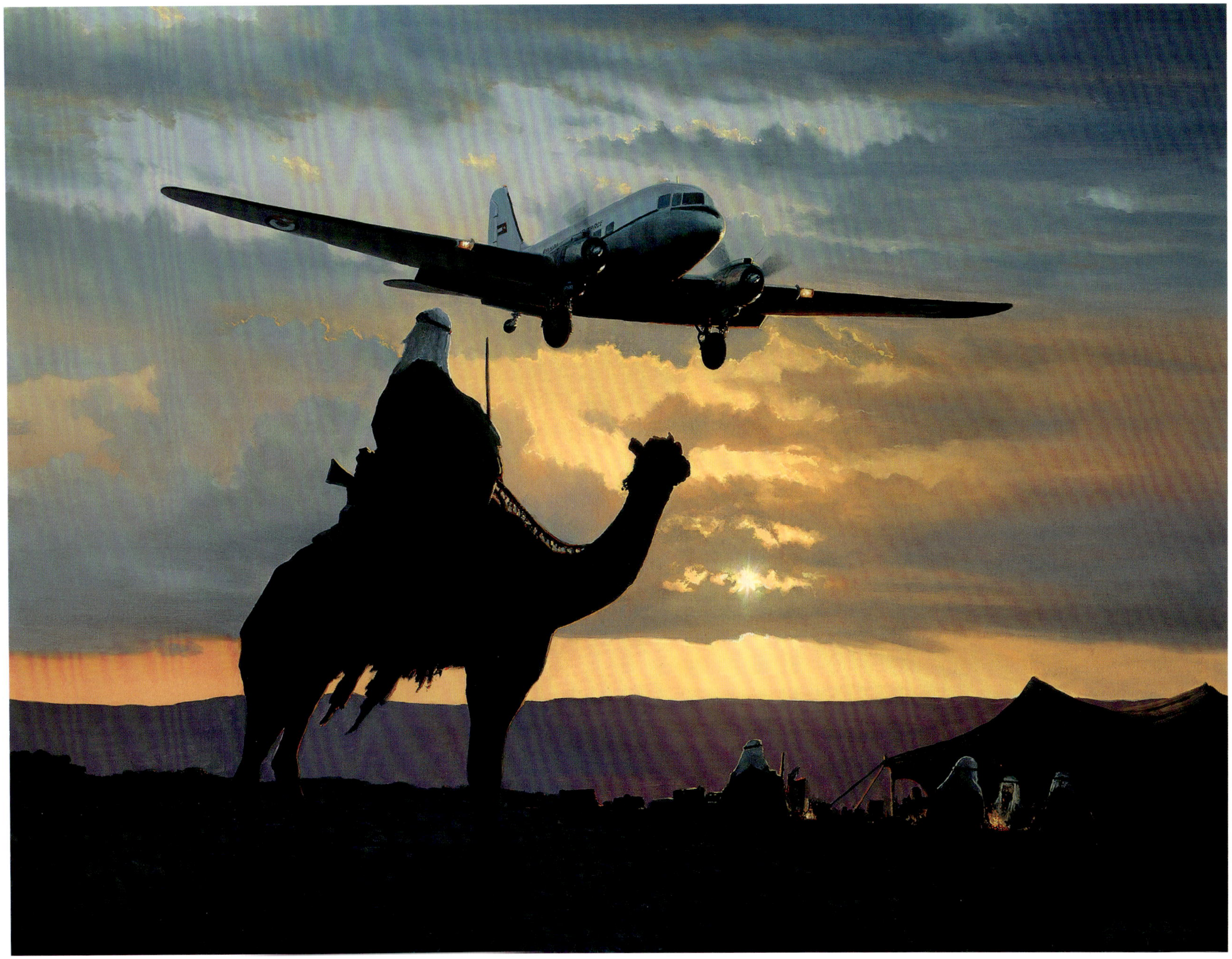

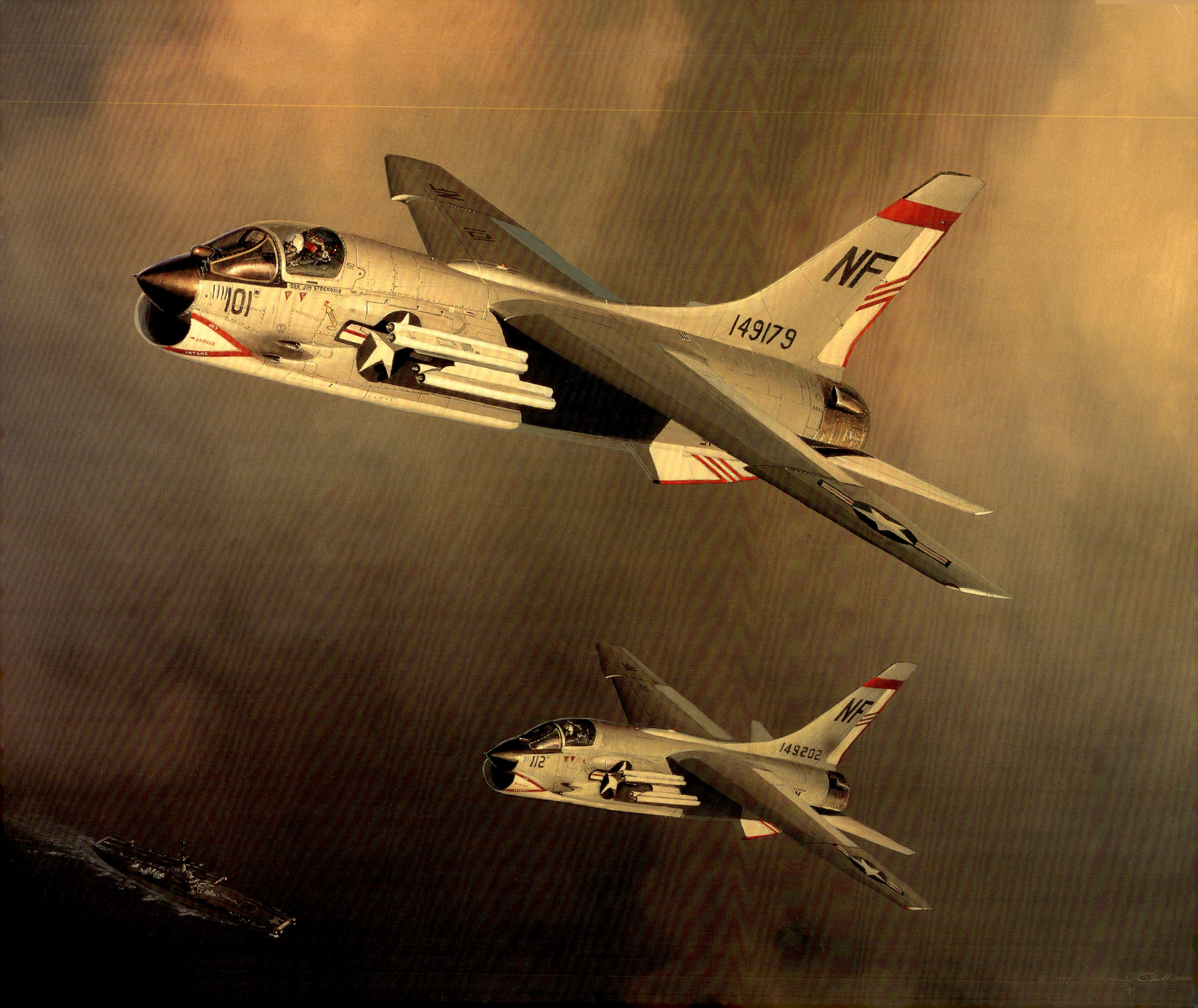
NF
149179
101
NF
149202
112

## Breaking Out on Top

A flight of Vought F-8E Crusaders break out above USS *Ticonderoga* (CVS-14) as they begin another mission against well-defended North Vietnamese targets. In the foreground, then-Commander and later Vice Admiral James B. Stockdale, a true American hero, took the lead. A graduate of Annapolis in 1947, he spent the major portion of his thirty-seven-year career as a carrier-based fighter pilot. He led the first American air strike against the North Vietnamese in August 1964. He was shot down and damaged his knee in the ejection on his third tour to the war zone, in September 1965. Captured and incarcerated in Hoa Lo prison in Hanoi, Commander Stockdale was destined to spend seven-and-one-half-years in brutal captivity. Yet, his leadership prevailed. He used the "tap code" to communicate and, for encouraging defiance by his fellow prisoners, he suffered four of his years in solitary confinement and two in leg irons.

When the Vietnam conflict ended in 1973, Stockdale and his prison mates finally came home. His damaged knee and a hearing loss caused by repeated prison beatings made flying difficult and risky. He was promoted and reassigned as President, Naval War College. Admiral Stockdale, the only 3- or 4-star officer to wear both the gold naval aviator wings and the Medal of Honor, retired from the Navy in 1979. This hero's naval career, his leadership and courage as a prisoner of-war and his contributions to U.S. higher learning are the stuff of legends.

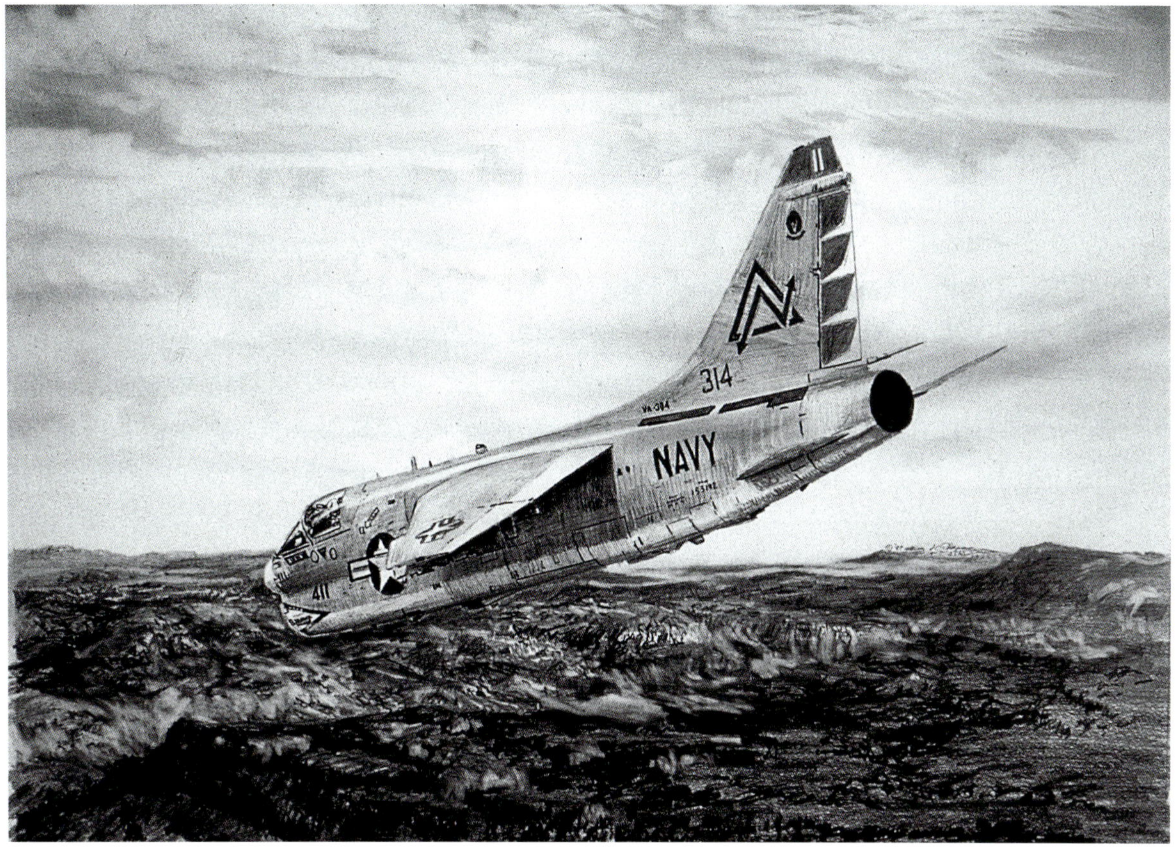


The A-7 Corsair II ground attack aircraft rolls in on target. Bill Phillips flew on just such a mission with the Arizona Air National Guard.

# On the Tail of the Whale

Two thirsty U.S. Navy "Red Rippers" approach a "whale," their KA-3 aerial tanker, to refuel. It is the 1960s and the two Vought F-8E Crusaders of the U.S. Navy's VF-11 have departed the USS *Franklin D. Roosevelt* (CVN-71). In 1952, the Navy Department announced a need for a rugged day fighter capable of supersonic speed and sustained carrier operations. Eight manufacturing companies rushed to their drawing boards and competition fell to Grumman's F-11F Tigercat and Chance Vought's XF8U-1 Crusader.

The Crusader was chosen. The prototype first flew at Edwards AFB in the Mojave Desert in 1955 and, impressively, exceeded MACH 1 in level flight, the first shipboard aircraft to achieve this milestone. It met the Navy's requirement for speed and, with a variable incidence wing and an interconnection of ailerons and flaps with the leading edge of the swept wing, introduced advanced technology able to slow the craft to a 100-knot landing speed, considered ideal for carrier operations. Following successful test flights at Edwards, carrier trials were conducted aboard *Forrestal* (CV-59) and the smaller *Bon Homme Richard* (CV-31) with similarly good results. The Crusader captured outstanding awards including the Collier Trophy for design and development, and the Thompson Trophy for attaining a speed in excess of 1,000-mph. More than 1,200 Crusaders had been built by the time production of the last of nine variants ended in late 1969.

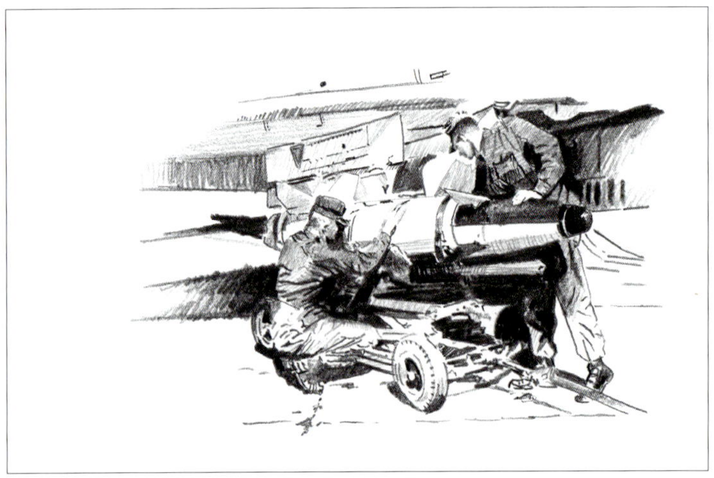

Setting the ordinance.

The Douglas A3D Skywarrior first flew in 1952. Powered by two Pratt and Whitney J57 engines, the Skywarrior was designed to serve as a nuclear weapons carrier and was the largest and heaviest aircraft intended for carrier operation. Fifty A3D-1s were built to fully investigate the nuclear weapons carry-and-deliver concept and a follow-on run of 164 aircraft permitted the development of three versions. The A-3B continued the original concept, the EA-3B was for Electronic Countermeasures (ECM) with four operator positions in the weapons bay, and the KA-3B fulfilled the role of refueling tanker. Officially the Skywarrior, the A-3 was more popularly known as "the Whale." It was big, heavy and it tended to bob up and down on an arrested landing.

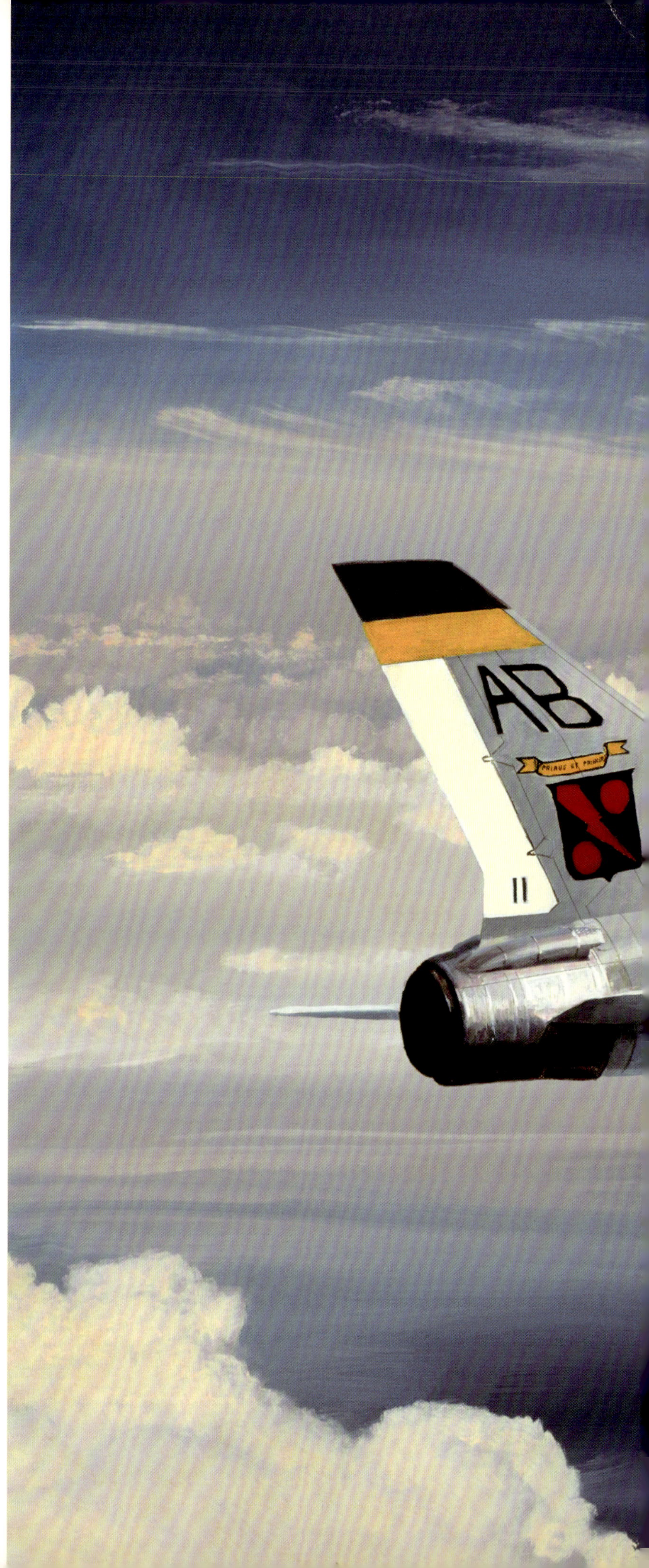

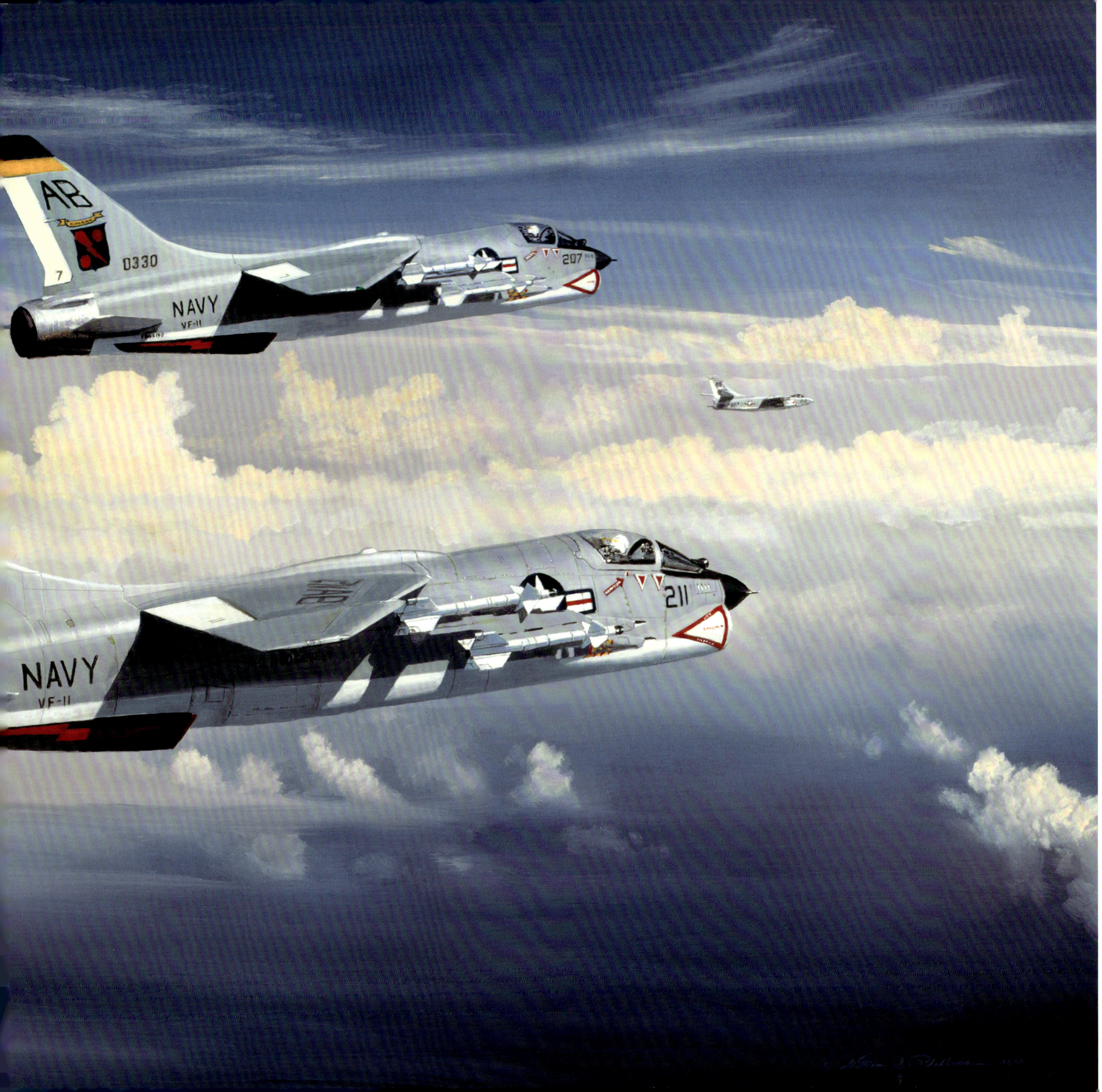
AB
0330
NAVY
VF-11
207
NAVY
VF-11
211

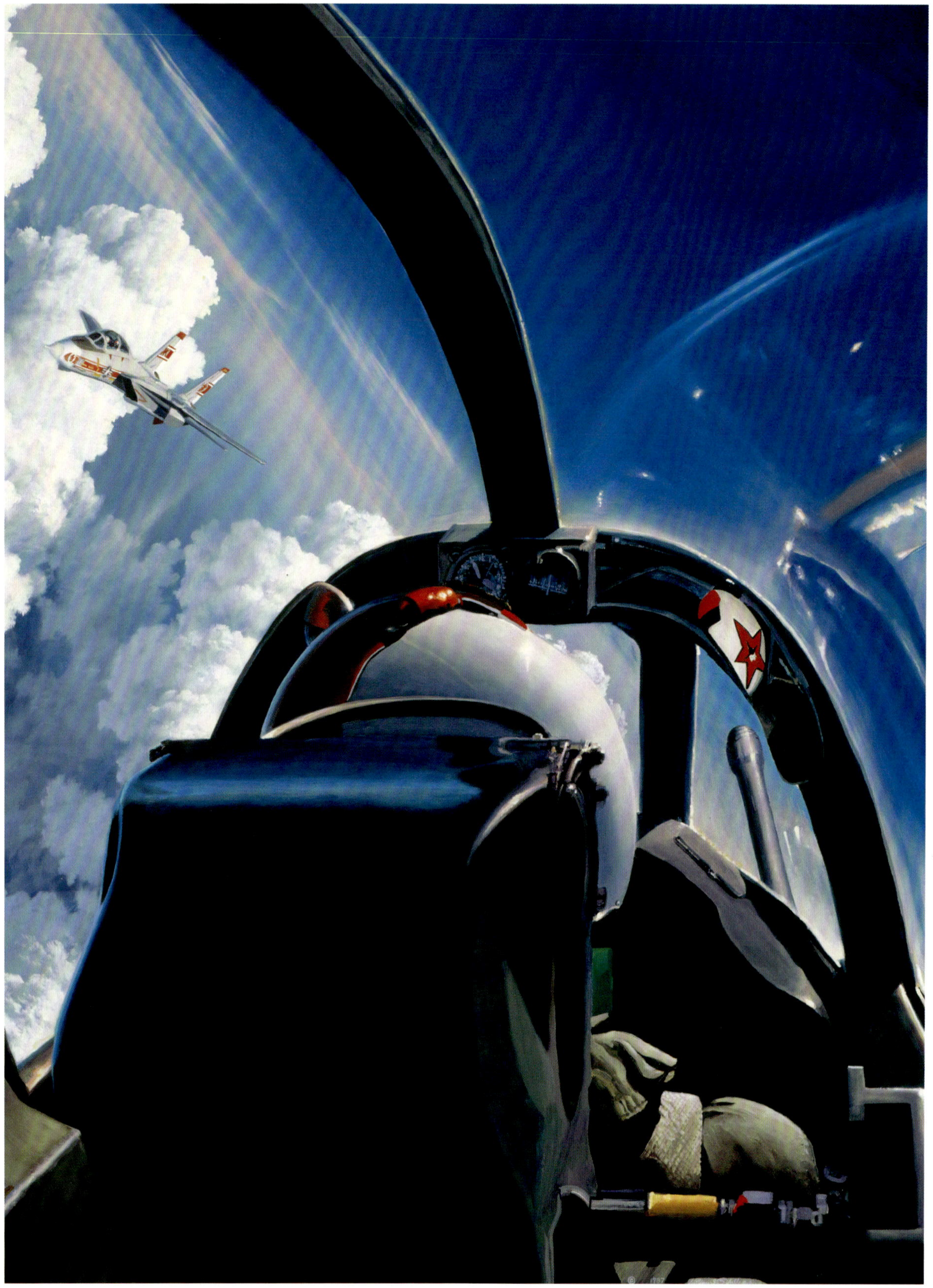

## Heads Up, Heads On

Top Gun, Naval Fighter Weapons School, a graduate school in Air Combat Maneuvering (ACM) training, utilizes McDonnell Douglas TA-4J Skyhawks of VF-126 as adversaries—bandits. As a backseater, Bill lived this head-to-head confrontation pitting the F-14 Tomcat against his TA-4J. Thanks to the Heads Up Display (HUD) his pilot's vigilance can remain outside the cockpit while momentarily referring to essential information projected directly in his field of vision.

## Into the Sunlit Splendor

Mount Hood towers over the Columbia River and the Willamette Valley of Oregon while this Oregon Air National Guard McDonnell Douglas F-4C Phantom II breaks through into the sunlight and rockets to altitude. A full moon adds to a charade of tranquility, but the mountain's cap of lens-shaped clouds portends high winds and strong downdrafts as the 123rd Fighter Interceptor Squadron, 142nd Fighter Interceptor Group's Phantom scrambles out of Portland into the possible turbulence of national defense.

Serving on twenty-four-hour alert as part of North American Air Defense Command, its crew is charged with interception, identification and, if necessary, destruction of an unidentified intruder. Then, as now, primary responsibility for the air defense of the United States falls to the Air National Guard. Currently a Fighter Wing, the 142nd was launched in 1941 as the 123rd Observation Squadron. It was later federally activated to perform observation and reconnaissance in the China-Burma-India Theater and in Europe during WWII, and mobilized to support the war in Korea. Now flying the F-15 A/B Eagle, the "Redhawks" continue to provide air defense and air superiority. As with all Air National Guard units, the 142nd maintains a high degree of skill and readiness as part of the All-Volunteer Total Force.

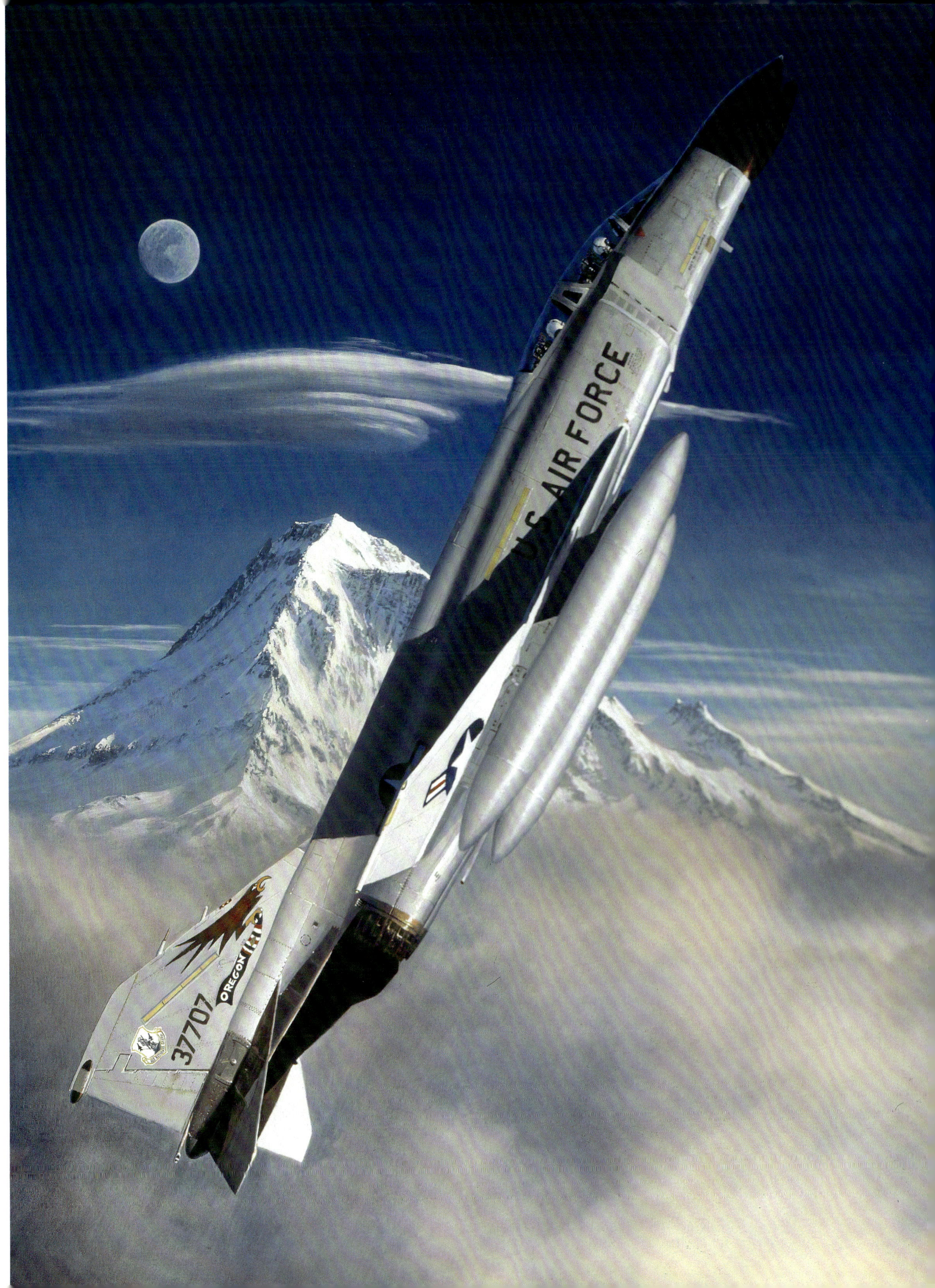

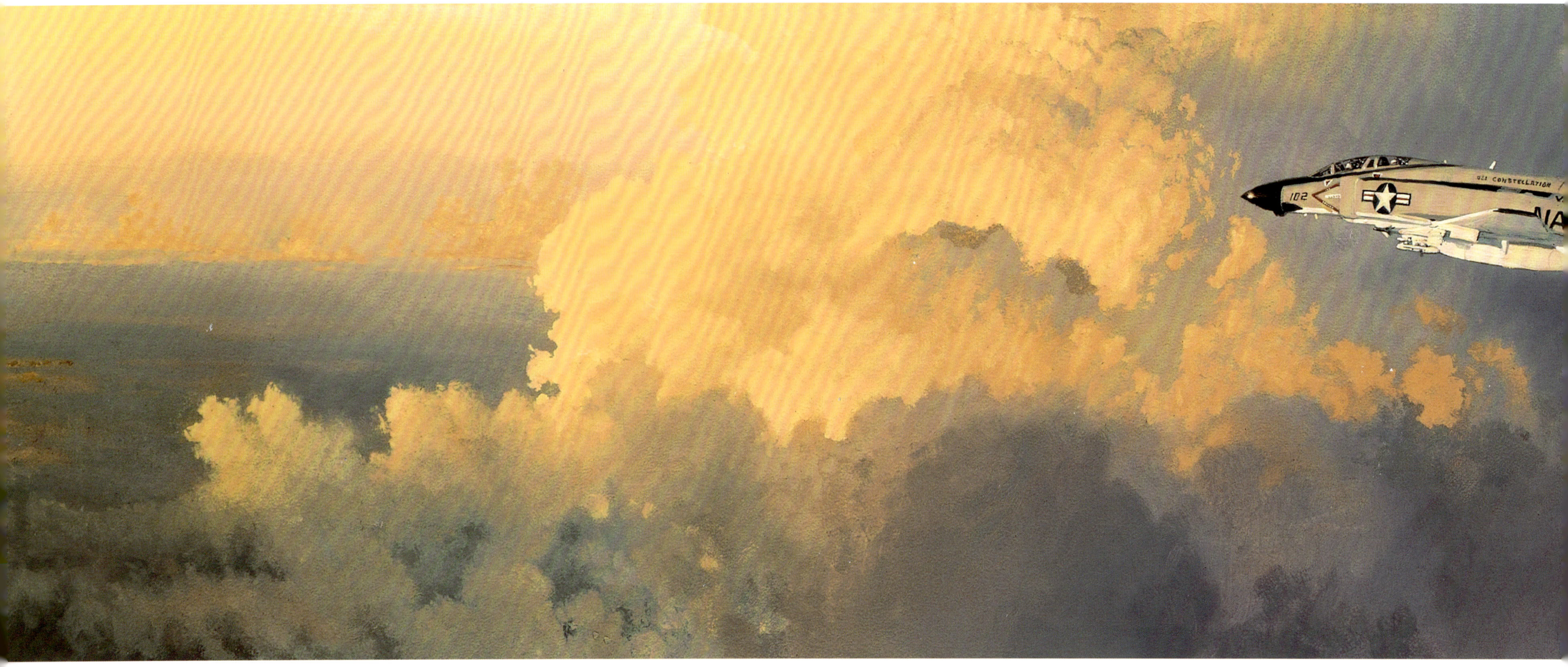

## Phantom Thunder

On May 10, 1972, Lt. Randy "Duke" Cunningham, his Radar Intercept Officer (RIO) Lt. (jg) "Willie" Driscoll, and their wingman Commander Dwight Timm, were jumped by twenty-two MiG-17, MiG-19, and MiG-21 fighters. A frenzied dogfight erupted as they came off a flak suppression strike against the Hai Dong railyards near Hanoi, North Vietnam. In downing three MiGs, Cunningham and Driscoll, flying Number 5800, the Carrier Air Group Commander's F-4 off the USS *Constellation* (CV-64), were credited with saving Timm's life. Adding to two previous victories, they became the first American aces of the Vietnam conflict and first to down the enemy with missiles.

En route to the carrier, their Phantom was struck by a surface-to-air missile. Cunningham and Driscoll ejected over the Gulf of Tonkin and were rescued by helicopter. The cruise of the *Constellation* was Duke Cunningham's second combat tour with VF-96. After an earlier cruise on USS *America* (CV-66), he completed the Navy's "Top Gun" Fighter Weapons Course before returning to action over Southeast Asia where he flew a total of 300 combat missions over North Vietnam and Laos. He later commanded VF-126, an adversary squadron designed to provide realistic combat training for aircrews.

The movie, *Top Gun*, was based on many of Duke and Willie's experiences in combat and at what the Navy dubbed Fightertown, USA. Highly decorated, Duke Cunningham retired from the Navy in 1987. Elected to the U.S. House of Representatives from California's 51st Congressional District in 1998, Congressman Cunningham has strongly supported military strength and preparedness.

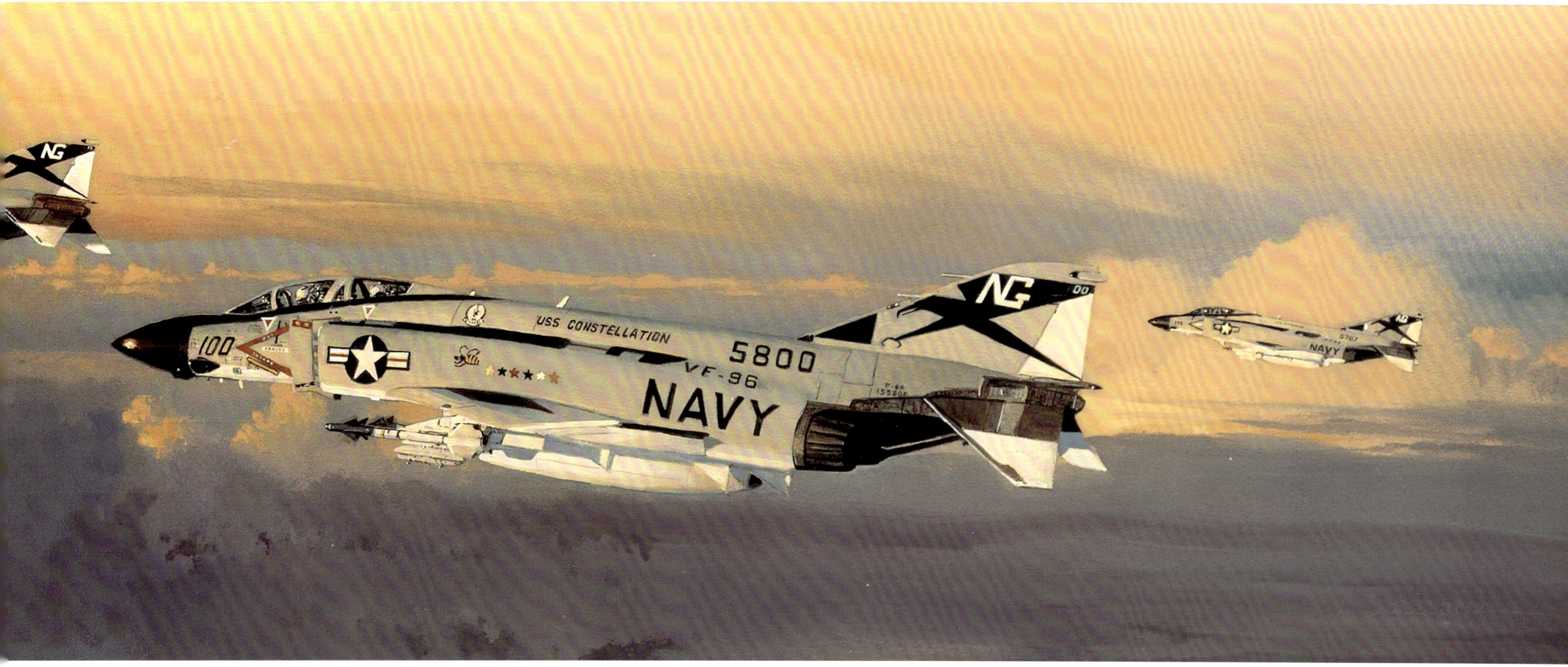
NG
USS CONSTELLATION
5800
VF-96
NAVY
100

# Sierra Hotel

The curtain was lowered on the Sundowners of VF-111, after fifty-three years of service, when it was officially disestablished following ceremonies on February 16, 1995 at Naval Air Station Miramar, California.

VF-111 is one of only two squadrons in U.S. Naval history to have fought in three wars—World War II, Korea and Vietnam. Their insignia, two F-4F Wildcats shooting down the Rising Sun, dates to WWII. In Korea, the Sundowners completed three combat tours in three years. Artist Bill Phillips had the opportunity to observe the Sundowners in action while flying against them in an F-106 during an Air Combat Maneuvering (ACM) mission at McChord AFB in 1977. This painting is his tribute to their McDonnell F-4N Phantom of the Vietnam era.

The F-4N substantially upgraded its predecessor, the B-model. F-4Bs served as defense fighters with secondary ground attack capability and an impressive ability to carry a 16,000-pound ordnance load, four Sparrow or four Sidewinder missiles. Upgrading to F-4N meant structurally rebuilding the B-model and refitting with new equipment, systems and avionics. The first, of 228 built, was delivered in February 1973 and remained in service into the 1980s.

With a proud service to its country during the Vietnam conflict, VF-111 made seven combat cruises and flew more than 12,500 missions over North and South Vietnam, Cambodia, Thailand, and Laos. It had transitioned from F-8 Crusaders to the shark-toothed F-4B Phantoms in 1971 and took to the USS *Coral Sea* deploying on what proved to be their last combat cruise to Southeast Asia. Their dramatic Phantoms could boast an impressive list of individual Battle Efficiency "Es" for the entire range of F-4 operations. It was the first jet aircraft purchased for the Navy from McDonnell Aircraft Company and the first purchase of a U.S. Navy fighter by the U.S. Air Force.

Blurring the foreground to demonstrate the speed and power of the Phantom, Phillips depicts this proud representative of VF-111 skimming the desert sand in a low-level dash. This highly recognizable F-4 Phantom, bearing the colors, markings, and insignia of VF-111, pays homage to a proud naval squadron.

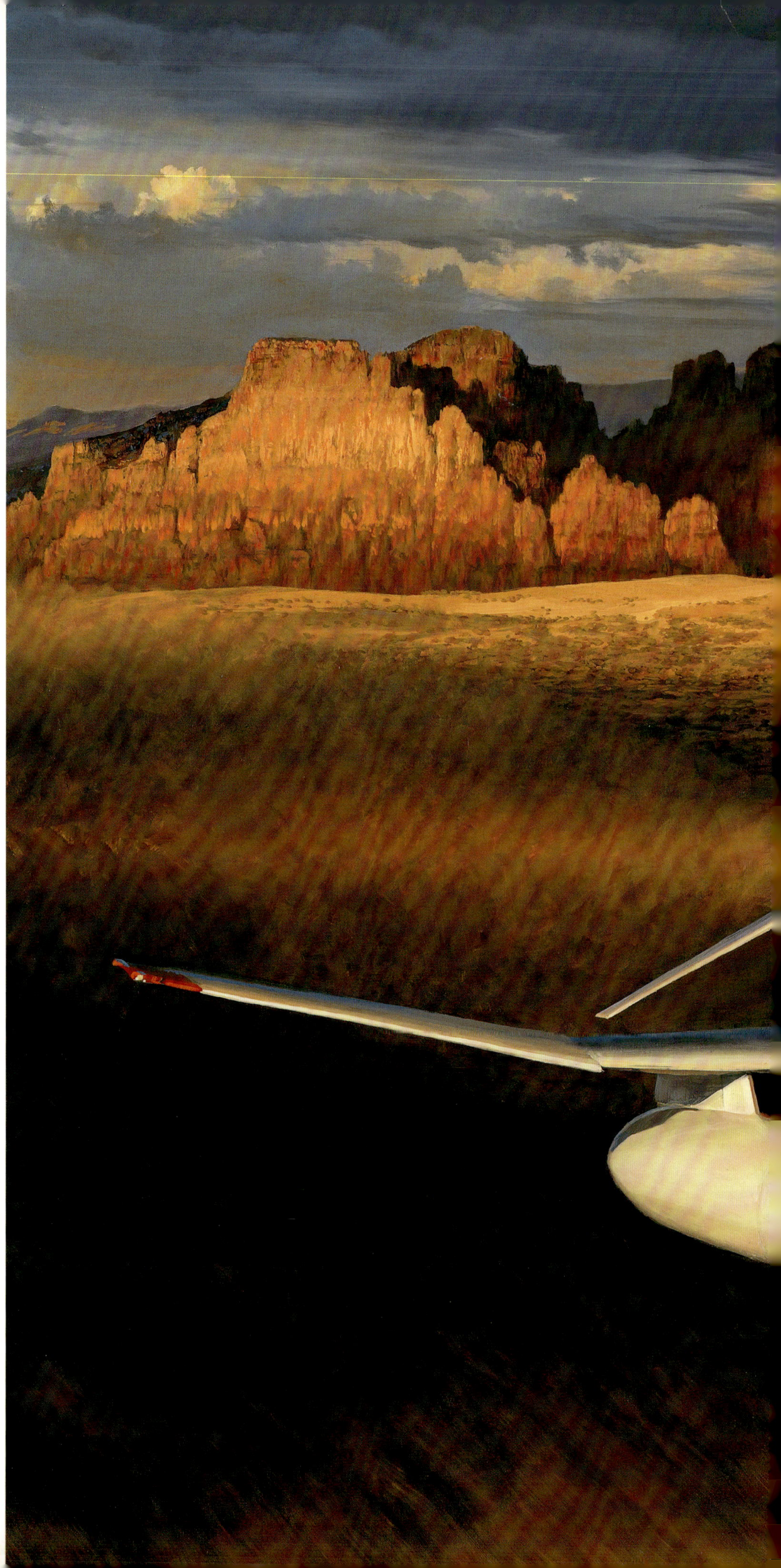

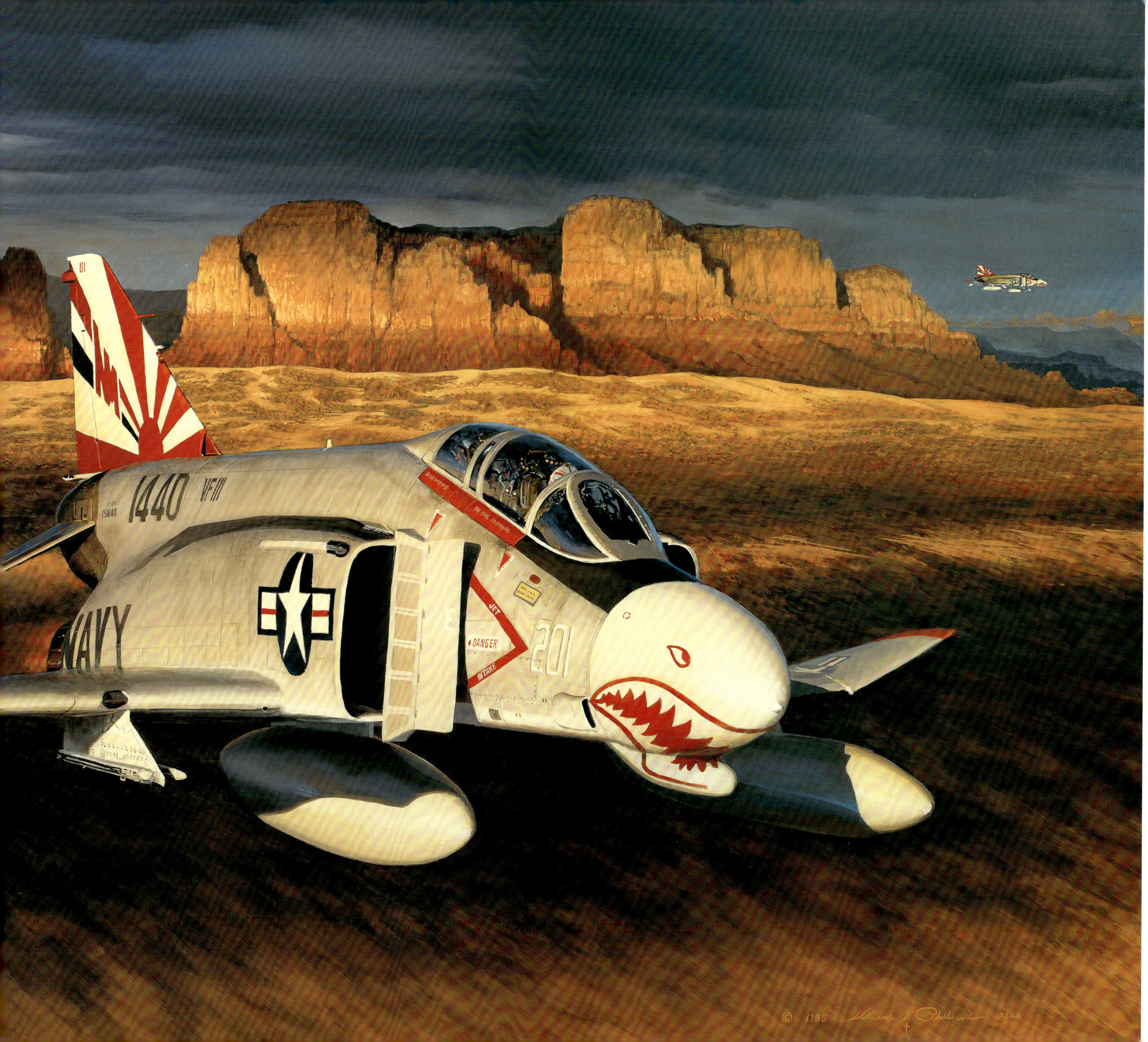
1440
VF111
NAVY
201
DANGER
JET
INTAKE
© 1988
ASAA

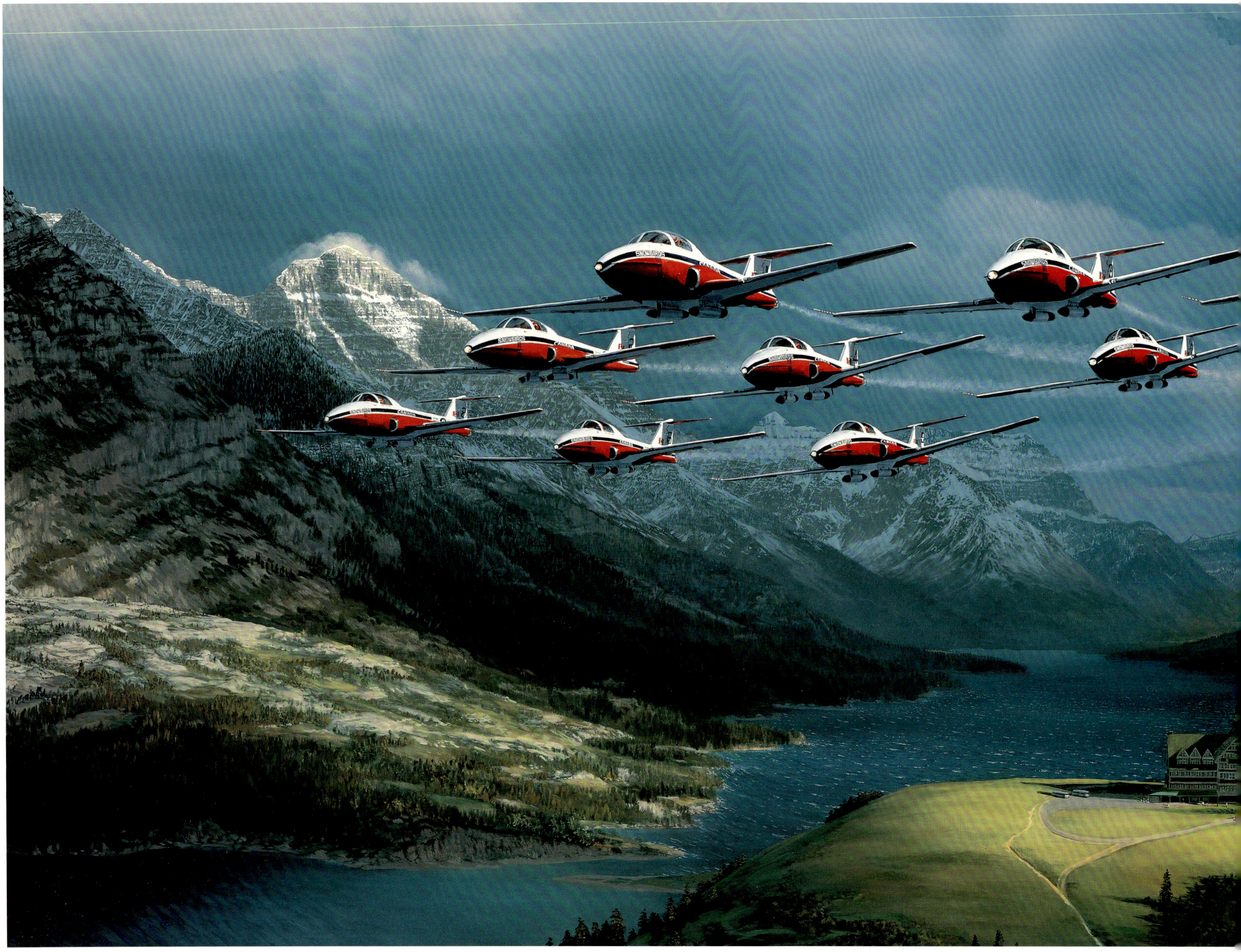
SNOWBIRDS
CANADA

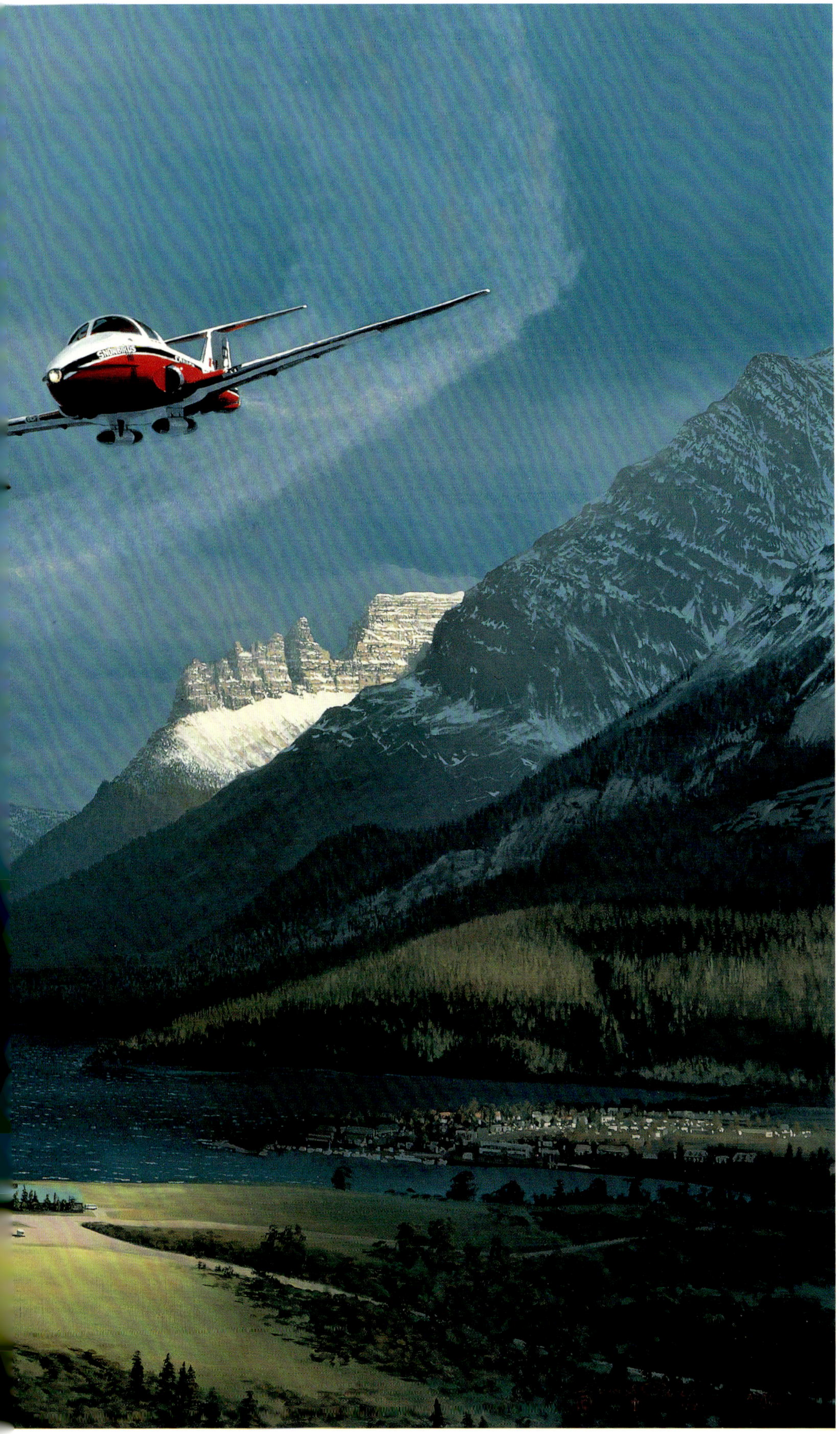

## Sunward We Climb

Swinging smoothly through the bottom of its nine-ship loop, one of the world's premier military flight demonstration teams performs over the Prince of Wales Hotel, Waterton Lakes National Park, Alberta. As the Canadian Forces Snowbirds slip in and out of the filtered sunlight, their performance of precision aerobatics is made even more spectacular when showcased within the natural bowl formed by the majestic Canadian Rockies and the International Peace Park, a jewel in Alberta's southwest corner.

Based at Moose Jaw, Saskatchewan, the Snowbirds brightly painted Canadair CT-114 Tutor aircraft vary from the Canadian Forces primary trainer only in their unique paint scheme. An agile craft, the Tutor lends itself well to the graceful, flowing aerial choreography for which the Snowbirds are known.

Now called the 431 Air Demonstration Squadron, the Snowbirds date to the November 11, 1942 formation of the 431 Bomber Squadron at Burn, England. With the motto, "Warriors of the Air" and the profile of a First Nations Iroquois on the squadron badge, squadron pilots flew against Axis targets in Europe prior to returning to RCAF Station Dartmouth, Nova Scotia and disbanding on September 5, 1945. On January 18, 1954, 431 Fighter Squadron equipped with the Canadair/North American F-86 Sabre was introduced at RCAF Station Bagotville and displayed the Sabre to the North American public at air shows. Disbanded in December 1954, it was reactivated in 1978 as the 431 Air Demonstration Squadron. It was presented with squadron colors in 1999 in recognition of twenty-five years of service.

Members of the team are highly skilled and experienced. All have served as Canadian Forces personnel and have accumulated years of military service prior to joining the squadron. They are selected on the basis of skill, capability, and performance trials, and take enormous pride in performing as extraordinary ambassadors for Canadian Forces. More than sixteen million spectators have thrilled at their showmanship and excellence.

## Cape Neddick Dawn

Dawn breaks sunny and cold, but the lighthouse keeper's house glows with warmth. You can almost smell coffee brewing and Maine blueberry pancakes on the griddle. The bomber's crew must be relieved to be nearly home.

Cape Neddick Lighthouse is built on a nubble island, a geographic term for a small rock island jutting into the ocean. The lighthouse, built in 1874 only 100 yards off the mainland, was dubbed the "Nubble Light." A cable trolley was erected to transport supplies to the island and, in the 1960s, the lighthouse keeper used the trolley's basket to carry his son back and forth to the mainland to attend school. Bad publicity forced the U.S. Coast Guard to put a stop to it.

Today, this quaint tourist site belongs to the town of York, Maine. According to legend, in 1842, two crewmembers from the bark, *Isidore*, both dreamed that they would not survive their next voyage. One refused to sail; the other sailed only to perish when the *Isidore* went down off Cape Neddick. It is whispered that a phantom crew in a ghostly vessel sometimes sails past the Nubble Light.

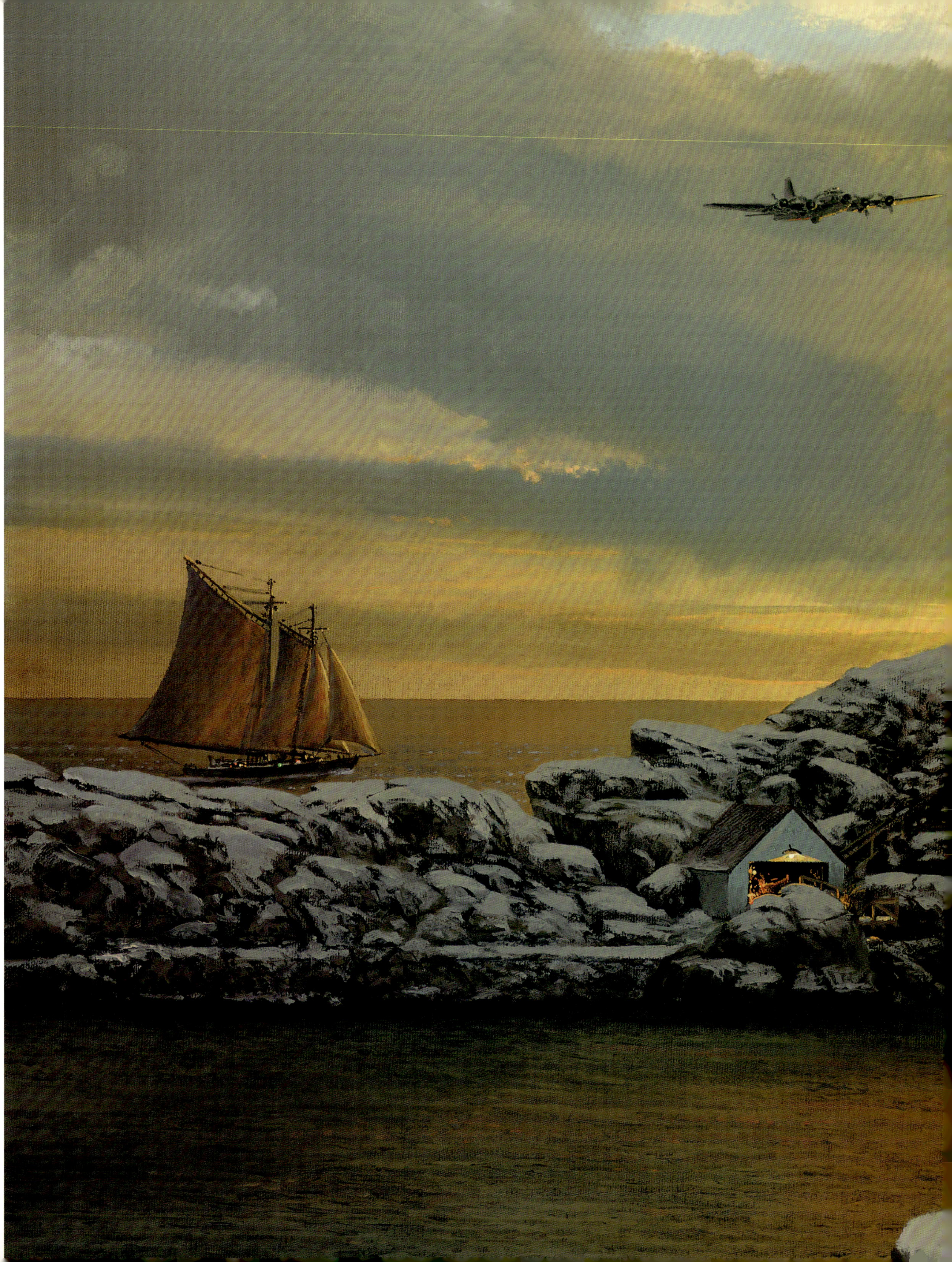

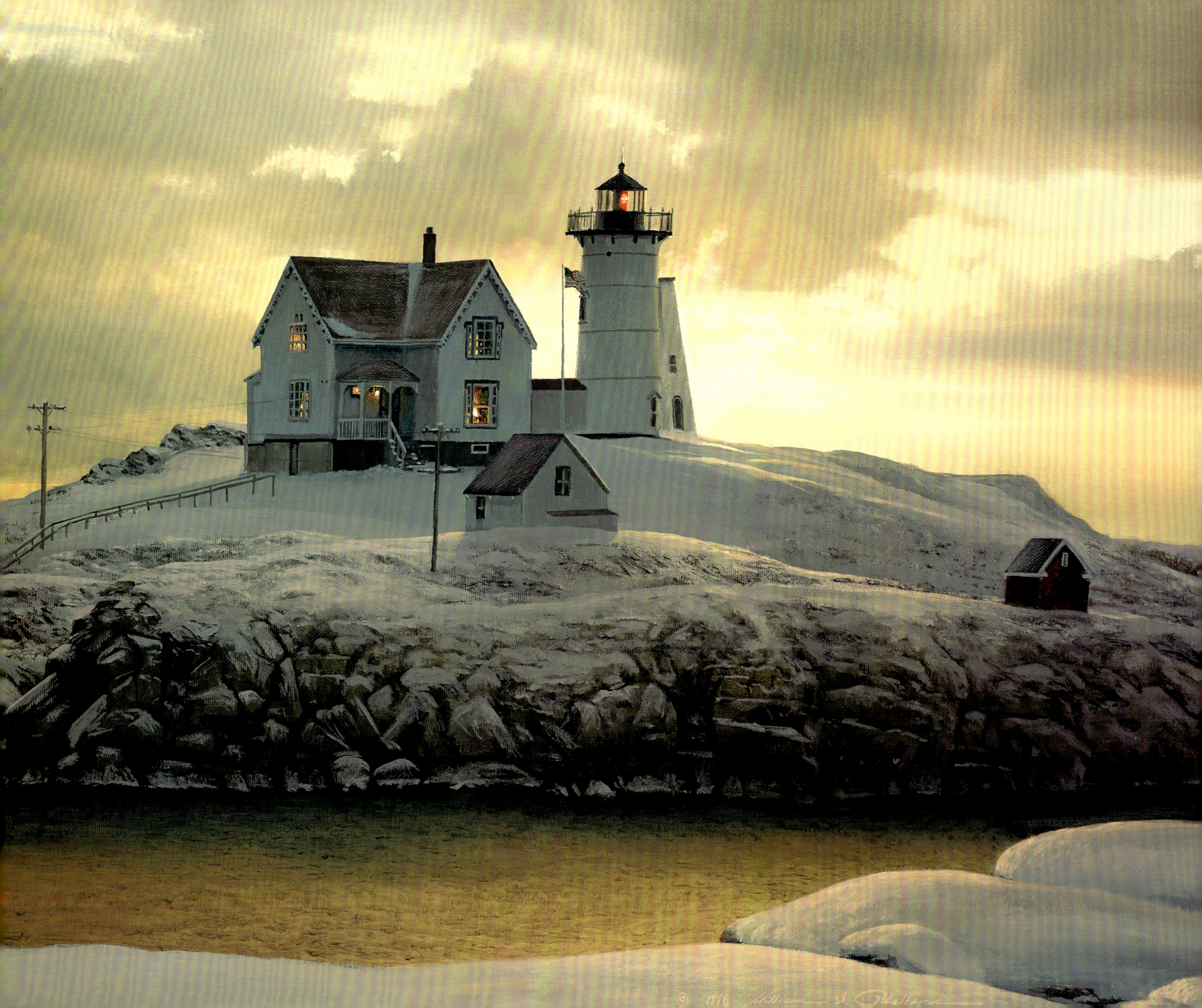

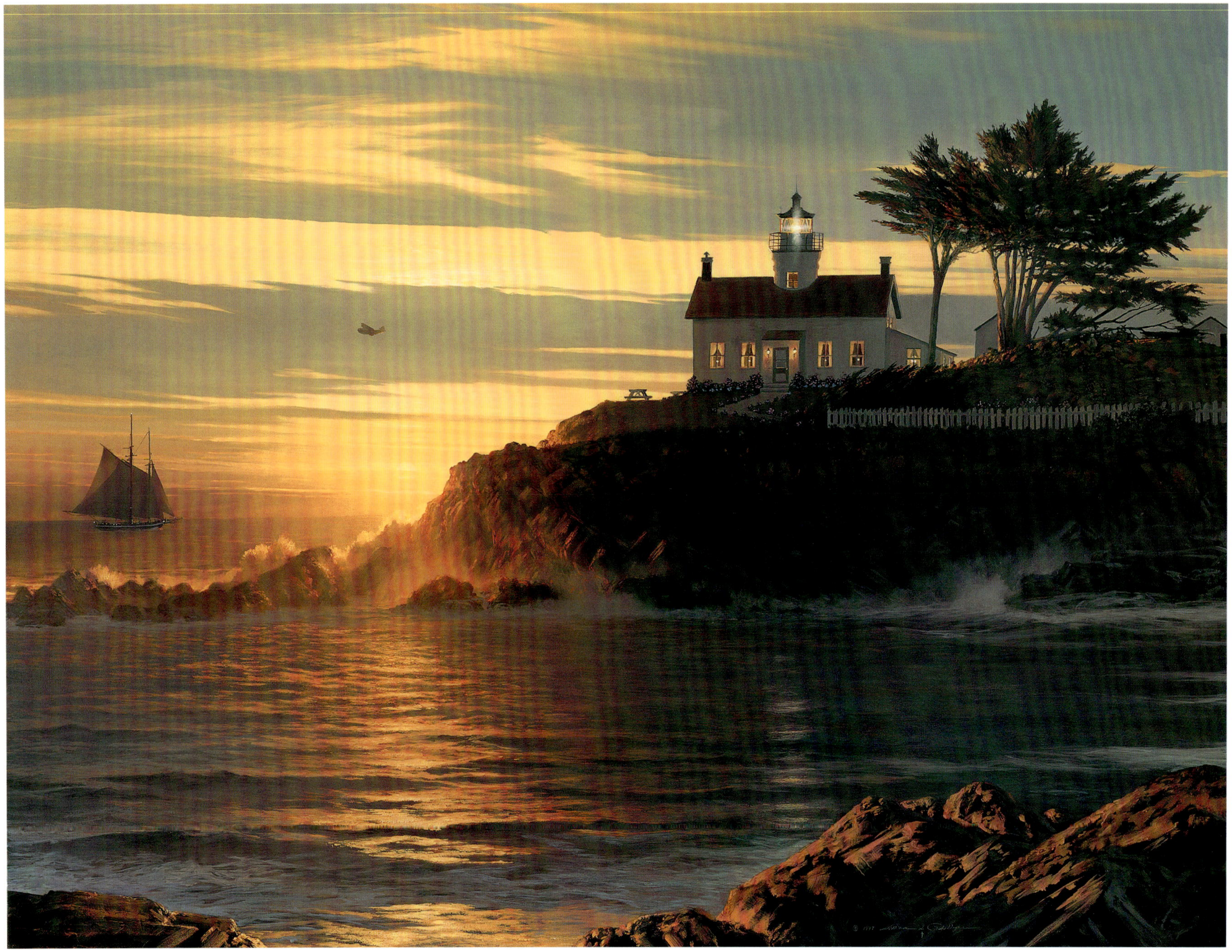

## Sunset Sentinel

Battery Point light, on the northern California coast near Crescent City, shined for the first time in 1856. The light has served to protect coastal shipping for nearly 150 years, except for a period from 1965 to 1982. No doubt the skipper of the schooner passing offshore, and the crew of the westbound airplane above, find the light a comforting navigational aid to their travels.

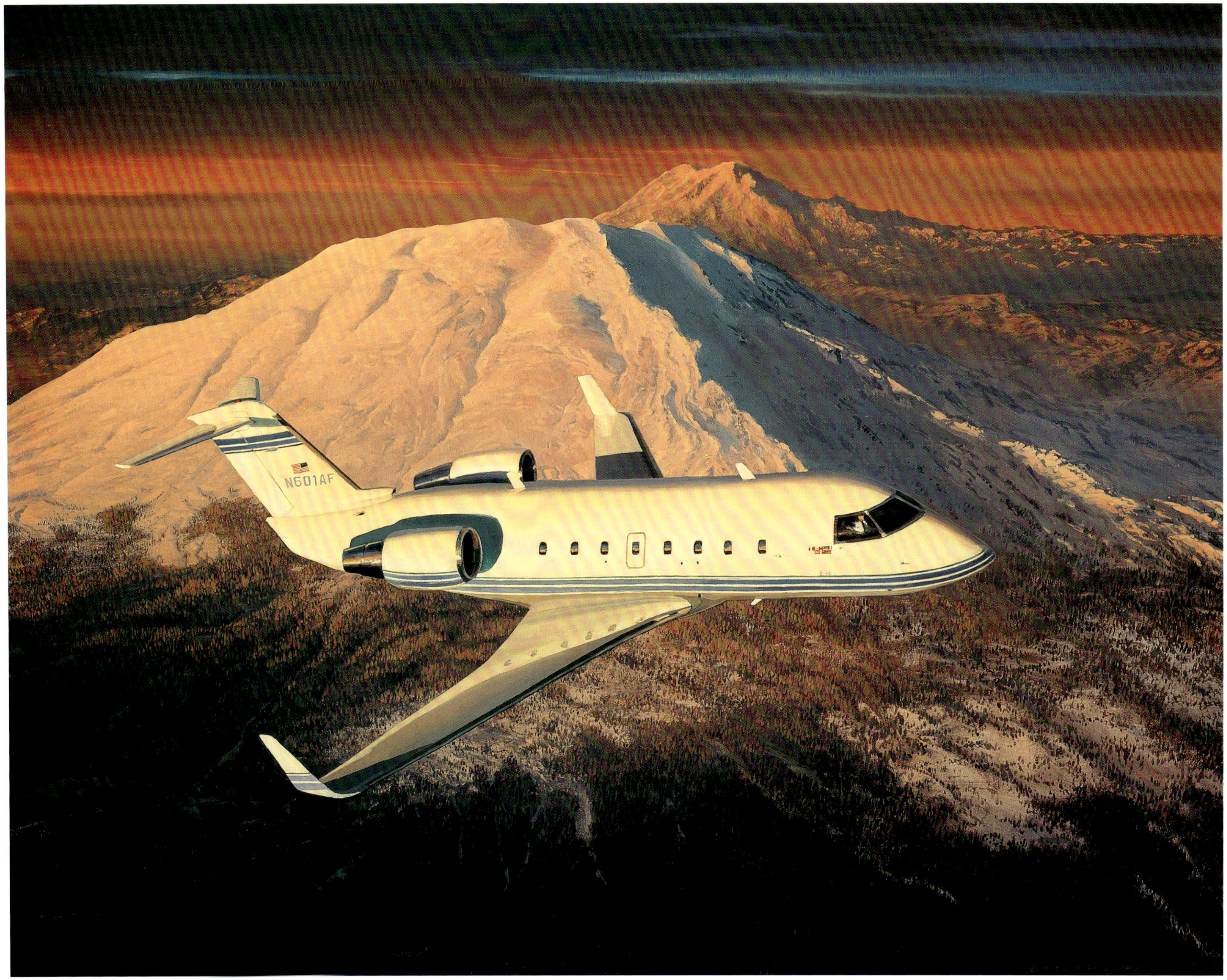


## Falcon Sunset

Against the snow-covered peak of Mount St. Helens, Washington, a Canadair CL-600 Challenger heads east bathed in a dramatic sunset, with Mount Rainier in the distance. The Challenger, designed in 1974 by Bill Lear as the LearStar 600, was sold to Canadair, and the U.S. and Canada both certified the type in 1980. Earliest models were powered by Arco Lycoming turbofans. Eighty-three were built, of which seventy were improved with drag reduction and increased range. The Challenger has become the father of a successful family of medium to long-range wide-body business jets.

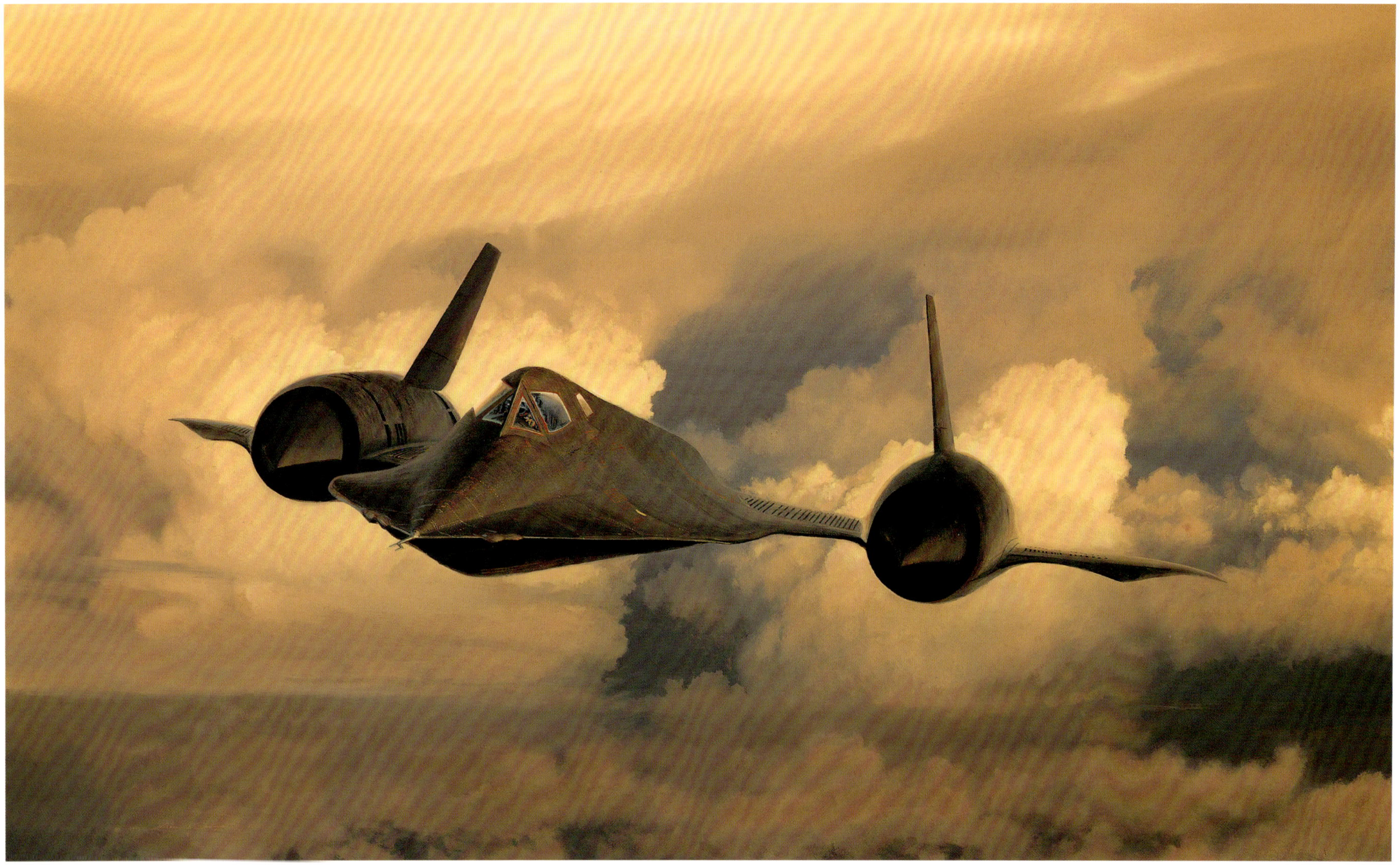

## Riding the Thunder

The Lockheed SR-71 Blackbird is attributed to the genius of Lockheed's famous "Skunk Works" led by Chief Research Engineer Kelly Johnson. Much about the SR-71 remains classified, but its records for speed and altitude are public. In "speed over a recognized course," the Blackbird flew at over 1,800 mph from New York to London in 1974. In 1976, it exceeded an "altitude in horizontal flight" of 85,000 feet. Its true value was its ability to photograph large portions of the earth's surface and to monitor the actions of real and potential enemies.

## Dawn, Desert Storm (opposite)

Dawn comes quickly to the desert as ground crews ready Air National Guard F-16 Fighting Falcons to launch their day's first combat missions. During Operation Desert Shield in 1990, following the Iraqi invasion of Kuwait, F-16s deployed to Al Kharj Air Base in Saudi Arabia directly from their home bases. Tanker support was critical since each fighter refueled as many as fifteen times during the long over-water flight. In the war phase, forty-two F-16s from South Carolina and New York flew 3,645 missions without an aircraft loss.

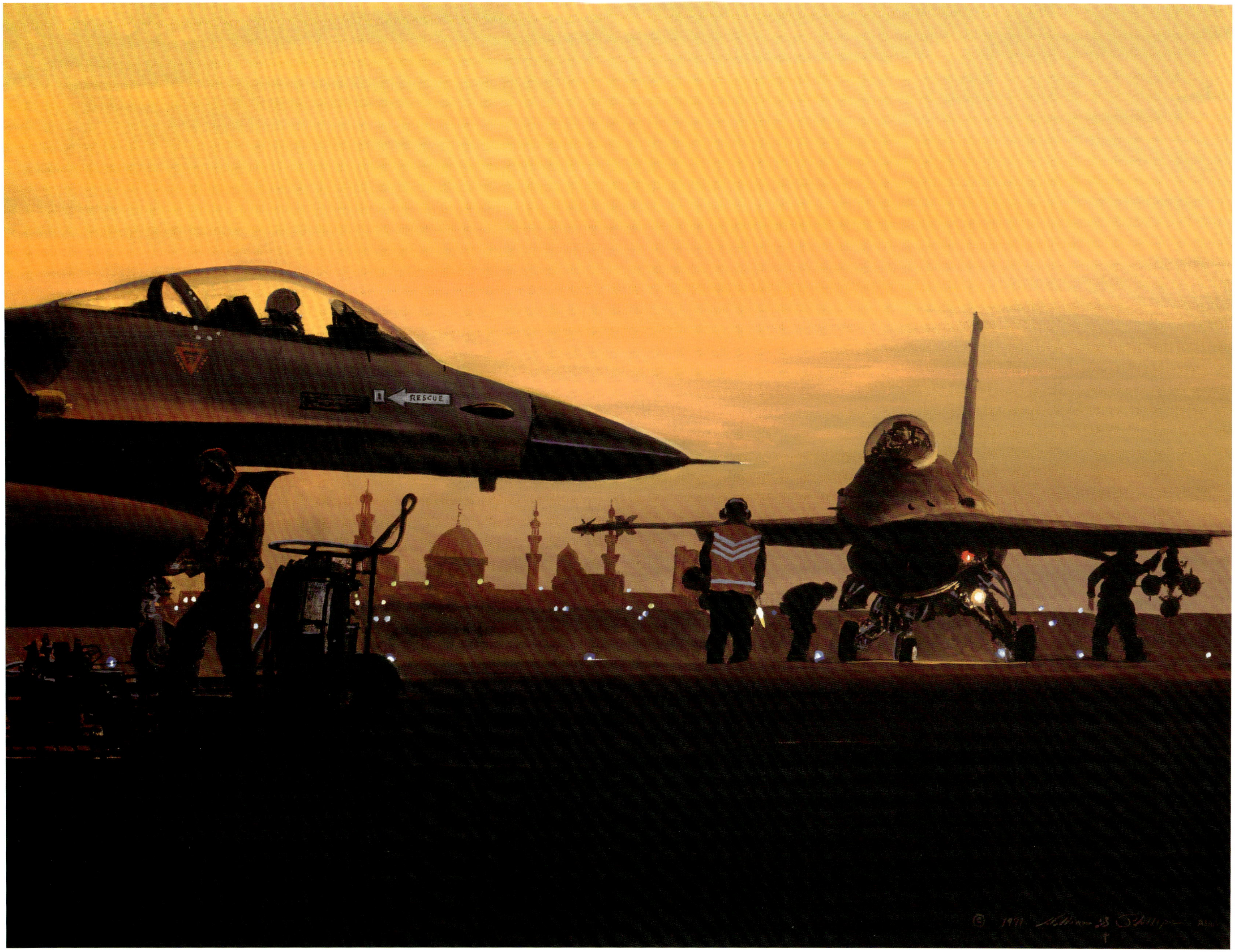
RESCUE

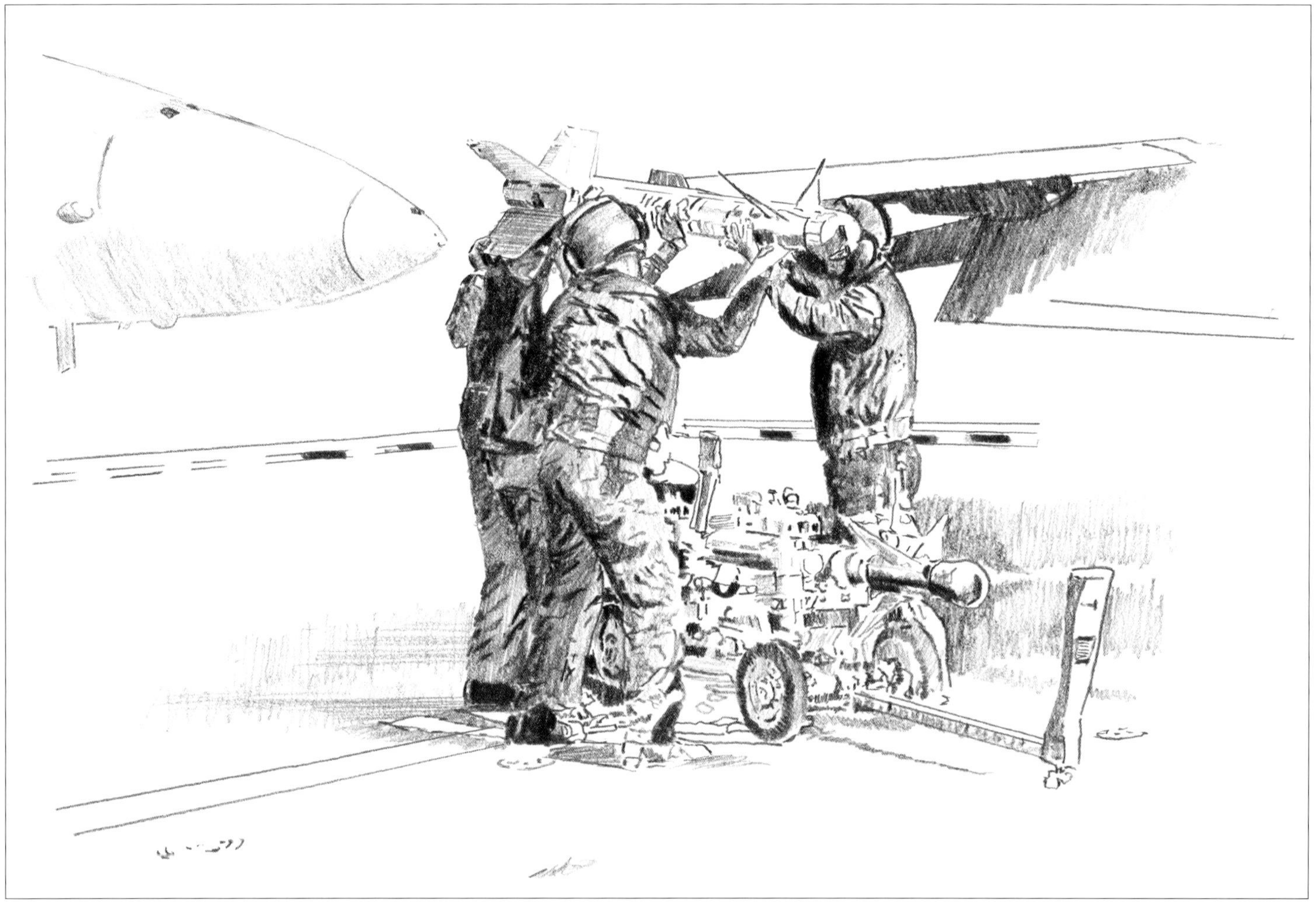

On-scene sketch of ordinance handlers uploading an F18 aboard the USS *Midway*.

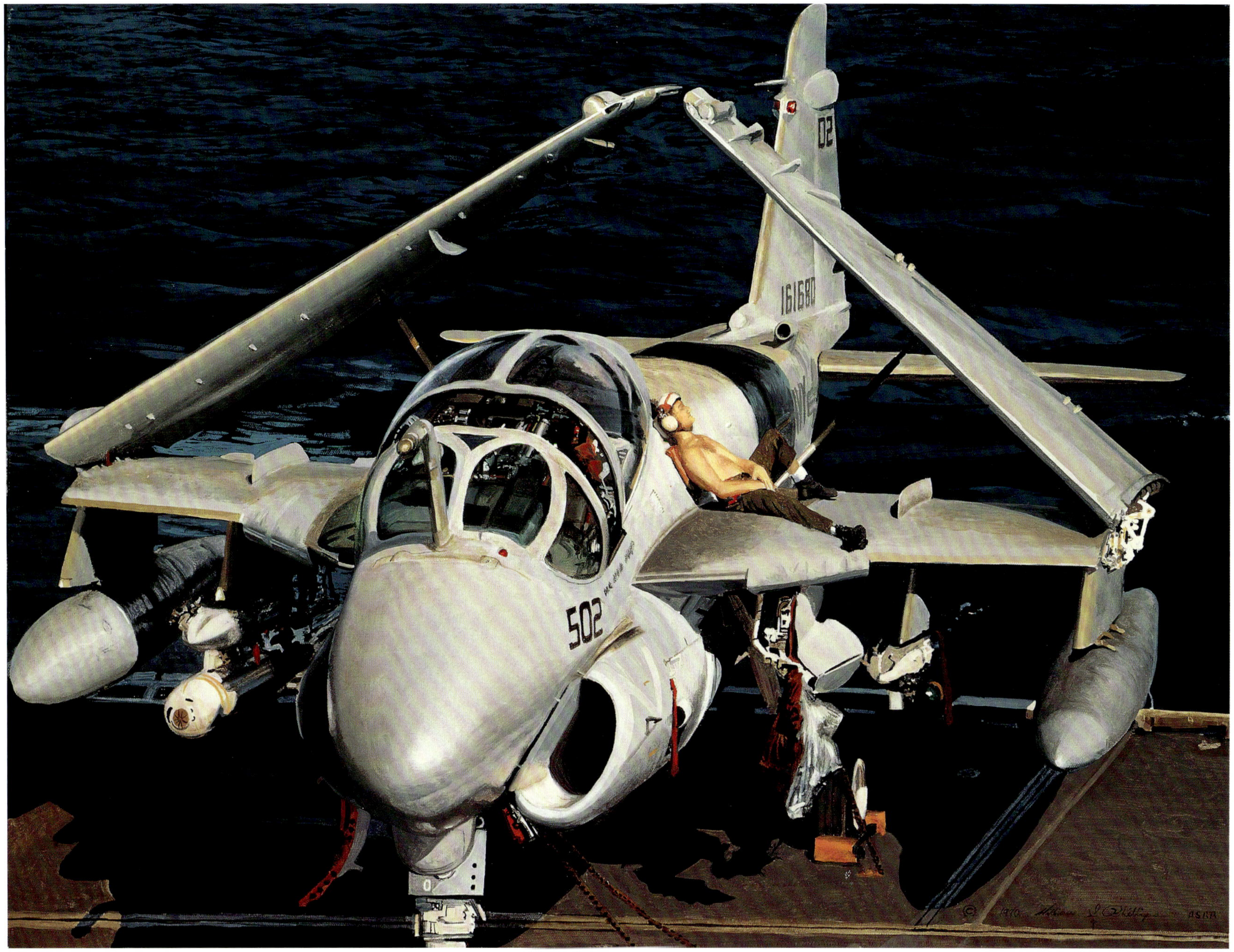

## Catching Some Rays

Breaks in the action on an aircraft carrier flight deck are few. The USS *Midway* deployed to the Gulf of Oman to guard oil tankers and, in a respite from action, this A-6 Intruder awaits its next sortie. The crewman, with his "ears" on, takes advantage of the sun and grabs a quick nap.

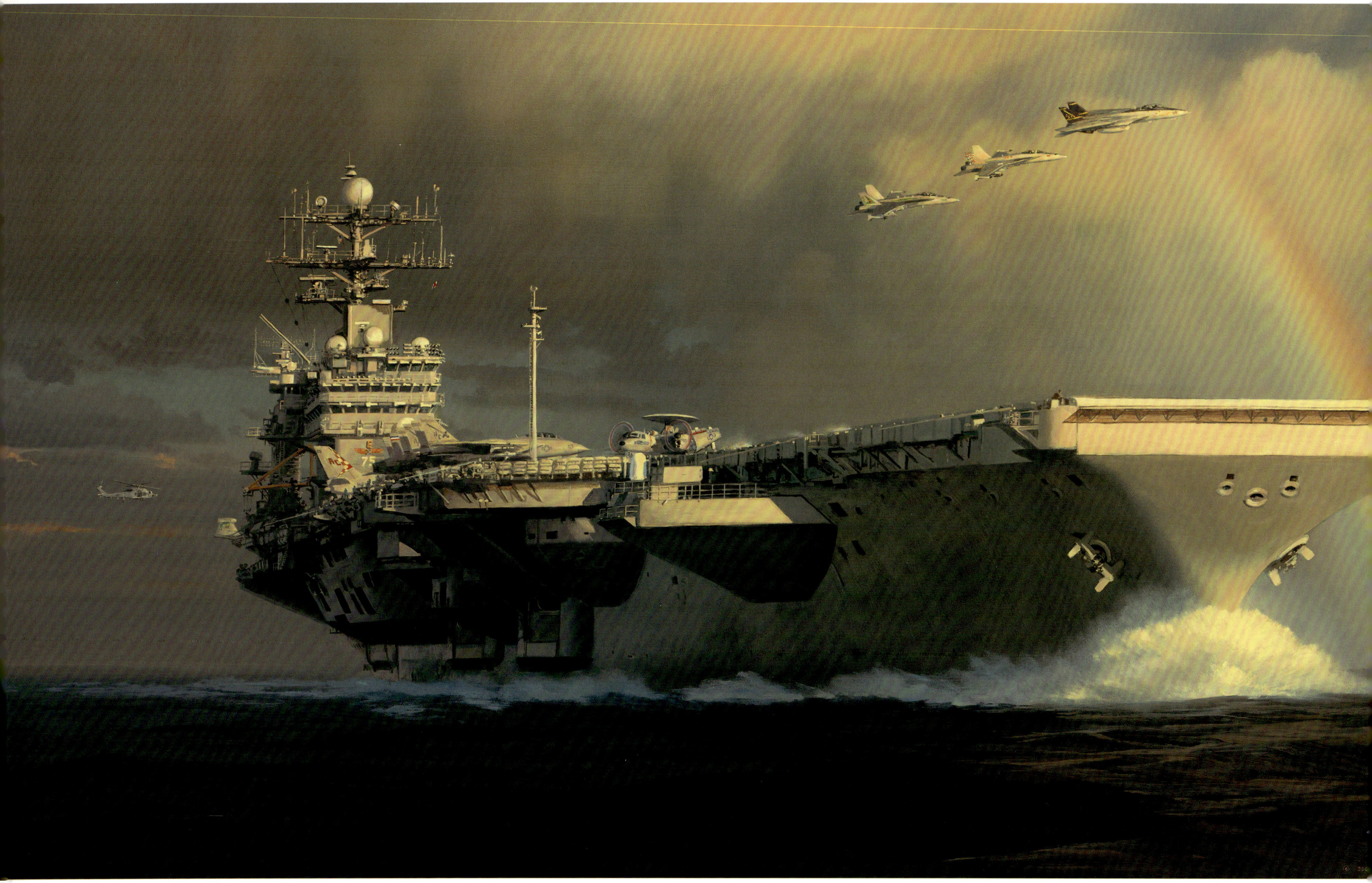

## Out of the Squall Line

This painting of the USS *Harry S. Truman* (CVN-75) is the most recent installment in the artist's carefully researched and meticulously depicted documentation of America's aircraft carriers. During a ten-day visit, he carefully observed details of every aspect of the ship and its flight operations.

Launched at Newport News, Virginia in 1996, the *Harry S. Truman* is truly an awesome ship. With its four catapults and four arresting cables, the flight deck covers four-and-one-half acres, the length of three football fields and the width of one football field. Powered by two nuclear reactors running four main engines that drive four propellers, the *Truman* can exceed thirty knots. A full complement of crew—60 officers and 3,200 sailors—maintains operating conditions that are prepared for contingency action throughout the world. Recently, the *Truman* participated in Operations Enduring Freedom, in Afghanistan, and in Iraqi Freedom, flying a traditional battle flag that pays tribute to its namesake and displays the ship's motto with Truman's famous quote, "Give 'em hell!"

An aircraft carrier's purpose—in war and in peace—is to provide a stable and ready platform from which its air wing can conduct flight operations in a wide variety of aircraft. Made up of almost 2,500 personnel, an air wing maintains and controls approximately 85 airplanes and helicopters. On a recent cruise with the *Truman*, Carrier Air Wing Three provided the aerial punch: Fighter Squadron 32 (VF-32) flying F-14B Tomcats; Strike Fighter Squadron 37 (VFA-37) flying F/A-18 Hornets; Strike Fighter Squadron 105 (VFA-105) flying F/A-18 Hornets; Marine Fighter Attack Squadron 115 (VMFA—115) flying F/A-18A Hornets; Carrier Airborne Early Warning squadron 126 (VAW-126) flying E-2C Hawkeyes; Electronic Attack Squadron 130 (VAQ-130) flying EA-6B Prowlers; Sea Control Squadron 22 (VS-22) flying 5-3B Vikings; Helicopter Anti-Submarine Squadron 7 (HELASRON HS-7) flying SH-60F and HH-60H Seahawks; and Fleet Logistics Support Squadron 40 (VRC-40) flying the C-2A Greyhound. With its supporting cast of armed warships and the logistics fleet essential to sustained operations at sea, the *Truman* and its air wing are significant contributors to America's war on international terrorism.

## List of Paintings